Algorithms and Computation in Mathematics • Volume 12

Springer
Berlin
Heidelberg
New York
Hong Kong
London
Milan
Paris
Tokyo

John R. Shackell

Symbolic Asymptotics

Springer

John R. Shackell
Institute of Mathematics,
Statistics and Actuarial Science
University of Kent
Canterbury
Kent CT2 7NZ
United Kingdom
e-mail: j.r.shackell@kent.ac.uk

Mathematics Subject Classification (2000): 68W30, 41A60

Library of Congress Cataloging - in - Publication Data

Shackell, John R., 1943 -
Symbolic asymptotics / John R. Shackell.
p. cm. – (Algorithms and computation in mathematics, ISSN 1431-1550 ; v. 12)
Includes bibliographical references and index.
ISBN 3-540-21097-0 (acid-free paper)
1. Asymptotic expansions. 2. Algorithms. I. Title. II. Series.
QA297.S435 2004
511'.4–dc22 2004045315

ISSN 1431-1550

ISBN 3-540-21097-0 Springer-Verlag Berlin Heidelberg New York

Springer-Verlag is a part of Springer Science+Business Media
springeronline.com

Typesetting: Camera-ready copy from the author
Cover design: *design & production GmbH*, Heidelberg

Printed on acid-free paper 46/3142at - 5 4 3 2 1 0 -

This book is dedicated to the memory of the author's parents.

Ralph Ronald Shackell 1899 - 1976
Jean Shackell (née Hatten) 1912 - 1994

Preface

This book is about calculating limits and asymptotic expansions of real functions. The subject has a long history and a vast literature, almost all of it concerning power series. There are already many books about asymptotic power series, and I am not the person best qualified to write another. So I have concentrated on the areas where power series are no longer the right tool. Even there I have given the most coverage to the developments that I have personally been involved in, not because they are necessarily the most important, but just because they are what I know most about. However I have tried to give a reasonably comprehensive bibliography.

I have taken the stance that the book should be accessible to anyone with a good general background in mathematics. This has caused the omission of some topics, for example the proofs in [102]. Very noticeably I have given little coverage to Joris Van Der Hoeven's thesis and related preprints. This work is undoubtedly important but it is also difficult, and the precise status of some of it remains unclear at the time of writing. Those who wish to find out more about Dr. Van Der Hoeven's work are referred to his website (currently http://www.math.u-psud.fr/~vdhoeven/).

The reader who wants only a quick introduction to symbolic asymptotics might suitably follow Chapter 1, glance through 2.1 and 2.2 as far as Algorithm 2, and then read 3.1, 3.2 3.3, 4.1, 4.2 and 5.1.

I have occasionally suggested exercises in the text. In many cases, an answer may be found in the reference cited. I have tried (mostly succeeding!) to avoid setting as exercises, problems which I have shrunk from trying myself but nonetheless believe to be soluble.

There are many areas of mathematics which are to a greater or lesser extent concerned with asymptotics, and sometimes there is some overlap with symbolic asymptotics. In particular this is the case with o–minimality, which is briefly alluded to in Chapter 8, and non–standard analysis, which is not hereafter mentioned at all. The book has already taken too long to write!

That it has been written at all owes much to the encouragement of colleagues both at the University of Kent and elsewhere around the world. I would like to take this opportunity to thank them all, together with my family. I would also like to thank the staff at Springer-Verlag. There were substantial periods when I made little progress with the book while I devoted myself to mind numbing administration of the sort that is ever more taking over the lives of British academics. Ms. Ruth Allewelt, Dr. Peters, and the series editor Manuel Bronstein showed considerable patience during these times.

John Shackell,
Canterbury,
11 May 2004.

Contents

1

Introduction

This book is about the automatic computation of asymptotic behaviour of functions of a real variable, x. This means that we shall wish to compute limits, and also to give some measure of how rapidly a given function approaches its limit. Throughout, we shall assume that our interest is in the behaviour as $x \to \infty$. This is for convenience only, since other cases can be considered by using a transformation of one of the forms $x \to \pm(x - x_0)^{-1}$. These give separate treatments of the cases when x tends to x_0 from above and below. We shall want our methods to be at least potentially implementable, and we shall, from time to time, give some consideration to actual implementations. However this book is really about algorithms and the mathematical principles that underlie them, rather than details of implementation.

Several questions present themselves:

(I). What sort of functions will be considered, and how will they be represented?
(II). How will the asymptotic behaviour be expressed?
(III). What are the key algorithmic problems which need to be overcome?
(IV). What mathematical techniques are to be employed?

This book represents an attempt to answer these questions from a particular perspective, and to report on the progress that has been achieved in various cases.

We take the view that a function is something which is built from constants and the identity function, x, by means of certain operations including arithmetic operations and others. Thus a function is given by an expression, which may be represented by a tree; for example the function $\exp(x^2 + 1)$ is represented by the tree below.

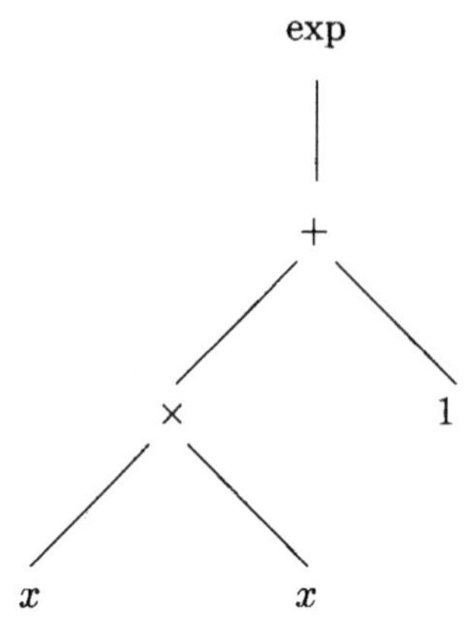

It is reasonable to ask which constants are to be allowed as leaves of the trees, and which operations are to be allowed as nodes. The obvious answer to the first question is to only allow integers. However if $\log x$ appears as a subexpression of our input function, it may be very inconvenient if $\log 2$ is not in our field of constants. We could add such constants as needed, but it is sometimes better for theoretical purposes to start from the set of real numbers as constants. Of course, this does not solve the problems as far as implementation is concerned. In fact there are substantial difficulties connected with constants in our area of study, as we shall see in the next chapter.

In addition to the arithmetic operations we may want to add real roots of algebraic equations and also consider the application of exponentials, logarithms, integrals, and solutions of more general differential equations.

As well as using trees as indicated above, it will be useful to employ a concept from differential algebra (see [45, 107] for example). A *tower* of sets of functions is a finite chain

$$\mathbb{K} = \mathcal{F}_0 \subset \mathcal{F}_1 \subset \cdots \subset \mathcal{F}_n. \tag{1.1}$$

Normally, the sets, $\mathcal{F}_i$, will be fields, or at least rings. Each $\mathcal{F}_i$, $1 \le i \le n$, will typically be generated from $\mathcal{F}_{i-1}$ by the addition of a single element, f_i, satisfying a differential equation over $\mathcal{F}_{i-1}$ of first order and degree. $\mathbb{K}$ will often be a field of constants. We shall frequently want to use induction on the index i, thereby reducing a problem about a function in $\mathcal{F}_i$ to one in $\mathcal{F}_{i-1}$.

One can define various classes of functions using towers (1.1), by placing restrictions on the differential equations which define the f_i. So if each f_i is a logarithm or an exponential, then the totality of functions definable using towers of function fields (1.1) is the field of *exp–log* functions. If f_i is also allowed to be a real root of an algebraic equation over $\mathcal{F}_{i-1}$ (which we could think of as a differential equation of order zero) then Hardy's class of *L-functions* is obtained. At the next stage of generalisation, one could allow f_i to be any integral of an element in $\mathcal{F}_{i-1}$, rather than just a logarithm, while retaining the possibility of f_i being an exponential or a real root of an algebraic equation; the class of functions so defined is the class of *Liouvillian functions.* Finally if one allows the f_i to be defined by any differential equations of first order and degree, one obtains the class of *Pfaffian functions.* The theory of Hardy fields suggests that this as a natural barrier. However there are some circumstances in which one can do more.

Having looked at the form which our input might take, we now turn our attention to the output. In other words, we look at question (II). The traditional approach here has been to use asymptotic power series, and it has led to many successes. However, as we shall see in Chapter 4 such series are frequently insufficient for our purposes, and may bring other problems. We shall present three more general type of expansion that avoid these difficulties; they are *multiseries, nested expansions* and *star products.* By using any of these it is possible to express the asymptotic behaviour of arbitrary exp-log functions, Liouvillian functions, inverse functions and implicit functions.

There is one more point that needs to be made here. The traditional stance in asymptotics is to regard asymptotic series as the prime objects of study. Most of the work is carried out in terms of the series, and the functions possessing these expansions often only appear quite late in the process. By contrast, we shall regard functions as the prime objects of study. One reason for this is that the convergence and summability theories that are used with asymptotic power series are less well developed for the more general expansions mentioned above. There are also more technical reasons concerning zero equivalence. As a result of our viewpoint, we shall not generally be concerned with proving the existence of functions having a particular expansion.

As regards question (III), many of the algorithmic problems will be best described when they are encountered in later chapters. However there are two which are worth mentioning now. We have seen that we can represent our input functions by expressions using either trees or towers of the form (1.1). Then each expression corresponds to a unique function, but unfortunately the representation is not one-to-one. In practice we have to work with the expressions, rather than directly with the functions. It is therefore vital that we have some way of determining when two different expressions represent the same function. This is the problem of *zero equivalence*, which will be treated in Chapter 2.

The second problem is related to zero equivalence. Suppose that we have a tower (1.1), with an expression $f \in \mathcal{F}_n$ for which we want to determine a nested expansion, for example. By induction, we may assume that we can do this for elements of $\mathcal{F}_{n-1}$. Then if f_n is given by a simple differential equation over $\mathcal{F}_{n-1}$, say f_n is an exponential or an integral, it is not too difficult to obtain an expansion for f_n. The trouble is that f_n may partially cancel with elements of $\mathcal{F}_{n-1}$ in the expression f. Here is an example from [97]. Let

$$f = \exp\left\{\frac{x}{x-1} + e^{-x}\right\} - \exp\left\{\frac{x}{x-1}\right\}. \tag{1.2}$$

If we proceed naively by expanding $x/(x-1)$ within the two exponentials and using the exponential series at 1, the terms will perpetually cancel out and the algorithm will fail to terminate. The problem is that e^{-x} is smaller than any constant power of x. We can recognise the existence of the cancellation by replacing e^{-x} by zero in the expression for f, obtaining an expression which is functionally equivalent to zero. (Note that we need to make explicit use of a zero-equivalence procedure here.). Then we can rewrite the right-hand side of (1.2) as an analytic function of the two 'variables', x^{-1} and e^{-x}. Since the terms not involving e^{-x} cancel We look successively at the positive powers, and discover that the coefficient of e^{-x} is $\exp\{1/(1-x^{-1})\}$, which is not equivalent to zero. Hence $f \sim e \cdot e^{-x}$. In general, we need to rewrite the given expression in terms of a finite number of sub-expressions which are all of a different 'order of growth', like x^{-1} and e^{-x}. Then we can expand in a similar way to the above. Of course, it is the rewriting that is the difficult part. In Chapter 5, we show how to do this for exp–log functions, and extend it to arbitrary Liouvillian functions. We can also uses the same ideas to handle extensions given by composition of a meromorphic solution of an algebraic differential equation with a function in the existing field.

In addition, we want to consider some cases which do not conform to the pattern of (1.1). Here there may be difficulties in obtaining the expansion of the newly-introduced object as well as in handling cancellations with existing objects. For example in Chapter 6 we look at algebraic differential equations of general order. We cannot give an algorithm to describe the asymptotics of an arbitrary solution of such equations. However we are able to obtain results for solutions that belong to some (a priori undetermined) Hardy field, and this at least has the merit of not restricting attention to a particular type of asymptotic growth in the way that searching for power-series solutions does.

Inverse functions are another case. We point out in Section 4.1 that inverse functions present formidable difficulties for asymptotic series expansions. Nested expansions on the other hand can just be inverted, [90], and multiseries can be used to give a full answer to the problem, [88]. We look in detail at inverse functions in Chapter 7.

In Chapter 8, we consider functions of several variables. This leads on to our results on implicit functions. Chapter 9 looks at functions which grow more rapidly than any iterated exponential, and related matters.

We have not yet given any answer to question (IV). Much of the theoretical underpinning for the algorithms we describe comes from the theory of Hardy fields, which is covered in Chapter 3. Beyond that, we shall need to use some differential algebra, some ideas from asymptotic series and of course many of the standard techniques of computer algebra, such as polynomial calculations, gcds etc. In Chapter 10 we look at what can be done with oscillating functions, and here some basic ideas from measure theory come into play.

In a number of cases, for example the algebraic extensions of Chapter 5, solutions of differential equations in Chapter 6 and implicit functions in Chapter 8, our techniques do not immediately give us the existence of solutions. Instead we demonstrate that any solutions that do exist (and lie in a Hardy field in the case of solutions of differential equations of arbitrary order) must have nested expansions of one of a number of types, which we list. Of course this exactly parallels the classical situation for asymptotic series. For algebraic extensions, we are able to adapt Sturm sequences to a Hardy-field setting in order to determine the number of solutions with the given nested form. A similar technique works for implicit functions in most cases.

Before closing this chapter, we want to briefly discuss the following question: What is the natural class of functions for the analysis side of symbolic computation? Of course one cannot hope to get a single answer to such a question, since there will always be special domains for special problems, but it seems worth giving the matter some thought. For the domain of functions very much determines the kind of theory that one obtains. It is instructive to contrast the analysis of Euler, Cauchy and Weierstrass in this way.

Thus far, in what is still a young subject, much of the attention has focussed on elementary functions. Sometimes these have been augmented by certain special functions, such as the error function, and sometimes Liouvillian functions have been considered. However the basic flavour has been that of elementary functions. It is natural to try and handle these first, and some very fine work has been done. Indeed the theory of integration in finite terms can make a good claim to be the jewel in the crown of symbolic computation. Nonetheless, I believe that the concentration on elementary functions and their near relatives is ultimately too limiting. Too often the only answer that an integration package can give to a user's demand is to assert that the input is not integrable in finite terms. From one point of view this represents a mathematical triumph. The engineer may be less impressed than we might hope however! In [109], David Stoutemyer points out the need to give qualitative information about functions. In particular he argues that the presentation of a result as a complicated elementary function may be insufficient. We would add that it often cannot be done! Thus computer-algebra packages need to give information about zeros, singularities, and of course limits, of functions, rather than always looking for elementary solutions.

So what is the correct domain in which to try to do this? One can make a case that it should ultimately be the domain of solutions of algebraic differential equations, augmented by one or two other special functions such as the gamma function. I am not saying that this can be achieved now, nor anything like it, but I do believe that mostly it will be achievable, and perhaps within a reasonable time span. Surely there will always be things we do not know about solutions of algebraic differential equations, just as there are about elementary functions, but this should not deter us.

For a view on the importance of algebraic differential equations from a different perspective, the reader is referred to [84], and the other papers of Lee Rubel on this subject.

2

Zero Equivalence

In this chapter we are concerned with algorithms to determine whether a given expression of the type discussed in the introduction represents the zero function or not. The expression may be built using roots, exponentials, logarithms, integrals or solutions of differential equations.

Because it is often obvious that an expression does, or does not, represent the zero function, one might get the idea that it it was a trivial matter to decide this in general. The example

$$\exp(\sin^2 x(\log x + 1)) \exp(\sinh^2 x^2) - \\ x \exp(-\cos^2 x) \exp\left(\frac{e^{2x^2}}{4} + \frac{1}{2}\right) \exp\left(-\left(\log\cos^2 x - \frac{e^{-2x^2}}{4}\right)\right)$$

is perhaps sufficiently complicated to convince otherwise. But of course this example only involves elementary functions, and one can envisage far worse when extensions by solutions of complicated differential equations are present. Moreover there are relatively simple function classes where the question of zero-equivalence is known to be undecidable, [65]. Even for the classes we shall be considering, the matter is not yet totally resolved.

Not only is zero-equivalence far from easy; it is also vitally important. The main point has already been mentioned in Chapter 1, but it is worth reiterating here. In order to compute with functions, we have to represent them by expressions and compute with these in the first instance. Unless we have a method of deciding when two expressions represent the same function, we cannot claim to be computing with functions at all. For example, if we have an expression in the form of a quotient, we shall not know whether it defines a function unless we can determine whether the denominator is functionally equivalent to zero.

We start by looking at L–functions. We may think of any given L–function as belonging to the top field of a tower (1.1). One can use induction on the number of fields in the tower to show that the number of zeros of a non-zero element of the top field is bounded, and this is essentially what Hardy did to prove that the L–functions form what we now call a Hardy field, [40]. By following similar lines, Dan Richardson, [66], and independently Angus Macintyre, [54], showed how to come up with an actual bound for a given non-zero L–function. As they pointed out, this gives a potential method of determining whether a given expression is zero as follows. One evaluates the expression at a number of points one larger than the bound obtained. If the function evaluates to zero at all of these, it must be the zero function; otherwise not, of course. However this leaves one question unanswered, namely how to decide whether a constant expression is zero. This problem is related to some old conjectures in transcendental number theory, which look formidably difficult.

2.1 Zero-equivalence of Constants

We recall some definitions and results concerning transcendence; see [125]. Let k and K be two fields with $k \subset K$. An element, x of K is said to be *algebraic* over k is there exists a non-zero polynomial, P with coefficients in k such that $P(x) = 0$. If no such P exists, x is said to be *transcendental* over k. Similarly we say that elements $x_1, \ldots, x_n$ of K are *algebraically independent* over k if there is no non-zero polynomial, P, in n indeterminates with coefficients in k such that $P(x_1, \ldots, x_n) = 0$. An infinite set $L \subset K$ is said to be algebraically independent over k if this is true of every finite subset of L. A *transcendence basis* of K over k is a maximal algebraically independent subset of K. Thus $\{x, e^x\}$ is a transcendence basis of $\mathbb{R}(x, e^x)$ over $\mathbb{R}$; the set $\{x, e^x + \sqrt{x}\}$ is another. It can be shown, [125], that transcendence bases always exist, and that any two have the same cardinal number of elements. That number is called the *transcendence degree* of the extension $K : k$.

We shall make use of the following conjecture. Our treatment here is based on [49] and [94].

SCHANUEL CONJECTURE *Let $\alpha_1, \ldots, \alpha_k$ be complex numbers which are linearly independent over the rational numbers $\mathbb{Q}$. Then the transcendence degree of*

$$\mathbb{Q}(\alpha_1, \ldots, \alpha_k, \exp(\alpha_1), \ldots, \exp(\alpha_k)) : \mathbb{Q}$$

is at least k.

The conjecture implies in particular that e and π are algebraically independent. For $2\pi i$ and 1 are linearly independent over $\mathbb{Q}$ and so the conjecture tells us that the transcendence degree of $\mathbb{Q}(1, 2\pi i, e, e^{2\pi i}) : \mathbb{Q}$ is at least 2. Thus the transcendence degree of $\mathbb{Q}(\pi, e)$ over $\mathbb{Q}$ is 2 and hence e and π are algebraically independent. In point of fact, even this very special case appears not to be known! However the case when $\alpha_1, \ldots, \alpha_k$ are algebraic numbers was proved by Lindemann, [51].

The first algorithm to decide zero-equivalence of constants involving exponentials and logarithms based on Schanuel's conjecture was given by Caviness and Prelle in [24], although some of the ideas were already present in [49]. This line of development subsequently played a part in the solution, on basis of the Schanuel conjecture, of Tarski's conjecture on the decidability of the reals with exponentiation, [55]. In [70, 71], Dan Richardson treats the problem of deciding zero equivalence of constants defined by systems of exponential equations from a computer-algebra point of view.

In zero-equivalence computations there is no particular difficulty in considering complex-valued constants and functions, and so we may as well allow the extra generality that this gives. Suppose then that we have a tower of fields

$$\mathbb{Q}(i) = \mathcal{F}_0 \subset \mathcal{F}_1 \subset \cdots \subset \mathcal{F}_n \tag{2.1}$$

in which for each $j = 1, \ldots, n$, $\mathcal{F}_j = \mathcal{F}_{j-1}(c_j)$ with c_j a constant which is either

(i) algebraic over $\mathcal{F}_{j-1}$ with given minimal polynomial, m_j,
(ii) an exponential, $c_j = \exp b_j$ where $b_j \in \mathcal{F}_{j-1}$, or
(iii) a logarithm, $c_j = \log b_j$ where $b_j \in \mathcal{F}_{j-1}$.

We assume that in (i), c_j is somehow specified as a particular root of m_j, and in (iii), c_j is a specified complex logarithm.

An element of $\mathcal{F}_n$ will be given as a rational function of c_n with coefficients in $\mathcal{F}_{n-1}$. Our task is then to determine whether an expression of the form $P(c_n)$ is equal to zero, where P is a given polynomial over $\mathcal{F}_{n-1}$. We may assume inductively that we can perform such a task for elements of $\mathcal{F}_{n-1}$. If c_n is given as algebraic over $\mathcal{F}_{n-1}$, then $P(c_n)$ will be zero if and only if P is divisible by m_n, and this can be determined by employing a division algorithm using the zero-equivalence procedure in $\mathcal{F}_{n-1}$. Of course if c_n is transcendental over $\mathcal{F}_{n-1}$ then $P(c_n)$ cannot be zero! So if c_n is given as an exponential or a logarithm, we need to know whether it is actually transcendental over $\mathcal{F}_{n-1}$ or is an element of the algebraic closure of $\mathcal{F}_{n-1}$ in disguise.

Suppose that among $c_1, \ldots, c_{n-1}$, only $c_{j_1}, \ldots, c_{j_{k-1}}$ are actually transcendental over $\mathcal{F}_{j_1-1}$, ..., $\mathcal{F}_{j_{k-1}-1}$ respectively, where $1 \le j_1 < j_2 < \cdots < j_{k-1} \le n-1$. We consider first the case when c_n is an exponential, $c_n = \exp b_n$. We wish to use the Schanuel conjecture with $\alpha_k = b_n$, so that $\exp(\alpha_k) = c_n$. The trick is to arrange things so that we know that only $k-1$ of $\alpha_1, \ldots, \alpha_k, \exp(\alpha_1), \ldots, \exp(\alpha_{k-1})$ are algebraically independent. Then if $\alpha_1, \ldots, \alpha_k$ are linearly independent over $\mathbb{Q}$ the Schanuel conjecture will imply that $c_n = \exp(\alpha_k)$ is transcendental over $\mathcal{F}_{n-1}$. On the other hand if we have a $\mathbb{Q}$-linear relation between $\alpha_1, \ldots, \alpha_k$, we can write c_n as a monomial in $\exp(\alpha_1), \ldots, \exp(\alpha_{k-1})$.

We organise matters as follows. Consider each c_{j_i}, $i = 1, \ldots, k-1$. If c_{j_i} is an exponential, $c_{j_i} = \exp b_{j_i}$, we put $\alpha_i = b_{j_i}$ so that $c_{j_i} = \exp(\alpha_i)$. On the other hand if c_{j_i} is a logarithm, $c_{j_i} = \log b_{j_i}$, we put $c_{j_i} = \alpha_i$, giving $\exp \alpha_i = b_{j_i}$. Now as an element of $\mathcal{F}_{n-1}$, α_k is algebraic over $\mathbb{Q}(\alpha_1, \ldots, \alpha_{k-1}, \exp \alpha_1, \ldots, \exp \alpha_{k-1})$ and it is a matter of deciding whether it is linearly dependent (over $\mathbb{Q}$) on $\alpha_1, \ldots, \alpha_{k-1}$. We write each α_i, $i = 1, \ldots, k$ as a rational function, $R_i(c_1, \ldots, c_{n-1})$, and seek solutions in integers of the equation

$$\sum_{i=1}^{k} z_i R_i(c_1, \ldots, c_{n-1}) = 0. \tag{2.2}$$

By cross-multiplying, we may reduce to the case when the R_is are polynomials. Moreover if c_{a_i} is algebraic over $\mathcal{F}_{a_i-1}$ for $i = 1, \ldots, n-k$, with minimal polynomial m_i of degree d_i, we may use the relations $m_i(c_{a_i}) = 0$ to ensure that no R_j contains powers of c_{a_i} of degree greater than $d_i - 1$. That done, we may equate coefficients in (2.2) to zero giving a set of linear equations for $z_1, \ldots, z_n$ with coefficients in $\mathbb{Q}$. We can then determine whether non-trivial solutions of (2.2) exist.

In the case when c_n is a logarithm, $c_n = \log b_n$, we take $\alpha_k = c_n$ giving $\exp \alpha_k = b_n$ again. For each $j = 1, \ldots, n-1$ we can express $\exp \alpha_j$ as a rational function $S_j(c_1, \ldots, c_{n-1})$. As before, we can arrange that each numerator and denominator only contains powers of c_{a_i} of degree less that d_i, for $i = 1, \ldots, n-k$. We then use the fact that $\alpha_1, \ldots, \alpha_k$ are linearly independent if and only if $\exp \alpha_1, \ldots, \exp \alpha_k$ are multiplicatively independent. Thus we seek solutions of

$$\prod_{i=1}^{k} S_i^{z_i}(c_1, \ldots, c_{n-1}) = 1 \tag{2.3}$$

in integers, $z_1, \ldots, z_k$. We use gcd computations to identify common factors between pairs of different numerators and denominators of S_js, ending up with expressions of the form

$$S_i = T_1^{r_1^i} T_2^{r_2^i} \cdots T_M^{r_M^i},$$

where the $r_1^i, \ldots, r_k^i$ are integers, $i = 1, \ldots, k$, and $T_1, \ldots, T_M$ are pairwise coprime. Then we substitute these relations into (2.3) and equate exponents of T_is to zero, giving a set of linear equations for $z_1, \ldots, z_n$ to be solved as before.

Once the algebraic dependencies between $c_1, \ldots, c_n$ are established, we can write an element of $\mathcal{F}_n$ as a rational function of $c_1, \ldots, c_n$ with each c_{a_i} only appearing with degree less than d_i. This element is then zero if and only if each coefficient of the numerator is zero.

Example

We apply the method described above to the expression

$$36 \log^2 \left(\exp\left(\frac{\sqrt{3}}{2}\right) \exp\left(\frac{2}{\sqrt{3}}\right) \right) - 147.$$

We write $c_1 = \sqrt{3}$, $c_2 = \exp(\sqrt{3}/2)$, $c_3 = \exp(2/\sqrt{3})$ and we set c_4 equal to $\log(\exp(\sqrt{3}/2)\exp(2/\sqrt{3}))$. Then we take $\mathcal{F}_0 = \mathbb{Q}$, and $\mathcal{F}_i = \mathcal{F}_{i-1}(c_i)$ for

$i = 1, \ldots, 4$. The minimal polynomial for c_1 is of course $m_1 = x^2 - 3$. We check algebraic dependencies, starting with c_2. In this particular case, we can use the Lindemann Theorem, and it tells us that the transcendence degree of

$$\mathbb{Q}\left(\frac{\sqrt{3}}{2}, \exp\left(\frac{\sqrt{3}}{2}\right)\right) : \mathbb{Q}$$

is at least one. So $\exp(\sqrt{3}/2)$ is indeed transcendental over $\mathbb{Q}(\sqrt{3}/2)$.

Next we look at c_3. We discover that $2/\sqrt{3} = (4/3)(\sqrt{3}/2)$, and hence $c_3 = c_2^{4/3}$. Now

$$c_4 = \log(c_2 c_3) = \frac{7}{3} \log c_2.$$

Hence

$$36 \log^2(c_2 c_3) = 196 \log^2 c_2 = 196(3/4) = 147,$$

and so our expression does represent zero.

In practice, the algorithms we shall be describing in later chapters will make frequent use of zero-equivalence computations for functions, and as we shall see later in this chapter, each such a computation typically requires many zero-equivalence calculations for constants. But it is clear that once the expression being tested becomes at all complicated, the above method will be quite long, and so it may be infeasible to use it for all such calculations. However most expressions are not zero and in the vast majority of cases this can be established very quickly by numerical methods. Thus the use of the Schanuel conjecture should be regarded as an ultimate arbiter when floating-point computations fail to yield an answer in a reasonable time.

Another point is that we shall sometimes consider function classes more general than the class of L-functions, for example Liouvillian functions, and there the Schanuel conjecture does not generally resolve zero-equivalence of constants. At the present state of development, floating-point methods are then all that is available. Until recently these could only be used in an ad hoc way, but a new conjecture given independently by Van Der Hoeven, [114, 117] and Richardson, [72] provides a standard method, at least for exp–log functions. We follow Richardson's version here as being perhaps slightly simpler.

2.1.1 Richardson's Uniformity Conjecture

The conjecture relates the length of a constant exp–log expression to the size of the number it represents. The essential idea is that very large numbers can only be represented by long expressions, and similarly for very small non-zero numbers. However exponentiation can ruin this. Hence expressions of the form e^A are only allowed if $|A| \leq 1$. Similarly A^n with $n \in \mathbb{N}^+$ is disallowed, although $A^{1/n}$ is not. Expressions which conform to these rules are said to be in *expanded form*. It is easy to rewrite a given expression into expanded form. For example A^3 becomes $A * A * A$ and $e^{\sqrt{2}}$ becomes $e * e^{\sqrt{2}-1}$.

The length of an expression in expanded form may be defined as follows. Fix a numerical base b, for example $b = 10$. The *length* of a non-negative integer is the number of its digits in this base. Then

$$\begin{aligned}
length(A \pm B) &= length(A) + length(B) + 1 \\
length(A * B) &= length(A) + length(B) + 1 \\
length(A/B) &= length(A) + length(B) + 1 \\
length(-A) &= length(A) + 1 \\
length((A)) &= length(A) + 2 \\
length(A^{1/n}) &= length(A) + length(n) + 1 \\
length(e^A) &= length(A) + 1 \\
length(\log A) &= length(A) + 1
\end{aligned}$$

So essentially one counts the number of symbols.

Conjecture **(Richardson)** *If E is an expression in expanded form which represents a non-zero number then $|E| > 1/N(k)$, where k is the length of E and $N(k)$ is the number of syntactically correct expressions in expanded form of length no more than k.*

If S is the number of symbols for operators, brackets and digits, then $N(k) \leq S^k$. In base ten, we have $S = 19$ for example. If the conjecture is true then one can decide whether a constant expression represents zero by approximating to within $1/N(k)$. This would be a fast practical method.

The Uniformity Conjecture is related to the Witness Conjecture of Van Der Hoeven, [114]. The latter is much less specific in that an expression involving an unspecified parameter is used rather than $N(k)$. This makes the conjecture more likely to be true, and presumably easier to prove (if true), but perhaps less useful in practical computation.

It is interesting to speculate about the relationship between the Uniformity Conjecture and the Schanuel Conjecture. In [72], Richardson suggests that it might be possible to prove the Schanuel Conjecture from the Uniformity Conjecture using the method of Gelfond. However he is clearly some way from being able to do this.

[72] also contains some comparisons of known estimates in transcendental number theory with the corresponding deductions from the Uniformity Conjecture. The fact that in most cases the latter gives very much stronger results has caused some to doubt whether the conjecture can possibly be true. On the other hand Richardson has made an extensive search for unusually small numbers and has found none that contradict the conjecture. Inevitably he has only looked at specific types of numbers.

Transcendental number theory remains an area in which there are many unanswered questions. Conjectures such as the Uniformity Conjecture can act as a focus for research, and this particular conjecture has additional practical benefits. The dependence on the choice of base is perhaps a weakness of the conjecture, but probably not a serious one. Certainly this is a most interesting and important area for ongoing research.[1]

Finally, it should also be borne in mind that difficulties with the signs of constants have always been present in mathematical calculation involving transcendental functions, and this fact has not caused Numerical Analysis to be abandoned together with the associated parts of Physics, Chemistry, Engineering etc! It is a matter of coping with the problem as best one can until such a time as the relevant theory becomes sufficiently advanced. The example given in [65] shows that there are fundamental barriers beyond which this cannot be achieved.

2.2 Zero-equivalence of Functions

In this section, we show that zero-equivalence for functions can be handled in a number of ways, given a method for constants. However we do not have a full algorithm for most of the classes of constants we shall encounter. The usual device in this area is to postulate the existence of an *oracle* which can decide zero-equivalence of constants for us. The theorems proved then hold modulo the oracle, and the methods given in this section are to be understood to be subject to such a limitation. Of course for the exp–log constants, we can use the Schanuel conjecture and the methods of Caviness–Prelle and Richardson.

We have already mentioned the algorithm of Richardson and Macintyre for determining the equivalence of L–functions. Alas the number of constant evaluations needed would seem to be very large for even a mildly complicated function, and so one might doubt whether this method can be regarded as a practical one.

Another approach parallels the method given in Section 2.1 but with functions instead of constants. In fact Schanuel made a second conjecture which is the equivalent of that of 2.1 for formal power series with constant series excluded. This conjecture was proved by Ax, [3]. There are connections right back to the work of Liouville on transcendental functions, [52, 53]. The prob-

[1] While this book was in the final stages of preparation, Dan Richardson announced a counter–example to the Uniformity Conjecture. If $E(x)$ is an expression with just k occurrences of x which has a zero of order greater than k, then iteration of E will produce constants which are smaller and smaller relative to the size of their defining expressions. An example is given by

$$E(x) = \log(1 + x) - 2\log(1 + \log(1 + x/2)),$$

which contains only two occurrences of x but is $O(x^3)$ at the origin. Clearly the conjecture needs to be modified to take this and similar examples into account. It will be interesting to see how the matter develops.

lem of determining the algebraic dependencies between apparently transcendental functions is of fundamental importance in the theory of integration in finite terms. Here Risch, [74, 75], extended the work of Liouville. A substantial development then followed, and indeed continues; see [29, 21] for example. Using these methods and building on earlier work by Epstein and Caviness, Rothstein and Caviness were able to give a zero-equivalence algorithm for Liouvillian functions (modulo a suitable constant oracle of course), [83].

The methods of Risch and Rothstein-Caviness involve determining all the algebraic dependencies in the relevant field of functions. So if the tower

$$\mathbb{K} = \mathcal{F}_0 \subset \cdots \subset \mathcal{F}_n$$

is defined by $\mathcal{F}_i = \mathcal{F}_{i-1}(f_i)$, for $i = 1, \ldots, n$, and we are given an expression $E(f_1, \ldots, f_n) \in \mathcal{F}_n$, the method requires us to find all the polynomials $Q(X_1, \ldots, X_n)$ for which $Q(f_1, \ldots, f_n) \equiv 0$. For the purposes of integration in finite terms, that may be necessary, but all we need to know here is whether this is true for the case when Q is equal to the numerator of E.

Thus we may write E as the quotient of two polynomials in $\mathbb{K}[f_1, \ldots, f_n]$ and we then seek to determine whether the numerator, P, is equivalent to zero. By induction, we may assume that we can do this for elements of $\mathbb{K}[f_1, \ldots, f_{n-1}]$; of course the case $n = 0$ may require the services of an oracle. We regard P as a polynomial in f_n with coefficients in $\mathcal{F}_{n-1}$. For the purposes of explanation, we shall frequently want to temporarily replace f_n by an indeterminate X. Suppose that f_n is define by the differential equation

$$\Omega(y)\frac{dy}{dx} = \Lambda(y) \qquad (2.4)$$

together with initial condition $y(x_0) = y_0$. We assume that $\Omega(y)$ does not vanish under the substitutions $x = x_0$, $y = y_0$, so in particular $\Omega(f_n) \not\equiv 0$. We define an operator

$$* : \mathbb{K}[f_1, \ldots, f_{n-1}][X] \longrightarrow \mathbb{K}[f_1, \ldots, f_{n-1}][X]$$

by

$$Q^* = \frac{\partial Q}{\partial x}\Omega(X) + \frac{\partial Q}{\partial X}\Lambda(X), \qquad (2.5)$$

where $\partial Q/\partial x$ is obtained by differentiating the coefficients of Q. We note that $Q^*(f_n)$ is the (total) derivative of $Q(f_n)$ multiplied by $\Omega(f_n)$. This gives the following method for determining whether or not $P(f_n) \equiv 0$.

Algorithm 1

To determine whether $P(f_n) \equiv 0$:

(1). Check that $P(X)$ vanishes under the substitutions $x = x_0$, $X = y_0$. If not, then of course $P(f_n) \not\equiv 0$. Otherwise let $\mathbf{G}$ be the Gröbner basis (see for example [23, 1]) consisting of the single polynomial P.

(2). For $j = 1, 2, \ldots$ compute $P^{*j}(X)$ and check that it vanishes under the substitutions $x = x_0$, $X = y_0$; if not then $P(f_n) \not\equiv 0$. If the substitutions do annihilate $P^{*j}(X)$, reduce P^{*j} with respect to the Gröbner basis $\mathbf{G}$*, obtaining Q. If Q is the zero polynomial then $P(f_n) \equiv 0$. Otherwise add Q to the Gröbner basis* $\mathbf{G}$*, and continue with the next j.*

Stage 2 must terminate after a finite number of steps, since $\mathbb{K}[f_1, \ldots, f_{n-1}][X]$ is a Noetherian ring. This means, [125], that we cannot for ever keep generating larger and larger ideals. So either we discover that some $P^{*j}(X)$ fails to vanish under the substitutions $x = x_0$, $X = y_0$, in which case $P(f_n) \not\equiv 0$, or else some P^{*j} reduces to zero with respect to $\mathbf{G}$. In the second case, P^{*j} must belong to the ideal generated by $P, P^*, \ldots, P^{*j-1}$. That is to say $P(f_n)$ satisfies a differential equation of the form

$$y^{(j)} = R(y, y', \ldots, y^{(j-1)})$$

where R is a polynomial over $\mathbb{K}[f_1, \ldots, f_{n-1}]$. Since we shall already have checked that $P(f_n)$ and its first $j-1$ derivatives vanish at $x = x_0$, the uniqueness theorem for differential equations guarantees that $P(f_n)$ must be the zero function.

Algorithm 1 also works when the f_ns are defined by higher-order differential equations. However it does have its drawbacks from the point of view of efficiency. For Gröbner-basis calculations can be quite lengthy, and we may have to perform a large number of them. The point can be illustrated by the following example from [100]. Suppose $P(f) = (f - \exp(x^2))(\exp x + x^M)$, with f defined by the differential equation $f' = 2xf$ and the initial condition $f(0) = 1$. Here M is a large integer. It is, of course obvious that $P(f) \equiv 0$. However Algorithm 1 will require $M+2$ differentiations because for $n < M+2$ the n–th derivative of $\exp x + x^M$ is not in the ideal generated by $\exp x + x^M$ and its first $n-1$ derivatives. Yet $P(f)$ is equivalent to zero entirely because of the first factor. Thus the algorithm requires a great deal of work sorting out the ideal structure in respect of a factor which has nothing to do with the zero equivalence. Obviously the same problem could occur in an example where the first factor was less transparently equivalent to zero. There is another approach.

If $P(f_n) \equiv 0$ then $P(X)$ and $P^*(X)$ have a common root at $X = f_n$. Hence $\gcd_X\{P, P^*\}$ will be a polynomial in X of degree at least one. It turns out that provided P is square free as a polynomial in X, this is nearly sufficient. Since the gcd can be computed using calculations in the coefficient field $\mathcal{F}_{n-1}$, we have the basis of an inductive method. Note that if P is not square free, P and P^* will inevitably have a root in common. To cover such an eventuality we compute $\delta = \gcd_X\{P, \partial P/\partial X\}$, and if this has degree greater than zero in X, we replace P by P/δ and clear any denominators of coefficients. This has the effect of removing multiple factors from P, which will not alter the zero-equivalence of $P(f_n)$ since $P(f_n) \equiv 0$ if and only if $P(X)$ has a factor $X - h$ with $h \equiv f_n$.

LEMMA 1 *Suppose that* $P(X) \in \mathbb{K}[f_1, \ldots, f_{n-1}][X]$ *is a square-free polynomial, and let* h *be a root of* $P(X)$ *in the algebraic closure of* $\mathbb{K}[f_1, \ldots, f_{n-1}]$. *Then* $X - h$ *is a factor of* $\gcd_X\{P, P^*\}$ *if and only if* h *satisfies the defining differential equation for* f_n. *That is to say, if and only if* $h'\Omega(h) = \Lambda(h)$.

Proof of Lemma 1
Let $P(X) = (X - h)Q(X)$. Then

$$P^*(X) = (\Lambda(X) - \Omega(X)h')Q(X) + (X - h)Q^*(X).$$

If $h'\Omega(h) = \Lambda(h)$, then we see that h is also a root of P^*. Conversely suppose that h is root of P^*. Now $Q(h) \neq 0$ since $P(X)$ is square free, and so we must have $\Lambda(h) - \Omega(h)h' = 0$. This completes the proof of Lemma 1.

It follows that if the degree of $g = \gcd_X\{P, P^*\}$ is at least one, g will be of the form $g = A\prod_{i=1}^{m}(X - h_i)$, where A belongs to $\mathbb{K}[f_1, \ldots, f_{n-1}]$ and each h_i belongs to the algebraic closure of $\mathbb{K}[f_1, \ldots, f_{n-1}]$ and satisfies (2.4). If one of the h_is also satisfies the initial condition $y(x_0) = y_0$ then it must be identical to f_n, by the uniqueness theorem for differential equations. However we have to be careful in the matter of checking the initial condition because we shall not have an explicit factorisation of g and there is the additional complication that $A(x_0)$ might be zero. Dividing g through by the leading coefficient might not solve the second problem since, with $x_0 = 0$, the leading coefficient might be $e^x - 1$ while another coefficient might be x. What we can do is to use the fact that A is analytic at x_0, and hence cannot vanish together with all its derivatives there.

In g, we substitute $x = x_0$ but leave X as an indeterminate. If this reduces g to zero, we must have $A(x_0) = 0$. We repeatedly differentiate g partially with respect to x until the substitution $x = x_0$ no longer reduces g to the zero polynomial. Then making both the substitutions $x = x_0$ and $X = y_0$ will reduce g to zero if and only if at least one of the h_is satisfies $h(x_0) = y_0$. To see this suppose that $A^{(k)}$ is the first derivative of A which does not vanish at x_0. Then the k-th partial derivative of g takes the form

$$\frac{\partial^k g}{\partial x^k} = A_0 G_0 + A' G_1 + \cdots + A^{(k-1)} G_{k-1} + A^{(k)} \prod_{i=1}^{m} (X - h_i),$$

where $G_j = {}^kC_j \partial^{k-j}\{\prod_{i=1}^{m}(X - h_i)\}/\partial x^{k-j}$. When we put $x = x_0$, all the terms on the right vanish except $A^{(k)}(x_0)\prod_{j=1}^{M}(X - h_i(x_0))$, so putting $X = y_0$ reduces the expression to zero precisely when at least one $h_j(x_0)$ is equal to y_0. Thus we have the following algorithm.

ALGORITHM 2
To determine whether $P(f_n) \equiv 0$:

(1). Check that $P(X)$ reduces to zero under the substitutions $x = x_0$, $X = y_0$. If not $P(f_n) \not\equiv 0$.

(2). Remove any coefficients with zero terms from $P(X)$. Assuming that this does not reduce P to the zero polynomial, compute the gcd of P and $\partial P/\partial X$. If the degree of the gcd in X is at least one, divide it into P and clear denominators to give the new P.

(3). Compute $g = \mathrm{gcd}_X\{P, P^*\}$. If the degree of g in X is zero then $P(f_n) \not\equiv 0$.

(4). Substitute $x = x_0$ in g leaving X as an indeterminate. If this reduces g to the zero polynomial, repeatedly replace g by its partial derivative with respect to x until this substitution no longer makes g zero.

(5). Then substitute $X = y_0$ in addition to $x = x_0$. $P(f_n)$ is equivalent to zero if and only if this final substitution makes g vanish.

It should be stressed that the calculations in steps 2 and 3 are performed over the ring $\mathbb{K}[f_1, \ldots, f_{n-1}][X]$ and not over a formal polynomial ring $\mathbb{K}[X_1, \ldots, X_{n-1}][X]$. This means in particular, that in gcd computations we have to apply the algorithm recursively to decide whether coefficients are zero or not.

The basic idea in this method seems to have occurred independently to a number of people at about the same time, [96, 67, 111]. The mathematical basis was known much earlier; see [122]. The treatment here is based on [100].

Examples

We consider again our example $P(f) = (f - \exp(x^2))(\exp x + x^M)$, with f defined by the differential equation $f' = 2xf$ and the initial condition $f(0) = 1$. We apply the new method to $P(X) = (X - e^{x^2})(e^x + x^M)$, giving $P^*(X) = 2x(X - e^{x^2})(e^x + x^M) + (X - e^{x^2})(e^x + Mx^{M-1})$, and hence $g = X - e^{x^2}$. Substituting $x = 0$ sends g to $X - 1$ and now putting $X = 1$ gives zero. Thus Algorithm 2 gives the result with one differentiation, one gcd computation and two substitutions.

Now consider the expression $\exp(\log x) - 3x$. Proceeding more formally, we take $f_1 = x$, defined by $f_1' = 1$, $f_1(1) = 1$. Then $f_2 = \log x$ is given by $f_1 f_2' = 1$, $f_2(1) = 0$, and $f_3 = \exp(\log x)$ by $f_1 f_3' = f_3$, $f_3(1) = 1$. We take $P(X) = X - 3f_1$, so that $P^*(X) = X - 3f_1$ also, and hence $g = X - 3f_1$. The substitution $x = 1$, sends g to $X - 3$, but now putting $X = 1$ gives -2. So $\exp(\log x) - 3x$ is not zero. In fact the algorithm as written would have discovered this at step 1.

Of course these are toy examples. Some more slightly serious ones are given in Section 2.2.1.

Algorithm 2 can also be adapted to cover cases where the f_is are defined by higher-order differential equations, but the generalisation is non-trivial. Firstly the extensions in our tower of fields will have to include derivatives. So if f_i satisfies

$$y^{(m_i)} \Omega_i(y, y', \ldots, y^{(m_i-1)}) = \Lambda_i(y, y', \ldots, y^{(m_i-1)}),$$

we shall need to take $\mathcal{F}_i = \mathcal{F}_{i-1}(f_i, f_i', \ldots, f_i^{(m_i-1)})$. From one point of view this is because we need our fields to be closed under differentiation. From another it is because even if derivatives are not present in the given expression, they will be generated as the algorithm progresses and will have to be handled at a later stage of the recursion.

In order to avoid unnecessary subscripting, we take the equation satisfied by f_n to be

$$y^{(m)}\Omega(y, y', \ldots, y^{(m-1)}) = \Lambda(y, y', \ldots, y^{(m-1)}), \tag{2.6}$$

and we suppose that the initial conditions are $f_n^{(i)}(x_0) = y_i$, $i = 0, \ldots, n-1$. We employ the notation $\langle y \rangle$ to stand for a list consisting of y and its relevant derivatives. Thus if y satisfies (2.6), we use $\mathcal{F}_{n-1}\langle y \rangle$ to stand for $\mathcal{F}_{n-1}[y, y', \ldots, y^{(m-1)}]$. Similarly if Q is a polynomial, $Q\langle f_n \rangle$ means $Q(f_n, f_n', \ldots, f_n^{(m-1)})$. So let P be an element of $\mathbb{K}\langle f_1, \ldots, f_n \rangle$ and suppose that P contains $f_n^{(i)}$, $0 \le i \le m-1$, but no higher derivative of f_n. Our task is to determine whether $P\langle f_1, \ldots, f_n \rangle$ is functionally equivalent to zero. We shall assume that we can do this for smaller values of i, or in the case $i = 0$ for smaller values of n. We replace f_n^i by an indeterminate, X, and $f_n, \ldots, f_n^{(i-1)}$ by $Y, \ldots, Y^{(i-1)}$. Thus we work over the ring

$$\mathbb{K}\langle f_1, \ldots, f_{n-1} \rangle [Y, Y', \ldots, Y^{(i-1)}][X, X', \ldots, X^{(m-i-1)}].$$

This is a differential ring provided we take the derivative of $Y^{(i-1)}$ to be X, and that of $X^{(m-i-1)}$ to be

$$\frac{\Lambda(Y, \ldots, Y^{(i-1)}, X, \ldots, X^{(m-i-1)})}{\Omega(Y, \ldots, Y^{(i-1)}, X, \ldots, X^{(m-i-1)})}$$

with the understanding that denominators are always cleared. Thus for

$$Q \in \mathbb{K}\langle f_1, \ldots, f_{n-1} \rangle [Y, Y', \ldots, Y^{(i-1)}][X, X', \ldots, X^{(m-i-1)}],$$

we define

$$Q^* = \Omega \left\{ \frac{\partial Q}{\partial x} + \sum_{j=0}^{i-2} \frac{\partial Q}{\partial Y^{(j)}} Y^{(j+1)} + \frac{\partial Q}{\partial Y^{(i-1)}} X + \sum_{k=0}^{(m-i-2)} \frac{\partial Q}{\partial X^{(k)}} X^{(k+1)} \right\} + \Lambda \frac{\partial Q}{\partial X^{(m-i-1)}}, \tag{2.7}$$

where $\Lambda = \Lambda(Y, \ldots, Y^{(i-1)}, X, \ldots, X^{(m-i-1)})$ and similarly for Ω. We note that

$$[Q^*]_{\{Y = f_n, X = f_n^{(i)}\}} = \left[\Omega \frac{dQ}{dx} \right]_{\{Y = f_n, X = f_n^{(i)}\}},$$

where X and Y are taken as functions of x for the total differentiation.

The basic idea of our method is to eliminate $X, \ldots, X^{(m-i-1)}$ from the equations $P^{*j}|_{\{Y=f_n\}} = 0$, $j = 0, 1, \ldots, m-i$. We begin by noting that, because P does not contain $X', \ldots, X^{(m-i-1)}$, the polynomial P^{*j} is of degree one in $X^{(j)}$, for $j = 1, \ldots, m-i-1$.

LEMMA **2** *Let $P\langle Y\rangle(X)$ be an element of the ring*

$$\mathbf{K}_{Y,X} =_{def} \mathbb{K}\langle f_1, \ldots, f_{n-1}\rangle[Y, Y', \ldots, Y^{(i-1)}][X],$$

such that $P|_{Y=f_n}$ is square free. Let S be the polynomial obtained from P^{*m-i} by eliminating $X', \ldots, X^{(m-i-1)}$ using the equations $P^{*j}|_{Y=f_n} = 0$, $j = 1, \ldots, m-i-1$, and clearing denominators. Let g be the gcd with respect to X of $P|_{Y=f_n}$ and $S|_{Y=f_n}$. Then $X - h$ is a factor of g if and only if h satisfies the defining differential equation for $f_n^{(i)}$. That is to say, if and only if

$$h^{(m-i)} = \frac{\Lambda(f_n, \ldots, f_n^{(i-1)}, h, \ldots, h^{(m-i-1)})}{\Omega(f_n, \ldots, f_n^{(i-1)}, h, \ldots, h^{(m-i-1)})}.$$

Proof of Lemma 2
Let $P = (X - h)Q$ with $Q \in \mathbf{K}_{Y,X}$. Then for $1 \leq j \leq m-i-1$,

$$P^{*j} = (X^{(j)} - h^{(j)})Q + (X^{(j-1)} - h^{(j-1)})Q_{j,1} + \cdots + (X - h)Q_{j,j}, \quad (2.8)$$

where $Q_{j,1} \ldots, Q_{j,j} \in \mathbf{K}_{Y,X}$. For the case $j = m-i$, we get

$$P^{*m-i} = \{\Omega - h^{(m-i)}\Lambda\}Q + \{(X^{(m-i-1)} - h^{(m-i-1)})Q_{m-i,1} + \cdots + (X - h)Q_{m-i,m-i}\}\Omega, \quad (2.9)$$

where $\Omega = \Omega\langle Y, X\rangle$ and $\Lambda = \Lambda\langle Y, X\rangle$. If we set P^{*j} to zero for $j = 1, 2, \ldots, m-i-1$, we get the equations

$$X^{(j)} = h^{(j)} + (X^{(j-1)} - h^{(j-1)})\frac{Q_{j,1}}{Q} + \cdots + (X - h)\frac{Q_{j,j}}{Q}. \quad (2.10)$$

We use these in reverse order to eliminate $X^{(m-i-1)}, \ldots, X'$ from the right–hand side of (2.9) to obtain S. Now if T is any polynomial in $X, \ldots, X^{(j)}$, the resulting of making the above substitution for $X^{(j)}$ in T is of the form

$$T(X, \ldots, X^{(j-1)}, h^{(j)}) + \{(X^{(j-1)} - h^{(j-1)})T_1 + \cdots + (X - h)T_j\}Q^{-r},$$

for some natural number r. Hence from (2.9), S may be written

$$S = \{\Omega - h^{(m-i)}\Lambda\}Q^R + (X - h)W, \quad (2.11)$$

where $W \in \mathbf{K}_{Y,X}\langle h\rangle$ and $R \in \mathbb{N}$. We substitute the function f_n for the indeterminate Y and note that $X = h$ is not a zero of $Q|_{Y=f_n}$, because

$P|_{Y=f_n}$ is square free. Thus $X = h$ is a root of $S|_{Y=f_n} = 0$ if and only if $\Omega\langle f_n, h\rangle - h^{(m-i)}\Lambda\langle f_n, h\rangle = 0$, which proves Lemma 2.

As in the case $m = 1$, Lemma 2 gives us a method of deciding the zero equivalence of P by performing gcd computations over the ring $\mathbb{K}\langle f_1, \ldots, f_{n-1}\rangle [Y, Y', \ldots, Y^{(i-1)}]$. Again the algorithm needs to be used recursively to decide whether coefficients obtained are equivalent to zero. The case when $i = 0$ is identical to the general one except that Y and its derivatives do not occur.

If we obtain a gcd which depends non-trivially on X, we know that it must be of the form

$$g = A\prod_{j=1}^{M}(X - h_j). \tag{2.12}$$

As before, we have to discover whether one of the h_js satisfies the same initial conditions as $f_n^{(i)}$, but now there is more than one condition to check. The idea is to substitute for the derivatives of X one at a time, having differentiated sufficiently to ensure that g does not reduce to zero under the previous substitutions. What is not so easy, is to see why this works. We know that g is of the form (2.12). We assume for convenience of notation that the h_i are ordered in a particular way which will become apparent. We start by substituting $x = x_0$ but leaving X and its derivatives as indeterminates, and as in the first-order case, we keep differentiating until this no longer yields zero. Let p be the first integer such that $A^{(p)}(x_0) \neq 0$. Suppose that $h_1(x_0) = h_2(x_0) = \cdots = h_{r_0}(x_0) = y_i$ but $h_j(x_0) \neq y_i$ for $j > r_0$. Substituting $x = x_0$ into $g^{(p)}$ alone will not give zero since there will be just one summand in the derivative of the product, namely $A^{(p)}(x_0)(X - h_1(x_0)) \cdots (X - h_M(x_0))$ which fails to vanish. Now substituting $X = y_i$ makes r_0 of the factors vanish. The next $r_0 - 1$ differentiations successively reduce the number of factors vanishing but the situation remains essentially unchanged until we reach $g^{(p+r_0)}$. Then the substitutions $x = x_0$, $X = y_i$ leave precisely one summand non-zero, namely

$$A^{(p)}(x_0)(X' - h_1'(x_0)) \cdots (X' - h_{r_0}'(x_0))(X - h_{r_0+1}(x_0)) \cdots (X - h_M(x_0)).$$

The point is that all other summands either contain a derivative of A less than the p-th or some $X - h_i(x_0)$ with $i \leq r_0$, and so must vanish.

Now suppose that $h_j'(x_0) = y_{i+1}$ for $1 \leq j \leq r_1$, with $r_1 \leq r_0$, but $h_j'(x_0) \neq y_{i+1}$ for $r_1 < j \leq r_0$. Here our notation assumes that the h_is are ordered so that this is possible. Then substituting $x = x_0$, $X = y_i$ and $X' = y_{i+1}$ into $g^{(p+r_0+k)}$ will yield zero for $k = 0, \ldots, r_1 - 1$, but $g^{(p+r_0+r_1)}$ will contain just one summand,

$$\begin{aligned} &A^{(p)}(x_0)(X'' - h_1''(x_0)) \cdots (X'' - h_{r_1}''(x_0))(X' - h_{r_1+1}') \\ &\quad \cdots (X' - h_{r_0}'(x_0))(X - h_{r_0+1}(x_0)) \cdots (X - h_M(x_0)), \end{aligned}$$

which does not vanish under these substitutions. Once again the point is that any other summand contains a factor which is differentiated more than in this summand, and hence there must be a factor which is differentiated less; this factor will evaluate to zero under the substitutions. So we can see that the method works generally.

As remarked, all the h_i satisfy the same differential equation as f_n. If one of them also satisfies the same initial conditions as f_n, we may apply a uniqueness theorem to deduce that this h_i is equal to f_n. However we need some care because there is no a priori reason to believe that the h_i are analytic at the evaluation point x_0. However as elements of the algebraic closure of $\mathbb{K}\langle f_1, \ldots, f_n\rangle$ they will certainly be analytic in a region whose boundary includes x_0. Fortunately a theorem given in Chapter V of [8] then guarantees that the h_i in question will be analytic at x_0 and so we must indeed have $h_i \equiv f_n$ for such a value of i. See [100] for further details.

ALGORITHM **3**
Let $P \in \mathbf{K}_{Y,X}$. To determine whether $P\langle f_n\rangle(f_n^{(i)}) \equiv 0$:

(1). Render $P\langle f_n\rangle(X)$ square-free in X by dividing by

$$\gcd_X\{P\langle f_n\rangle(X), \partial P\langle f_n\rangle(X)/\partial X\}.$$

(2). Compute P^{*^j} for $j = 1, \ldots, m-i$ and substitute the expressions for $X^{(m-i-1)}, \ldots, X'$ obtained respectively from $P^{*^{m-i-1}} = 0, \ldots, P^* = 0$ into $P^{*^{m-i}}$. Then clear denominators to form S. Next compute $g = \gcd_X\{P, S\}$. If g is independent of X then $P\langle f_n\rangle(f_n^{(i)}) \not\equiv 0$.

(3). Otherwise substitute $x = x_0$ into g leaving X and any of its derivatives as indeterminates. If the result is zero, keep replacing g by g^* until these substitutions do not yield zero. Now make the additional substitution $X = y_i$. If this fails to make the expression vanish then $P\langle f_n\rangle(f_n^{(i)}) \not\equiv 0$.

(4). Otherwise keep replacing g by its derivative until all the above substitutions no longer make g vanish, and then make the additional substitution $X' = y_{i+1}$. Once again, if this fails to make the expression vanish then $P\langle f_n\rangle(f_n^{(i)}) \not\equiv 0$.

(5). Continue in this way. The j-th stage begins by checking that the extra substitution $X^{(j-1)} = y_{i+j-1}$ makes the current expression for g zero. Then differentiate until this is no longer the case, and check that the additional substitution $X^{(j)} = y_{i+j}$ makes g zero once again. If not then $P\langle f_n\rangle(f_n^{(i)}) \not\equiv 0$.

(6). If the stage when $j = m - i - 1$ is reached and the last substitution, $X^{(m-1)} = y_{m-1}$ does make the current g zero, then indeed $P\langle f_n\rangle(f_n^{(i)}) \equiv 0$. Otherwise $P\langle f_n\rangle(f_n^{(i)}) \not\equiv 0$.

One can see from the algorithm that in some sense it is quite difficult for an expression to be zero–equivalent. This accords with practical experience, where most expressions are found to correspond to non–zero functions, and very frequently this is 'obvious'. Thus in a practical implementation it would be well worth making some simple tests which might prove that the expression is not zero-equivalent, such as evaluation of the function and perhaps its first few derivatives at some selected points.

The gcd computations in steps 1 and 2 are carried out over the ring $\mathbb{K}\langle f_1, \ldots, f_{n-1}\rangle[Y, Y', \ldots, Y^{(i-1)}]$. Hence, as already remarked, the algorithm has to be called recursively to decide whether certain coefficients occurring in the gcd calculations are equivalent to zero.

2.2.1 Examples

Our first example indicates how the algorithm handles a moderately complicated expression involving trigonometric functions. Let

$$\begin{aligned} f(x) = {} & \sin^4 2x + \sin^3 2x \sin^2 x - \sin^3 2x \cos^2 x - 2\sin^3 2x \sin x \cos x \\ & +2\sin^3 2x \sin x - 2\sin^2 2x \sin^3 x \cos x + 2\sin^2 2x \sin x \cos^3 x \\ & +2\sin^2 2x \sin^3 x - 2\sin^2 2x \sin x \cos^2 x - 4\sin^2 2x \sin^2 x \cos x \\ & + \sin^2 2x \sin^2 x - 4\sin 2x \sin^4 x \cos x + 4\sin 2x \sin^2 x \cos^3 x \\ & + \sin 2x \sin^4 x - \sin 2x \sin^2 x \cos^2 x - 2\sin 2x \sin^3 x \cos x \\ & -2\sin^5 x \cos x + 2\sin^3 x \cos^3 x. \end{aligned}$$

In fact Maple's *simplify* handles this quite well using knowledge of trigonometric identities. However the point about Algorithm 3 is that it uses only the differential equations defining sine and cosine, and so the method works just as well with other transcendental functions given by differential equations.

We parameterise the expression for f by setting $\sin 2x = X_2$ and $\sin x = X_1$, so that $X_1' = \cos x$ and $X_2' = 2\cos 2x$. Then X_1 is defined by the differential equation $y'' = -y$ with initial conditions $y(0) = 0$, $y'(0) = 1$, and X_2 is similarly given by $y'' = -4y$ with conditions $y(0) = 0$ and $y'(0) = 2$. Our expression f is then represented by

$$\begin{aligned} P = {} & X_2^4 + X_2^3 X_1^2 - X_2^3 (X_1')^2 - 2X_2^3 X_1 X_1' + 2X_2^3 X_1 - 2X_2^2 X_1^3 X_1' \\ & +2X_2^2 X_1 (X_1')^3 + 2X_2^2 X_1^3 - 2X_2^2 X_1 (X_1')^2 - 4X_2^2 X_1^2 X_1' + X_1^2 X_2^2 \\ & -4X_2 X_1^4 X_1' + 4X_2 X_1^2 (X_1')^3 + X_2 X_1^4 - X_2 X_1^2 (X_1')^2 - 2X_2 X_1^3 X_1' \\ & -2X_1^5 X_1' + 2X_1^3 (X_1')^3. \end{aligned}$$

We treat X_2 as the 'top variable'; that is to say, we are using a tower $\mathbb{Q} \subset \mathbb{Q}(X_1, X_1') \subset \mathbb{Q}(X_1, X_1')(X_2, X_2')$. The gcd of P and $\partial P/\partial X_2$ turns out to be $X_2 + X_1$, and dividing through by this, to render P square free, gives

$$\begin{aligned} P = {} & -2X_1^4 X_1' - 2X_2 X_1^3 X_1' + X_2 X_1^3 + X_2^2 X_1^2 + 2X_1^2 (X_1')^3 \\ & -2X_2 X_1^2 X_1' + 2X_2 X_1 (X_1')^3 + X_2^2 X_1 - 2X_2^2 X_1 X_1' \\ & -X_2 X_1 (X_1')^2 - X_2^2 (X_1')^2 + X_2^3. \end{aligned}$$

A mildly complicated calculation for $g = \gcd\{P, S\}$ results in

$$g = X_2^2 + X_2X_1^2 - X_2(X_1')^2 - 2X_2X_1X_1' - 2X_1^3X_1' + 2X_1(X_1')^3.$$

We commence substitutions by putting $X_1 = 0$ and $X_1' = 1$, these being the values at $x = 0$. This gives $X_2^2 - X_2$, which is clearly not the zero polynomial. However putting $X_2 = 0$ does give zero, and so we have passed step 3 of Algorithm 3. Next we differentiate, to obtain

$$g' = 2X_2'X_2 + X_2'X_1^2 - X_2'(X_1')^2 - 2X_2'X_1X_1' + 4X_2X_1X_1' - 2X_2(X_1')^2 \\ - 12X_1^2(X_1')^2 + 2(X_1')^4 + 2X_2X_1^2 + 2X_1^4.$$

In this expression, we substitute $X_1 = 0$, $X_1' = 1$ and $X_2 = 0$, giving $2 - X_2'$. Then putting $X_2' = 2$ makes this polynomial zero, and so we are entitled to conclude that the expression $f(x)$ represents the zero function.

It is worth having a closer look at what is going on in this example. Although the algorithm never uses the fact, $f(x)$ factorizes as

$$f(x) = (\sin 2x + \sin x)^2(\sin 2x - 2\sin x\cos x)(\sin 2x + \sin^2 x - \cos^2 x),$$

and so

$$P = (X_2 + X_1)^2 * (X_2 - 2 * X_1 * X_1') * (X_2 + X_1^2 - (X_1')^2).$$

Dividing by the gcd of P and $\partial P/\partial X_2$ removes the square factor. As regards the three remaining factors, $-X_1$ does not satisfy the differential equation defining X_2 but $2 * X_1 * X_1'$ and $(X_1')^2 - X_1^2$ do. Hence

$$g = (X_2 - 2 * X_1 * X_1') * (X_2 + X_1^2 - (X_1')^2).$$

Now both $2 * X_1 * X_1'$ and $(X_1')^2 - X_1^2$ satisfy the first initial condition for X_2, namely vanishing at $x = 0$. Only $2 * X_1 * X_1'$ satisfies the second initial condition, that the derivative shall evaluate to 2 at $x = 0$, but of course that is enough to give f a zero-equivalent factor.

Our second example is taken from [100] and shows the working in more detail on a simpler expression. Let

$$f(x) = e^{2ix} - 2ie^{ix}\sin x - 1.$$

We take $X_1 = \sin x$, so that X_1 satisfies $X_1'' = -X_1$ with initial conditions $X_1(0) = 0$, $X_1'(0) = 1$. Similarly, we define $X_2(x) = \exp(ix)$ by $X_2' = iX_2$ with initial condition $X_2(0) = 1$. Our polynomial is then

$$P = X_2^2 - 2iX_2X_1 - 1. \tag{2.13}$$

We begin by checking that $P|_{X_1=\sin x}$ is a square-free polynomial in X_2. We have

$$\frac{\partial P}{\partial X_2} = 2X_2 - 2iX_1$$

and hence

$$P - \left(\frac{X_2}{2} - \frac{iX_1}{2}\right)\frac{\partial P}{\partial X_2} = X_1^2 - 1.$$

But $(X_1^2 - 1) \not\equiv 0$, since substituting $x = 0,\ X_1 = 0$ gives -1. Thus $P|_{X_1=\sin x}$ is indeed square free. In this case, we have

$$S = P^* = 2iX_2^2 - 2iX_2X_1' + 2X_2X_1. \tag{2.14}$$

Our next task is to compute $g = \gcd_{X_2}\{P, S)$. If we let $Q = iP - P^*/2$, then

$$Q = X_2(iX_1' + X_1) - i.$$

We observe that $iX_1' + X_1 \not\equiv 0$, since substituting $x = 0,\ X_1 = 0,\ X_1' = 1$ gives a non-zero result. Next

$$-(iX_1' + X_1)P + X_2Q = X_2\{2iX_1(iX_1' + X_1) - i\} + (iX_1' + X_1).$$

On multiplying through by $iX_1' + X_1$, we obtain

$$\begin{aligned} -(iX_1' + X_1)^2P + X_2(iX_1' + X_1)Q = \\ X_2(iX_1' + X_1)(-2X_1'X_1 + 2(X_1')^2 - i) + (iX_1' + X_1)^2. \end{aligned} \tag{2.15}$$

Also

$$\begin{aligned} (-2X_1'X_1 + 2iy_1^2 - i)Q = X_2(iX_1' + X_1)(-2X_1'X_1 + 2(X_1')^2 - i) \\ -i(-2X_1'X_1 + 2iX_1^2 - i). \end{aligned} \tag{2.16}$$

On subtracting (2.16) from (2.15), we obtain on the right-hand side

$$(iX_1' + X_1)^2 + i(-2X_1'X_1 + 2iX_1^2 - i) = -(X_1')^2 - X_1^2 + 1.$$

Let $W = -(X_1')^2 - X_1^2 + 1$. To determine whether $W \equiv 0$, we call the algorithm recursively. We have $W^* = 2X_1'X_1 - 2X_1X_1' = 0$ and so $\gcd_(W, W^*) = W$. This establishes that W contains a factor of the form $X_1' - h$, $h \in (\mathbf{Q}[x][X_1])_{alg}$, where h satisfies the differential equation for X_1', i.e. $(X_1')' = -X_1$. In fact, as we can see, there are two choices for h, namely $\pm\sqrt{1 - X_1^2}$. However the algorithm proceeds automatically without using this knowledge. We make the substitutions $x = 0,\ X_1 = 0$, but keep X_1' as an indeterminate; under these, $W \mapsto -(X_1')^2 + 1$. Now putting $X_1' = 1$ reduces W to zero, and hence $W \equiv 0$.

However W was the remainder obtained on reducing P modulo Q, and hence $Q = \gcd(P|_{X_1=\sin x}, S|_{X_1=\sin x})$. Then $i(iX_1' + X_1)^{-1}$ must satisfy the differential equation $X_2' = iX_2$. (The reader may verify that $i(iX_1' + X_1)^{-1} = \exp(ix)$, though of course the algorithm does not use this fact.) Finally, the substitutions need to be applied to Q. First, on putting $x = 0,\ X_1 = 0,\ X_1' = 1$, we see that $Q \to iX_2 - i$. Putting $X_2 = 1$ then does give zero, and we are now entitled to conclude that $P \equiv 0$.

We close this section by taking a closer look at some of the algebra underlying Algorithm 3. The results will be of relevance in Section 2.4.

2.2.2 The sets $\mathcal{S}_i$

We write ζ_j for the homomorphism from $\mathbb{K}[X_1, \ldots, X_j]$ into $\mathbb{K}$ generated by evaluation at the initial-condition point, x_0.

Next we define the subset $\mathcal{S}_1$ of $\mathbb{K}[X_1]$ to be $\{0\}$, and for $2 \le j \le n$ we suppose that $\mathcal{S}_{j-1}$ has already been defined. We write $\mathcal{I}_{j-1}$ and $\mathcal{J}_{j-1}$ for the ideals generated by $\mathcal{S}_{j-1}$ in $\mathbb{K}[X_1, \ldots, X_{j-1}]$ and $\mathbb{K}[X_1, \ldots, X_j]$ respectively. Then we define

$$\mathcal{S}_j = \{P \in \mathbb{K}[X_1, \ldots, X_j] \; ; \; \zeta_j(P) = 0 \;\; \& \;\; \exists S \in \mathcal{J}_{j-1} : \; P^* = PQ + S\},$$

where $Q \in \mathbb{K}[X_1, \ldots, X_j]$. Thus $\mathcal{S}_j$ may be thought of as the set of polynomials in $\mathbb{K}[X_1, \ldots, X_j]$ which divide their own star derivative modulo $\mathcal{J}_{j-1}$ and vanish under the application of ζ_j. Functions which divide their own derivatives seem to play an important role in a number of areas of differential algebra. See for example [20, 69, 106, 100] and also [122].

LEMMA **3** *$\mathcal{S}_j$ is a semigroup under multiplication.*

Proof of Lemma 3
Let P_1 and P_2 be two elements of $\mathcal{S}_j$. Then there exist, $Q_1, Q_2 \in \mathbb{K}[X_1, \ldots, X_j]$ and $S_1, S_2 \in \mathcal{J}_{j-1}$ such that $P_i{}^* = P_i Q_i + S_i$, $i = 1, 2$. We have

$$(P_1 P_2)^* = P_1 P_2{}^* + P_1{}^* P_2 = P_1 P_2 (Q_1 + Q_2) + S_1 P_2 + S_2 P_1.$$

Also $\zeta_j(P_1 P_2) = \zeta_j(P_1)\zeta_j(P_2) = 0$. Hence $P_1 P_2 \in \mathcal{S}_j$, which establishes the lemma.

Our present interest in the sets $\mathcal{S}_j$ stems from the following result; c.f. Lemma 2.

THEOREM **1** *Let $P \in \mathbb{K}[X_1, \ldots, X_n]$. Then $P \equiv 0$ if and only if $P \in \mathcal{I}_n$.*

Proof of Theorem 1
For $j = 1, \ldots, n$ we use the notation η_j for the differential homomorphism from $\mathbb{K}[X_1, \ldots, X_j]$ to $\mathcal{F}_j$ which sends a polynomial expression to the function it represents. We also write

$$\tilde{\eta}_{j-1} \; : \; \mathbb{K}[X_1, \ldots, X_{j-1}][X_j] \longrightarrow \mathcal{F}_{j-1}[X_j]$$

for the map obtained by applying η_{j-1} to the coefficients. The theorem is proved by induction.

First let $P \in \mathbb{K}[X_1]$. Then since f_1 is assumed to be transcendental over $\mathbb{K}$, $P \equiv 0$ if and only if P is the zero polynomial; i.e. if and only if $P \in \mathcal{I}_1$.

Now suppose that $n > 1$ and that the result is true for $n-1$. Then $\mathcal{I}_{n-1}$ is a prime differential ideal since it is the kernel of the differential homomorphism η_{n-1}. Similarly $\mathcal{J}_{n-1}$ is a prime differential ideal since it is the kernel of $\tilde{\eta}_{n-1}$. So $\mathbb{K}[X_1, \ldots, X_n]/\mathcal{J}_{n-1}$ is a differential domain. We overload the symbol * to denote also the derivation in $\mathbb{K}[X_1, \ldots, X_n]/\mathcal{J}_{n-1}$, and write ω_n for the natural differential homomorphism from $\mathbb{K}[X_1, \ldots, X_n]$ to $\mathbb{K}[X_1, \ldots, X_n]/\mathcal{J}_{n-1}$.

Now let $P \in \mathcal{S}_n$ and write $\tilde{P} = \omega_n(P)$. Then $\tilde{P}$ divides $\tilde{P}^*$, so there exists $\tilde{Q} \in \mathbb{K}[X_1, \ldots, X_n]/\mathcal{J}_{n-1}$ such that $\tilde{P}^* = \tilde{Q}\tilde{P}$. Hence

$$\frac{d\eta_n(P)}{dx} = \eta_n(P)\eta_n(Q),$$

and it follows that $\eta_n(P) = K \exp(\int \eta_n(Q))$ for some constant K. However $\eta_n(Q)$ is non-singular at the initial-condition point x_0 and $\zeta_n(P) = 0$. This implies that $K = 0$, and so $P \equiv 0$. Thus elements of $\mathcal{I}_n$ are functionally equivalent to zero.

Now suppose that $P \equiv 0$. The map $\tilde{\eta}_{n-1}$ induces an isomorphism

$$\rho : \frac{\mathbb{K}[X_1, \ldots, X_n]}{\mathcal{J}_{n-1}} \longrightarrow \mathcal{F}_{n-1}[X_n]. \tag{2.17}$$

This means that if $\tilde{\eta}_{n-1}(P) = 0$, then $P \in \mathcal{J}_{n-1} \subseteq \mathcal{I}_n$.

Suppose then that $\tilde{\eta}_{n-1}(P) \neq 0$ and write $P_\eta = \tilde{\eta}_{n-1}(P)$. Since we are going to show P_η is a multiple of an element $\tilde{\eta}_{n-1}(S)$ with $S \in \mathcal{S}_n$, we may remove any multiple factors from P_η and suppose that P_η is square free. Since $P_\eta(f_n) = 0$, the factorization of P_η over the algebraic closure of $\mathcal{F}_{n-1}$ must contain the factor $X_n - f_n$. Since ${X_n}^* - f_n'$ must vanish when we replace X_n by f_n, it follows that $X_n - f_n$ is also a factor of $g = \gcd_{X_n}\{P_\eta, {P_\eta}^*\}$.

Let m be the minimal polynomial of $X_n - f_n$ over $\mathcal{F}_{n-1}$, and let M be an element of $\mathbb{K}[X_1, \ldots, X_n]$ such that $\tilde{\eta}_{n-1}(M) = m$. Now m must divide g and hence it must divide P_η. Let $P_\eta = mh$ and let $H \in \mathbb{K}[X_n]$ be such that $\tilde{\eta}_{n-1}(H) = h$. Then $\tilde{\eta}_{n-1}(P - MH) = P_\eta - mh = 0$. Hence

$$P - MH \in \ker(\tilde{\eta}_{n-1}) = \mathcal{J}_{n-1}. \tag{2.18}$$

We show that $M \in \mathcal{S}_n$. Firstly m must divide m^* since m divides ${P_\eta}^* = mh^* + m^*h$ but does not divide h (as P_η is square free). Let $m^* = ml$ and let $L \in \eta_{n-1}^{-1}(\{l\})$. Then since η_{n-1} is a differential homomorphism,

$$\tilde{\eta}_{n-1}(M^*) = m^* = ml = \tilde{\eta}_{n-1}(ML).$$

Hence $M^* - ML \in \ker(\tilde{\eta}_{n-1})$ as required. Now (2.18) shows that $P \in \mathcal{I}_n$. This completes the proof of Theorem 1.

COROLLARY *The ideal $\mathcal{I}_n$ is a prime differential ideal.*

Proof of Corollary 1
This is immediate since $\mathcal{I}_n$ is the kernel of the differential homomorphism η_n.

COROLLARY *$\mathcal{I}_n/\mathcal{J}_{n-1}$ is a prime principal ideal of $\mathbb{K}[X_1, \ldots, X_n]/\mathcal{J}_{n-1}$.*

Proof of Corollary 2
The proof of the theorem shows that $\mathcal{I}_n/\mathcal{J}_{n-1}$ is generated by the image of M, which is irreducible since m is.

2.3 Modular Methods in Zero Equivalence

It might seem that after Algorithm 3 our problems are solved. However at the implementation stage, the algorithms run into the classic problems of gcd calculation, [47], namely massive growth of coefficients and of degrees. Of course the coefficients and degrees of the gcd cannot be so very large, but the problem arises from the necessity to multiply up by a leading coefficient when dividing one polynomial by another. The subresultant algorithm is designed to mitigate this problem. In implementation, it was found to help a little with the coefficient growth, but the problem remained, [91]. One answer to these problems in multivariate gcd calculation is to use homomorphic methods, but because our algorithm recurses during the gcd calculation it is not immediately clear that these are applicable to it.

The classical method is thus to perform gcd calculations in $\mathbb{Z}_q$ rather than $\mathbb{Z}$, where q is prime. Except for a few bad choices for q, it turns out that reducing modulo q commutes with the gcd operations; see for example [36]. Thus one can obtain the gcd mod q in this way. From this information, perhaps using different primes, one can recover the gcd itself if desired. As regards the interaction between mod-q calculations and differentiation, we have the following result.

Lemma 4 *Let $\tilde{\phi}_q$ be the homomorphism from $\mathbb{Z}[X_1, \ldots, X_n]$ to $\mathbb{Z}_q[X_1, \ldots, X_n]$ induced by the mod-q homomorphism, ϕ_q, from $\mathbb{Z}$ to $\mathbb{Z}_q$, and let $P \in \mathbb{Z}[X_1, \ldots, X_n]$. Then $(\tilde{\phi}_q(P))^{*}{}_q = \tilde{\phi}_q(P^*)$, where ${}^{*}{}_q$ denotes differentiation mod q; that is to say ${}^{*}{}_q = \tilde{\phi}_q \circ {}^{*} \circ i_q$, with i_q the injection of $\mathbb{Z}_q[X_1, \ldots, X_n]$ into $\mathbb{Z}[X_1, \ldots, X_n]$.*

Proof
This amounts to little more than the definition of ${}^{*}{}_q$. We have

$$P = i_q \circ \tilde{\phi}_q(P) + qP_1,$$

for some polynomial P_1. Hence

$$P^* = {}^* \circ i_q \circ \tilde{\phi}_q(P) + qP_1{}^*,$$

and so

$$\tilde{\phi}_q(P^*) = \tilde{\phi}_q \circ {}^* \circ i_q(\tilde{\phi}_q(P)) = (\tilde{\phi}_q(P))^{*}{}_q,$$

which completes the proof of Lemma 4.

Thus reduction of coefficients modulo q is a differential homomorphism, and so the use of these modular methods works well in our setting. Moreover we can take advantage of the fact that we do not have to know precisely what the gcd is, but only whether it depends on the variable X_n and whether it vanishes under initial-condition substitution, perhaps after some differentiation. Hence we can avoid the use of the Chinese Remainder Theorem.

2.3.1 Hensel Lifting

As an alternative to using a number of different primes, we can use just one and perform Hensel lifting.

The classical method of Hensel lifting may be summarized as follows, [36]. Let q be a prime and r a positive integer. A solution of an equation $F(v) = 0$ is sought, where v (unknown) and F (given) are polynomials over some ring $\mathbb{K}[X_1, \ldots, X_j]$. The assumption is that we have a solution, u_r modulo q^r. The method then generates a solution modulo q^{r+1}.

We write $u_{r+1} = u_r + q^r u$ and solve $F(u_r + q^r u) = 0 \bmod q^{r+1}$ for u. From the power series, we see that we want

$$F(u_{r+1}) = F(u_r) + q^r uF'(u_r) + O(q^{r+1}) = 0 \bmod q^{r+1}. \tag{2.19}$$

Hence

$$u = \frac{F(u_r)/q^r}{F'(u_r)} \bmod q.$$

Note that $F(u_r)$ is divisible by q^r since u_r is our mod–q^r solution.

In our present situation, we suppose that we have calculated an element, G_r of $\mathcal{S}_j$ modulo q^r which divides our polynomial P modulo $\mathcal{J}_{j-1}$ (and modulo q^r). We wish to discover whether an element, G, of $\mathbb{K}[X_1, \ldots, X_{j-1}]$ exists which satisfies the following properties:

I. G divides P mod $\mathcal{J}_{j-1}$;
II. G divides G^* mod $\mathcal{J}_{j-1}$;
III. $\zeta_j(G) = 0$.

We have an element, G_r, which satisfies these properties modulo q^r. Apart from the slight complication of working modulo $\mathcal{J}_{j-1}$, we can lift the first property in the normal way for gcd problems, [36]. Thus we have

$$P = G_r Q_r \mod \mathcal{J}_{j-1} \; (\text{mod } q^r),$$

which we lift to a similar relation with r replaced by $r + 1$, by defining u as above. Then we can check whether the resulting G_{r+1} satisfies II and III.

Alternatively, we might consider lifting property II in order to discover elements of the various $\mathcal{S}_j$. This would be especially useful if extensive calculations in the ring $\mathcal{F}_n$ are anticipated. For example if we had complete information about the $\mathcal{S}_j$s we could define canonical forms; see paragraph 2.4.1 below.

We suppose for the moment that q is large enough so that no power of exponent as large as q is encountered in our calculations. Our polynomial G_r satisfies a relation of the form

$${G_r}^* = G_r A_r + q^r B_r \mod \mathcal{J}_{j-1}, \tag{2.20}$$

where A_r and B_r belong to $\mathbb{Z}[X_1, \ldots, X_j]$. Let $G_{r+1} = G_r + q^r U$. We want to determine U such that ${G_{r+1}}^*$ divides G_{r+1} modulo $\mathcal{J}_{j-1}$ and q^{r+1}. So we want D_r in $\mathbb{Z}[X_1, \ldots, X_j]$ such that

$$G_{r+1}{}^* = G_{r+1}A_r + q^{r+1}D_r \mod \mathcal{J}_{j-1}.$$

Substitution for G_{r+1} leads to

$$U^* = UA_r - B_r \mod \mathcal{J}_{j-1} \pmod q.$$

This can be rewritten as

$$\bar{\Omega}_j \frac{\partial U}{\partial x} + \bar{\Lambda}_j \frac{\partial U}{\partial X_j} = UA_r + \frac{G_r A_r - G_r{}^*}{q^r} \mod \mathcal{J}_{j-1} \pmod q. \tag{2.21}$$

By using the isomorphism ρ of (2.17), we may interpret the coefficients in (2.21) as elements of $\mathcal{F}_{j-1}$. The standard approach to solving an equation such as (2.21) is to use the method of characteristics, [126]. One takes a new independent variable, s, and looks for solutions of the system

$$\frac{\partial x}{\partial s} = \Omega_j, \quad \frac{\partial X_j}{\partial s} = \Lambda_j, \quad \frac{dU}{ds} = UA_r + \frac{G_r A_r - G_r{}^*}{q^r} \pmod q. \tag{2.22}$$

Both sides of (2.21) will then be of the form dU/ds mod q.

In the present situation, we may suppose that q has been chosen so that modulo q, Λ_j and Ω_j have no common factor and G_r is square free. Then the fact that

$$\frac{\partial G_r}{\partial x}\Omega_j + \frac{\partial G_r}{\partial X_j}\Lambda_j = G_r A_r \mod q^r,$$

shows that G_r and Ω_j cannot have a common factor mod q either. It follows, using the Euclidean algorithm, that Ω_j is invertible as an element of $\mathcal{F}_{j-1}[X_j]/(G_r)$. Hence we may divide (2.21) through by Ω_j and consider the system corresponding to (2.22),

$$\frac{\partial x}{\partial s} = 1, \quad \frac{\partial X_j}{\partial s} = \frac{\Lambda_j}{\Omega_j}, \quad \frac{dU}{ds} = \frac{UA_r + (G_r A_r - G_r{}^*)/q^r}{\Omega_j} \pmod q. \tag{2.23}$$

The first equation has solution $x = s + k$ for some constant k, and the second then merely restates the definition of X_j. We already know a solution of this mod q; it is given by $G_1(x, X_j) = 0$. The third equation of (2.23) is a Risch differential equation over an algebraic extension of $\mathcal{F}_{j-1}$, and we would like a solution in this extension, if one exists. Much work has been done on this problem, and in many cases an answer can be guaranteed, [22, 4, 7]. However it is not currently soluble at our full level of generality, and even when it can be solved the computation is generally far from simple. Since our whole purpose here is to improve the efficiency of Algorithm 3, there is little point in using a time-consuming and difficult subalgorithm. In special cases we might find a polynomial solution of (2.23) quite easily. If we knew a good bound on the degree of such a polynomial, we could merely substitute and compare coefficients. However, Example 2 below shows that this will not always be possible.

If the degrees of the polynomials can be as large as q^r, there is a further complication in that any element of $\mathbb{Z}[X_1^{q^r},\ldots,X_j^{q^r}]$ has zero derivative mod q^r. The phenomenon will disappear once r is large enough, but it might lead to time being wasted with spurious solutions for low values of r.

Whether we use a number of different primes or Hensel lifting, we need to be able to bound the coefficients of the gcd so as to be sure that our answer modulo the appropriate $\bar{q}$ is equal to the real gcd. Example 2 below gives the depressing news that we cannot expect any sort of reasonable bound in general.

What can be done is to calculate modulo a number of suitably large primes chosen at random. If we take the probability of a coefficient being divisible by a prime p to be p^{-1}, and we use primes $p_1,\ldots,p_n$ then the probability of a non–zero coefficient being reduced to zero modulo all the primes is clearly $p_1^{-1}p_2^{-1}\cdots p_n^{-1}$. Thus we can make the probability of obtaining a false zero very small.

2.3.2 Examples

Our first example illustrates the use of Hensel lifting.

Suppose that X_1 and X_2 satisfy $X_1' = -2X_1$ and $X_2' = 2X_2 + X_1$. We let $G = 4X_2 + X_1$ and observe that $G^* = 2G$. If $q = 3$, then $G_1 = X_2 + X_1$ and $A_1 = 2$. We assume that this is the polynomial which we have computed, and that we wish to obtain G. We have $\Lambda_1 = -2X_1$, $\Lambda_2 = 2X_2 + X_1$ and $\Omega_1 = \Omega_2 = 1$. With $j = 2$, $\partial/\partial x$ may be taken as $-2X_1\partial/\partial X_1$ and (2.21) becomes

$$-2X_1\frac{\partial U}{\partial X_1} + (2X_2+X_1)\frac{\partial U}{\partial X_2} = 2U + \frac{2G_1 - G_1{}^*}{3} = 2U + X_1.$$

Using the fact that $-2X_1 = X_1 (\text{mod } 3)$ we see that here the system (2.23) is

$$\frac{\partial X_1}{\partial s} = X_1, \quad \frac{\partial X_2}{\partial s} = 2X_2 + X_1, \quad \frac{dU}{ds} = 2U + X_1 \quad (\text{mod } 3).$$

Since X_2 and U satisfy the same differential equation, we may take $U = X_2$. This gives $G_2 = G_1 + 3U = 4X_2 + X_1$ as required.

It is worth noting that a slightly different way of doing the calculation leads to $G_2 = X_2 - 2X_1$. This is a correct answer modulo 9, since

$$(X_2 - 2X_1)^* = 2X_2 + 5X_1 = 2(X_2 - 2X_1) \quad \text{mod } 9.$$

Example 2
Consider the function $y(x)$ defined as the solution of

$$\frac{dy}{dx} = \frac{(24x^2 + 62x + 14)y}{(35y^2 + 12y + 1)x} \tag{2.24}$$

satisfying the initial condition $y(0) = 0$. Note that any solution of (2.24) divides its star derivative when (2.24) itself is used as the defining differential equation.

Separation of the variables and the use of partial fractions shows that $y(x)$ satisfies the algebraic equation

$$7^{196}(5y+1)^{420}(4x+1)^{21} - (7y+1)^{300}(3x+7)^{196} = 0. \qquad (2.25)$$

This contains coefficients with nearly 500 digits. The left-hand side is also irreducible, as shown by substituting $x = -1$ and using the MAPLE *irreduc* command.

What the example shows is that we are not going to get good bounds for either coefficients or degrees of elements of the $\mathcal{S}_j$s. For if we take $\Lambda_2(x,y) = (24x^2+62x+14)y$ and $\Omega_2(x,y) = (35y^2+12y+1)x$ then the right-hand side of (2.24) is a generator of $\mathcal{S}_2$. There is nothing particularly unique about this example, and still larger ones could undoubtedly be constructed by considering expressions of the form

$$\sum_{i=0}^{M} \frac{a_i}{p_i x + b_i} + \sum_{i=M+1}^{N} \frac{a_i}{p_i y + b_i}$$

with the p_is pairwise coprime. The difficulty might be to determine if the resulting polynomial was irreducible, but presumably it would be in many cases.

2.4 Growth in Degrees

The modular methods described above address the problem of large coefficients. However that still leaves the problem of the growth of degrees. Here the usual strategy with gcd computation is to use point evaluations and rely on the principle that a polynomial of a given total degree is determined once its value at sufficiently many points is known. So one replaces all the variables except a single 'main' one by a set of integer values. Then, still working mod q, one computes the gcd. Since this is now a single-variable computation, there is no question of the degrees increasing, and the fact that one is working over $\mathbb{Z}_q$ takes care of coefficient growth. Once again the actual gcd can be reconstructed from the information obtained with different p and different sets of point evaluations. See [36], or a similar text, for details.

The problem for us is that reduction via point evaluations is *not* a differential homomorphism. Hence we need to use global data (i.e. from all the various point evaluations) when we differentiate. This reduces the scope for parallelization, and makes it difficult to see how something like Hensel lifting could be used here. Reconstructing the polynomial is likely to be one of the more expensive operations, and we can generally avoid this. As has been pointed out elsewhere, the results obtained from the calculations with the various moduli and point evaluations constitute a representation of the gcd, albeit an unfamiliar one. Thus, for the present problem, one of our main tasks is to show how the basic algorithm of section 2.2 can be recast in terms of this representation. The method for differentiating is as follows:

Consider a polynomial $Q(y)$ of degree $K-1$ over an integral domain D, and suppose that Q takes the value v_i at the point $\mathbf{c}_i$ for $i = 1, \ldots, K$, where the points $\mathbf{c}_1, \ldots, \mathbf{c}_K$ are all distinct. Then

$$Q(y) = \sum_{i=1}^{K} \left(\prod_{j \neq i} \frac{y - \mathbf{c}_j}{\mathbf{c}_i - \mathbf{c}_j} v_i \right).$$

Hence

$$Q'(y) = \sum_{i=1}^{K} \left\{ \sum_{l \neq i} \left(\prod_{j \neq i,l} \frac{(y - \mathbf{c}_j)v_i}{(\mathbf{c}_i - \mathbf{c}_j)(\mathbf{c}_i - \mathbf{c}_l)} \right) \right\}.$$

A small amount of manipulation, reveals that for $m = 1, \ldots, K$,

$$Q'(\mathbf{c}_m) = v_m \sum_{i \neq m} \frac{1}{\mathbf{c}_m - \mathbf{c}_i} + \sum_{i \neq m} \frac{v_i}{\mathbf{c}_i - \mathbf{c}_m} \left\{ \prod_{j \neq i,m} \frac{\mathbf{c}_m - \mathbf{c}_j}{\mathbf{c}_i - \mathbf{c}_j} \right\}. \tag{2.26}$$

Note that although (2.26) contains quotients, the denominators must divide exactly into the numerators after rearrangement simply because the coefficients of the derived polynomial do belong to D.

To obtain the homomorphic images of $\partial R^{n-1}/\partial X_\nu$, $1 \leq \nu \leq n-2$, we put $y = X_\nu$ and $D = \mathbb{Z}_{p_j}[X_{n-1}]$ for the appropriate j and carry out the computations above. We note that there is the possibility of pre-calculating the sub-expressions which only involve the $\mathbf{c}_i$s. From the images of the partial derivatives, it is easy to obtain those of $R^{n-1\,*}$ using (2.7).

Provided that pre-calculations can be used in a suitable way, it may be that this method of differentiating point-evaluation images will turn out to be fast enough not to cause problems in most cases. However Example 2 stands as a warning here.

2.4.1 Canonical Forms

As a consequence of Theorem 1 and its two corollaries, there exist polynomials $s_i \in \mathcal{S}_i$, $i = 1, \ldots, n$, which are irreducible or zero and which generate $\mathcal{I}_n$. Each s_i is unique modulo the ideal generated by $s_1, \ldots, s_{i-1}$. Of course for some i, f_i will be transcendental over $\mathcal{F}_{i-1}$ and so those $\mathcal{S}_i$ will only contain the zero polynomial. If we knew a maximal set of non-zero s_i we could use reduction modulo them to give canonical forms for elements of $\mathcal{F}_n$. For elementary functions and integrals, the method of Rothstein and Caviness, [83], could in principle be used to determine such a set. However the method is not fast, and not currently applicable to more general situations.

What our method allows us to do is to build up information about $\mathcal{S}_n$. When we test expressions for zero equivalence using Algorithm 2, we calculate a function of the form $g = \gcd(P, P^*)$. If our expression turns out to be equivalent to zero, and in other cases where g is non–trivial, g will belong to $\mathcal{S}_n$. We can then incorporate it into our list, and use it to reduce expressions encountered in the future. Unless we already have non-zero s_i for every i, which is generally unlikely, we will not be sure that our set is maximal. In many cases, relations between the transcendental functions of interest will already be known, and these could be input as s_i. For example, if $y'' + y = 0$ is used to define $\cos x$, then the relation $(y')^2 + y^2 = 1$ can be input. Any previously unknown relations between transcendental functions discovered during zero-equivalence determination would presumably be of considerable interest.

2.5 Systems of Partial Differential Equations

The use of algebraic methods in the study of differential equations forms part of differential algebra. This subject was founded by Ritt, and developed by Kolchin and others. The work of Ax, Risch and Rothstein-Caviness treated earlier in this chapter has strong connections with differential algebra. Recently Gröbner-basis techniques have been adapted to handle computational problems in differential algebra, in particular by F. Boulier [16, 17] and E. Mansfield [58]. A. Péladan-Germa applied the results of Boulier and others to give a zero-equivalence method for functions defined by systems of (possibly non-linear) partial differential equations, [63]. As one would expect, not every every such system can be handled by her algorithm, but in some sense most can.

We sketch the method of Péladan-Germa below together with some of the theory surrounding it. For further details, the reader is referred to [63, 16, 17].

2.5.1 Notation

$\mathbf{K}$ will denote a differential field of characteristic zero. $\delta_1, \ldots, \delta_m$ will be derivations acting on $\mathbf{K}$ and various differential rings containing it. $y_1, \ldots, y_n$ will denote differential indeterminates. We may think of $y_1, \ldots, y_n$ as unknown $\mathcal{C}^\infty$ functions of variables $x_1, \ldots, x_m$, and $\delta_1, \ldots, \delta_m$ as the operations $\partial/\partial x_1, \ldots, \partial/\partial x_m$, although there are other possibilities. Let

$$\Theta =_{def} \{\delta_1^{a_1} \cdots \delta_m^{a_m}; a_1, \ldots, a_m \in \mathbb{N}\}.$$

An element θ of Θ will be called a *derivation operator*; it will be a *proper* derivation operator if not all of the a_is are zero.

A *ranking* on the set of derivatives $\{\theta y_i; \theta \in \Theta,\ 1 \le i \le n\}$ is a total ordering such that

1. For all proper derivations δ_i, all $\theta \in \Theta$ and all y_j, we have $\delta_i \theta y_j > \theta y_j$.
2. For all derivations δ_i, all $\theta_1, \theta_2 \in \Theta$ and all y_j, y_k, we have

$$\theta_1 y_j > \theta_2 y_k \quad \Rightarrow \quad \delta_i \theta_1 y_j > \delta_i \theta_2 y_k.$$

We shall assume the existence of a fixed ranking on the set of derivatives.

$\mathbf{K}\langle y_1, \ldots, y_n \rangle$ will denote the ring of polynomials in the y_is and their derivatives. Let P be any differential polynomial in $\mathbf{K}\langle y_1, \ldots, y_n \rangle$. The *leader*, u, of P is the largest derivative (in the ranking) which appears in P. If the leader u is a derivative of y_j then j is the *class* of P. Let d be the degree of u in P. The coefficient, I_P, of u^d in P is called the *initial* of P. Similarly, $S_P = \partial P / \partial u$ is called the *separant* of P.

An important technique consists of eliminating derivatives between polynomials. Gröbner bases are generally used for this. However one may need to use the derivatives of some of the polynomials, as well as the polynomials themselves. Let u be the leader of P and let d be its degree, as above. Another differential polynomial, R, is said to be *partially reduced* with respect to P if R contains no proper derivative of u. Then R is said to be *reduced* with respect to P if in addition R is of degree less than d in u. Note that u need not be the leader of R, since R may contain derivatives higher in the ranking. Let P and Q be any two non–zero differential polynomials. By subtracting derivatives of P, we can calculate an R which is partially reduced with respect to P and which satisfies

$$S_P^b Q = \sum_{j_1, \ldots, j_m} B_{j_1, \ldots, j_m} \delta_1^{j_1} \cdots \delta_m^{j_m} P + R, \tag{2.27}$$

where $b \in \mathbb{N}$, the sum is finite and the $B_{j_1, \ldots, j_m}$ are differential polynomials. R is called the *partial remainder* of Q with respect to P, by analogy with the remainder in polynomial division. Similarly one can calculate an R' which is reduced with respect to P and such that

$$I_P^a S_P^b Q = \sum_{j_1, \ldots, j_m} B_{j_1, \ldots, j_m} \delta_1^{j_1} \cdots \delta_m^{j_m} P + R'. \tag{2.28}$$

R' is then called the *remainder* of Q with respect to P.

A set E of differential polynomials is said to be *autoreduced* if every element of E is reduced with respect to every other element. In the same way as above, one has the concept of a (partial) remainder of Q with respect to an autoreduced set of polynomials, E. Then Q multiplied by powers of all the separants and initials of elements of E may be expressed in terms of the derivatives of the elements of E plus a partial remainder, as in (2.27). Similarly (2.28) may be generalized.

If E is an autoreduced subset of a set F such that no non-zero element of $F \setminus E$ is reduced with respect to E, then E is called a *characteristic set* of F. A characteristic set for a differential ideal is thus an autoreduced set which 'generates' the ideal in the sense above. That is to say, each element of the ideal multiplied by appropriate powers of initials and separants is obtained as a sum of multiples of elements in the characteristic set.

From an algebraic viewpoint, the easiest way to handle the complication of needing to multiply by separants and initials is to introduce inverses for them. Let $\mathcal{J}$ be a differential ideal of $\mathbf{K}\langle y_1, \ldots, y_n\rangle$, with characteristic set C, and let $\mathcal{H}_C^\infty$ denote the set of all products of initials and separants of elements of C. Then we write $\mathcal{J} : \mathcal{H}_C^\infty$ to denote the ideal of all $Q \in \mathbf{K}\langle y_1, \ldots, y_n\rangle$ such that $QH \in \mathcal{J}$ for some $H \in \mathcal{H}_C^\infty$. To see that $\mathcal{J} : \mathcal{H}_C^\infty$ is closed under addition note that if Q_1H_1 and Q_2H_2 belong to $\mathcal{J}$, then so does $(Q_1 + Q_2)H_1H_2$. The effect of reducing modulo $\mathcal{J} : \mathcal{H}_C^\infty$ is to introduce inverses for elements of $\mathcal{H}_C^\infty$. Problems would arise if $\mathcal{H}_C^\infty$ were to contain zero divisors, but this will not occur in our situation.

As is well known, one of the fundamental ideas in Gröbner bases is that of the S-polynomial. It is interesting to note that the corresponding idea in Differential Algebra, found in [76], apparently pre-dates the theory of Gröbner bases. Let P_i and P_j be two elements of an autoreduced subset of $\mathbf{K}\langle y_1, \ldots, y_n\rangle$ whose leaders, $\theta_i y_k$ and $\theta_j y_k$ are derivatives of the same indeterminate y_k. Let ϕ_i and ϕ_j be elements of Θ of minimal order such that $\phi_i\theta_i = \phi_j\theta_j$. Then the Δ-polynomial of P_i and P_j is

$$\Delta_{ij} = S_{P_j}\phi_i P_i - S_{P_i}\phi_j P_j,$$

where S_{P_j} and S_{P_i} are the separants.

An autoreduced set of differential polynomials, E, is said to be *coherent* if each of its Δ-polynomials, Δ_{ij}, can be reduced to zero using derivatives of elements of E with leaders less than the corresponding $\phi_i\theta_i y_k$. (Obviously $\phi_i P_i$ and $\phi_j P_j$ reduce Δ_{ij}; coherence is about being able to reduce using lower derivatives).

Given a subset Φ of $\mathbf{K}\langle y_1, \ldots, y_n\rangle$, we write (Φ) for the ideal generated by Φ and $[\Phi]$ for the differential ideal generated by Φ. Similarly $\{<\Phi>\}$ will denote the radical differential ideal generated by Φ. That is to say $\{<\Phi>\}$ is the set of elements $Q \in \mathbf{K}\langle y_1, \ldots, y_n\rangle$ some power of which is a $\mathbf{K}\langle y_1, \ldots, y_n\rangle$ linear combination of derivatives of elements of Φ.

2.5.2 Radical Differential Ideals

The importance of the radical of an ideal is well known in algebraic geometry. The key theorem is of course the Nullstellensatz. To examine the situation in the differential case, we make the following definition.

Given an arbitrary subset of Σ of $\mathbf{K}\langle y_1, \ldots, y_n\rangle$, a *basis* for Σ is a **finite** subset Φ of $\mathbf{K}\langle y_1, \ldots, y_n\rangle$ such that $\Sigma \subset \{<\Phi>\}$.

THEOREM **2 (Ritt-Raudenbush)** *Every subset of* $\mathbf{K}\langle y_1, \ldots, y_n\rangle$ *has a basis.*

The theorem is encouraging, but note that we may need to take the *radical* differential ideal to generate from a finite subset. In fact there need not be a finite subset, Φ', such that the differential ideal, $[\Phi']$, generated by Φ' contains Σ. Examples may be found in [76].

The following formulation of the Nullstellensatz gives a method for determining membership of a radical ideal.

THEOREM **3 (Hilbert's Nullstellensatz)** *Let $P_1, \ldots, P_m$ be a finite set of differential polynomials. Then a differential polynomial, P, belongs to the radical differential ideal $\{< \{P_1, \ldots, P_m\} >\}$ if and only if the system of equations and inequation $P_1 = P_2 = \cdots = P_m = 0$, $P \neq 0$ is inconsistent.*

2.5.3 Computing Characteristic Sets

Given a finite subset $\Sigma \subset \mathbf{K}\langle y_1, \ldots, y_n \rangle$, we can compute a characteristic set for Σ as follows:

- Rank the elements of Σ according to their leaders (so two elements with the same leader have the same rank).
- Choose $A_1 \in \Sigma \setminus \{0\}$ of minimal rank. If $A_1 \in \mathbf{K}$ or if $\Sigma \setminus \{0\}$ contains no element reduced with respect to A_1, then $\{A_1\}$ is a characteristic set for Σ.
- Otherwise choose A_2 to be an element of $\Sigma \setminus \{0\}$ of minimal rank reduced with respect to A_1. Note that by definition of A_1, the class of A_2 must be greater than the class of A_1. If $\Sigma \setminus \{0\}$ contains no element reduced with respect to $\{A_1, A_2\}$ then $\{A_1, A_2\}$ is a characteristic set for Σ.
- Otherwise choose A_3 to be an element of minimal rank in $\Sigma \setminus \{0\}$ reduced with respect to $\{A_1, A_2\}$, and so on.

2.5.4 The Rosenfeld-Gröbner Algorithm

DEFINITION **1** *A* regular system, *Ω, of differential equations consists of a finite set of equations $P_i = 0$, $i = 1, \ldots, m$, and inequations $I_{P_i} \neq 0$, $S_{P_i} \neq 0$ $(i = 1, \ldots, m)$, $Q_j \neq 0$ $(j = 1, \ldots, J)$, such that*

1. *the set $A = \{P_1, \ldots, P_m\}$ is autoreduced and coherent,*
2. *each Q_j is partially reduced with respect to A,*
3. *I_{P_i} and S_{P_i} are the initial and separant of P_i for $i = 1, \ldots, m$.*

A differential ideal $\mathcal{J}$ is called regular *if there is a regular system, Ω, with $\mathcal{J} = [A] : \mathcal{H}_A^\infty$.*

LEMMA **5 (Rosenfeld)** *Suppose that A is an autoreduced and coherent subset of $\mathbf{K}\langle y_1, \ldots, y_n \rangle$. Then every polynomial in $[A] : \mathcal{H}_A^\infty$ which is partially reduced with respect to A also belongs to $(A) : \mathcal{H}_A^\infty$.*

This is of crucial importance because it sometimes allows one to convert a problem in differential algebra into a problem in (non-differential) ring theory. It provides a strengthening of Seidenberg elimination. The germ of the idea can be traced back to Ritt. The following theorem is a consequence.

THEOREM **4 (Rosenfeld)** *A regular system admits a differential model if and only if it admits an algebraic model.*

An algebraic model is one where the proper derivatives of a y_i are treated as new independent variables having no connection with y_i or each other. That is to say, one ignores the differential structure.

THEOREM **5 (Boulier et al.)** *Every regular ideal is radical.*

The Rosenfeld-Gröbner Algorithm below is due to F. Boulier, D. Lazard, F. Ollivier and M. Petitot. Given a system of equations and inequations, Σ, it produces a finite number of regular systems, $\Omega_1, \ldots, \Omega_s$ whose differential models form a partition of the differential models of Σ. So a set of functions satisfies Σ if and only if it satisfies precisely one Ω_i.

ALGORITHM **4 (Boulier et al.)**
Input: *A differential system, Σ of equations and inequations.*
Output: *A finite family $\{\Omega_1, \ldots, \Omega_s\}$ of regular systems whose differential models form a partition of the differential models of Σ.*
Function ObviouslyInconsistent *returns TRUE if a non-zero element of* **K** *appears as P on the left of an equation $P = 0$ or if zero appears on the left of an inequation.*

begin
if not *ObviouslyInconsistent(Σ)* **then**
  A := *a characteristic set of Σ_{eq} (the equations of Σ);*
  $\{h_1, \ldots, h_r\}$:= *the set of initials and separants of Σ not in* **K**;
  $R := ((\Sigma_{eq} \setminus A) \cup \Delta\text{-polys}(\Sigma))$ rem A;
  if $R = \emptyset$ *or* $R = \{0\}$ **then**
(Δ-polys reduce to zero, so A is coherent. Moreover A reduces Σ_{eq} to zero.)
    $\Omega_{eq} := A$;
    $\Omega_{ineq} := (\Sigma_{ineq}$ partial rem $A) \cup \{h_1 \neq 0, \ldots, h_r \neq 0\}$;
(Ω is now a regular system.)
    B:= *a Gröbner basis for* $(A) : \mathcal{H}_A^\infty$;
    if $B \neq \{1\}$ **then**
(The system is consistent.)
      output Ω and B;
    endif
  else
    $\Gamma_{eq,r+1} := A \cup R$;
    $\Gamma_{ineq,r+1} := \Sigma_{ineq} \cup \{h_1 \neq 0, \ldots, h_r \neq 0\}$;
    Rosenfeld-Gröbner(Γ_{r+1});
(The elements of R are reduced with respect to A; so the new char. set will be 'lower'.)
  endif
(If consistent, this gives a regular system whose inequalities include $h_1 \neq 0, \ldots, h_r \neq 0$.)
  for $i := r$ **down to** *1* **do**
    $\Gamma_{eq,i} := \Sigma_{eq} \cup \{h_i = 0\}$;
    $\Gamma_{ineq,i} := \Sigma_{ineq} \cup \{h_{i-1} \neq 0, \ldots, h_1 \neq 0\}$;

Rosenfeld-Gröbner(Γ_i)
(If consistent, gives a regular system with inequalities $h_1 \neq 0, \ldots, h_{i-1} \neq 0$ etc.)
od
endif
end

Note that the algorithm also computes a Gröbner basis, B_i, for each $\Omega_{eq,i}$. The algorithm terminates because after we have added R to A we must produce a characteristic set which is lower in the well ordering on chains given by Ritt.

The Rosenfeld-Gröbner algorithm requires a method for calculating a Gröbner basis for $(A) : \mathcal{H}_A^\infty$. This is classical:

ALGORITHM **5**

1. *For each $h_i \in \mathcal{H}_A^\infty$, introduce a new indeterminate, z_i, and a new equation $h_i z_i - 1 = 0$.*
2. *Compute a Gröbner basis using an ordering which has all the z_is greater than all the derivatives of the y_js.*
3. *Remove from the basis any polynomials which involve any of the z_is.*

2.5.5 Consequences, Applications

THEOREM **6 (Boulier et al.)** *With Σ and $\{\Omega_1, \ldots, \Omega_s\}$ as above, the radical differential ideal $\{<\Sigma>\}$ is the intersection of the regular differential ideals $[\Omega_{eq,i} : \mathcal{H}_{\Omega_{eq,i}}^\infty]$. i.e.*

$$\{<\Sigma>\} = [\Omega_{eq,1} : \mathcal{H}_{\Omega_{eq,1}}^\infty] \cap \cdots \cap [\Omega_{eq,s} : \mathcal{H}_{\Omega_{eq,s}}^\infty].$$

Using the Gröbner bases calculated by the Rosenfeld-Gröbner algorithm, we obtain a test for membership of the radical differential ideal $\{<\Sigma>\}$.

THEOREM **7 (Boulier et al.)** *With the above notation,*

$$P \in \{<\Sigma>\} \Leftrightarrow \forall i = 1, \ldots, s,\ P \text{ partial rem } \Omega_{eq,i} \in (B_i).$$

As indicated in Section 1.3, it has long been known how to calculate a characteristic set for a given **finite** set. The Rosenfeld-Gröbner algorithm allows us to compute a characteristic set for the radical $\{<\Sigma>\}$, provided that $\{<\Sigma>\}$ is known to be prime. Unfortunately, there is currently no algorithm for deciding this in general.

THEOREM **8 (Boulier et al.)** *Let $\{<\Sigma>\}$ be a prime differential ideal. Then*

$$\{<\Sigma>\} = [\Omega_{eq,1} : \mathcal{H}_{\Omega_{eq,1}}^\infty].$$

Moreover an algorithm exists to compute a characteristic set for $\{<\Sigma>\}$.

2.5.6 Zero Equivalence

A. Péladan-Germa adapted the work of Boulier and his collaborators to solve zero-equivalence problems involving functions defined implicitly by partial differential equations. We summarize the essentials of her method.

ALGORITHM **6 (A. Péladan-Germa)**
Input: *A set of differential polynomials, A, and a finite set of initial conditions, C, at an evaluation point, c, such that the equations* $A_i = 0,\ A_i \in A$ *and the conditions C define a set of functions* $\phi_1, \ldots, \phi_\nu$. *Also a differential polynomial, P.*
Output: *A decision as to whether* $P\langle\phi\rangle = P(\phi_1, \phi_1', \ldots, \phi_\nu, \phi_\nu', \ldots)$ *is functionally equivalent to zero.*
begin

1. *Express the system* $A \cup \{P\}$ *as a partition of regular systems using the Rosenfeld-Gröbner algorithm. If no regular system results, then* $P\langle\phi\rangle \not\equiv 0$.
2. *Let* $\bar{\Theta}$ *denote the finite set of derivatives involved in the initial conditions. For each regular system,* Ω_i, *if no separant or initial vanishes at c, check the vanishing of the derivatives* $\theta(P)(c)$, $\theta \in \bar{\Theta}$. *If all are zero then* $P\langle\phi\rangle \equiv 0$.
3. *Otherwise some separants and/or initials will vanish at c but not identically (since the system is regular).*
4. *Regard the derivatives in* $\theta(P)$, $\theta \in \bar{\Theta}$, *as independent variables (i.e. ignore the differential structure) and write W for the algebraic variety defined by the vanishing of the* $\theta(P)$. *Similarly write W′ for the variety defined by the vanishing of the product, H, of all the separants and initials. Then* $P \equiv 0$ *if and only if* $c \in \overline{W \setminus W'}$. *(The Zariski and metric closures agree here). This happens if and only if c belongs to the variety defined by* $(\{\theta(P),\ \theta \in \bar{\Theta}\}) : \mathcal{H}^\infty$, *which can be tested by computing a Gröbner basis for that ideal as above.*

end

2.6 Finding Symmetries

Determining whether a particular symmetry holds is a zero–equivalence problem. Thus f has reflective symmetry about α if $f(\alpha + x) - f(\alpha - x) \equiv 0$. Similarly f has period α if $f(x + \alpha) \equiv f(x)$. When α is given, the techniques of this chapter may be applied directly. However if we ask whether f has any reflective symmetry or whether f is periodic, then α has to be treated as a parameter to be determined by the zero equivalence. The method of Algorithms 2 and 3, for example, may be applied to an f in the appropriate class with the 'constants' now being functions of α. At various stages of the computation, different branchings of the algorithm will be taken according to whether a leading coefficient is zero or not, and the final answer will depend on whether one or more coefficients are zero. Thus the zero equivalence of say $f(\alpha + x) - f(\alpha - x)$ can be reduced to the zero equivalence of various 'constants'; in other words we need to discover values of α, if any, for which these exp–log functions are zero or non–zero as the case may be. The following is from [92].

Example
We seek reflective symmetries of the function

$$f(x) = x^2 - x + \sin \pi x,$$

where $\sin \pi x$ is defined as the solution, y_1 of the equation

$$y'' + \pi^2 y = 0$$

satisfying the initial conditions $y(0) = 0$, $y'(0) = \pi$. Let

$$F(\alpha, x) = f(\alpha + x) - f(\alpha - x).$$

We search for values of α which make $F(\alpha, x)$ the zero function. We have

$$\begin{aligned} F(\alpha, x) &= \{x^2 + (2\alpha - 1)x + \alpha^2 - \alpha + y_1(\alpha + x)\} \\ &\quad -\{x^2 - (2\alpha - 1)x + \alpha^2 - \alpha + y_1(\alpha - x) \\ &= (4\alpha - 2)x + y_1(\alpha + x) - y_1(\alpha - x).. \end{aligned}$$

So

$$\frac{dF}{dx} = 4\alpha - 2 + y'(\alpha + x) + y'(\alpha - x)$$

and

$$\frac{d^2F}{dx^2} = y''(\alpha + x) - y''(\alpha - x) = -\pi^2\{y_1(\alpha + x) - y_1(\alpha - x)\}.$$

We treat $y_1(\alpha + x)$ as the leading "variable" and take the gcd of F and F^{**}. We arrive at the identity

$$F + \frac{F^{**}}{\pi^2} = (4\alpha - 2)x,$$

and since this does not involve the leading variable, F can only be zero if $(4\alpha - 2)x \equiv 0$; this holds if and only if $\alpha = 1/2$. In that case F and F^{**} have a common factor $y_1(\alpha + x) - y_1(\alpha - x)$ and so we look at initial conditions. We have $F(1/2, 0) = 0$ and $F'(1/2, 0) = 2y_1'(1/2) = 2\cos\pi/2 = 0$, and so $1/2$ is indeed a point of reflective symmetry of f.

The simplicity of this example hides the fact that leading coefficients in the polynomial remainder sequence will in general depend on α, and the computation will then have to branch into cases according to whether the coefficient in question is zero or not. Thus a tree of different cases involving equations and inequations in α may develop, and the question arises as to the best way of handling this in implementation.

Another point is that the equations concerned will often involve transcendental functions. Methods for finding zeros of such equations, at least in the case of simple functions, are given in [71] and [57].

3

Hardy Fields

The origin of the theory of Hardy fields can be traced back to the work of Hardy in asymptotics, and beyond to the infinitärcalcül of P. du Bois-Reymond, [31]. Others including E. Landau and E. Borel also contributed to the development. We refer the reader to [39] for details on the antecedents of the subject.

The definition of a Hardy field first appeared in the Bourbaki series, [18], where a number of basic properties of general Hardy fields were established. After that the topic seems to have attracted little attention until 1972, when A. Robinson showed that the real algebraic closure of a Hardy field can itself be identified as a Hardy field, [77]. The use of the valuation in connection with Hardy fields was extensively studied in the book by Lightstone and Robinson, [50].

Then in the nineteen eighties the subject started to accelerate with a number of papers by M. Boshernitzan, [9, 10, 11, 12, 13, 14] and M. Rosenlicht, [78, 79, 80, 81]. The early papers of Boshernitzan do not use Hardy–field terminology, and are much concerned with the properties of the class E (a definition is given later). However they contain many important results about Hardy fields. A number of these were obtained independently by Rosenlicht, who also re–emphasised the importance of the valuation and added the key concepts of comparability class and rank.

Currently, the theory of Hardy fields is being applied in asymptotics, [15, 90, 103, 101, 98, 87], differential equations, [99, 102, 105], model theory [113], and also in theoretical automation, [34, 35]. Part of the message of this book is to stress the very important role played by Hardy fields in asymptotics.

3.1 Definitions and Examples

We recall the notion of the germ of a function. Let f_1 and f_2 be two $\mathcal{C}^\infty$ functions defined on intervals (a_1, ∞), (a_2, ∞) of $\mathbb{R}$ respectively. We write

$f_1 \approx f_2$ if there is an interval (a_3, ∞) on which $f_1(x) = f_2(x)$. The relation $\approx$ is easily seen to be an equivalence relation.

DEFINITION **2**
The germ *of a $\mathcal{C}^\infty$ function, f, is defined to be its equivalence class under the relation $\approx$.*

In practice it will cause no difficulty if we use the same notation for a function and its germ, and we do not generally distinguish between the two. This device allows us to avoid detailed consideration of the intervals where functions are defined. It is natural to do this since we are only interested in the behaviour of the function for large positive values of x, and so the domain of definition does not normally matter provided it includes *some* interval of the form (a, ∞).

We write $\mathcal{X}$ for the set of germs of $\mathcal{C}^\infty$ real–valued functions defined on intervals of the form (a, ∞). The set $\mathcal{X}$ is a ring under the algebraic operation inherited from the functions. Similarly, if $\mathbf{f} \in \mathcal{X}$ and f is an element of the equivalence class $\mathbf{f}$, we define $\mathbf{f}'$ to be the equivalence class of f'. Then $\mathcal{X}$ is closed under differentiation, and hence is a differential ring.

DEFINITION **3** ([18])
A Hardy field *is a differential subfield of $\mathcal{X}$.*

This definition is much more powerful than one might at first imagine. Let $\mathcal{F}$ be a Hardy field and let f be a non–zero element of $\mathcal{F}$. Then f must have an inverse, $1/f$, in $\mathcal{F}$, and so f (or strictly speaking any one of its representatives) cannot have arbitrarily large zeros. Hence either $f(x) > 0$ for all sufficiently large x or else $f(x) < 0$ for all sufficiently large x. We write $f > 0$ in the first case, and $f < 0$ in the second; similarly we write $f > g$ when $f - g > 0$. Thus any Hardy field carries a total order *which reflects the relative sizes of the underlying functions for arbitrarily large values of x.* Any Hardy field we introduce will always be assumed to be equipped with this order.

The fact that the asymptotic behaviour of two elements can be compared using this order is one of the main reasons for the importance of Hardy fields in asymptotics.

Another reason is closely related to this. If $f \in \mathcal{F}$ as above, then f' is either identically zero, or positive, or negative in the order. So f is eventually monotonic and must therefore tend to a limit, either finite or infinite.

Examples
The field of rational numbers, $\mathbb{Q}$, is a Hardy field, and is contained in every Hardy field. The field of rational functions with real coefficients, $\mathbb{R}(x)$, is also easily seen to be a Hardy field. A slightly more interesting example is $\mathbb{R}(x, \log x)$. Any non–zero element of this field may be represented in the form

$$f(x) = \frac{x^n (\log x)^m \sum_{i,j=0}^{N} c_{i,j} x^{-i} (\log x)^{-j}}{\sum_{i,j=0}^{M} d_{i,j} x^{-i} (\log x)^{-j}} \tag{3.1}$$

where $m, n \in \mathbb{Z}$, the $c_{i,j}$ and $d_{i,j}$ belong to $\mathbb{R}$ with $c_{0,0}$ and $d_{0,0}$ non–zero. Clearly $f(x)$ has the sign of $c_{0,0}/d_{0,0}$ for large values of x, and hence $\mathbb{R}(x, \log x)$ is indeed a Hardy field. Other examples of Hardy fields include $\mathbb{R}(e^x)$ and $\mathbb{R}(x, e^{x^2}, \int e^{x^2})$.

It is also true (though harder to prove) that the set of exp–log functions forms a Hardy field. Moreover one can add real–algebraic closure to the allowable operations, giving Hardy's class of *L–functions* - the eponymous Hardy field. Hardy's proof (in [40]) that the L–functions do indeed satisfy Definition 3 involved using induction to bound the number of zeros, and this was used for zero–equivalence purposes by Richardson and subsequently by Macintyre (see Chapter 2).

It is worth pointing out that if f belongs to a Hardy field, $\mathcal{F}$, containing a field of constants $\mathbb{K}$, then $\mathbb{K}(f, f', f'', \ldots)$, the set of all $\mathbb{K}$–rational functions in f and its derivatives, is also a Hardy field contained in $\mathcal{F}$. We recall our notation from the previous chapter.

NOTATION *If $\mathcal{F}$ is a differential field, we use $\mathcal{F}\langle y \rangle$ to denote the field $\mathcal{F}(y, y', y'', \ldots)$. If $Q \in \mathcal{F}\langle y \rangle$, we write*

$$Q\langle y \rangle = Q(y, y', y'', \ldots).$$

3.2 Building Hardy Fields

One of the difficulties with Hardy fields is that it is not generally true that the union of two Hardy fields is contained in a Hardy field . This was first shown in [10]. The following is a slightly different example, though on basically similar lines, given in [79]. Our treatment of it follows [101]. Let Φ be defined by

$$\Phi(x) = \frac{\Gamma'(x)}{\Gamma(x)} + x.$$

Since the gamma function satisfies no algebraic differential equation over $\mathbb{R}(x)$, [43], the same is true of Φ. Now for λ a fixed real number, we let

$$\Psi_\lambda(x) = \Phi(x) + \lambda e^{-x} \sin x,$$

and we define $\mathcal{G}_\lambda$ to be the field $\mathbb{R}\langle \Psi_\lambda \rangle$. Then each $\mathcal{G}_\lambda$ is a Hardy field, essentially because a non–zero rational expression in Φ and its derivatives cannot cancel. Hence it will be asymptotic to a constant times some power of x, which will "hide" any terms involving $e^{-x} \sin x$.

Looking in more detail, we recall that the Stirling series gives an asymptotic series (see Chapter 4.1) for Φ, namely

$$\Phi(x) \simeq x + \sum_{s=0}^{\infty} B_s x^{-1-s}, \tag{3.2}$$

where the B_s are the Bernoulli numbers. Hence

$$\Phi'(x) \simeq 1 - \sum_{s=0}^{\infty} (s+1) B_s x^{-2-s} \tag{3.3}$$

and for $n \geq 2$,

$$\Phi^{(n)}(x) \simeq \sum_{s=0}^{\infty} (-1)^n \frac{(s+n)!}{s!} B_s x^{-n-1-s}. \tag{3.4}$$

We regard the various infinite series above as elements of the formal power series field

$$\mathbb{R}((1/x)) = \left\{ \sum_{i=0}^{\infty} a_i x^{m-i}; m \in \mathbb{Z}, a_0, a_1, \ldots, \in \mathbb{R} \right\}.$$

By overloading the notation, we use $\Phi^{(n)}(x)$ to stand for the formal series on the right of (3.4) (or of (3.3) or (3.2) if n is 0 or 1). The relation $\Gamma(x+1) = x\Gamma(x)$ yields

$$\Phi^{(n)}(x+1) = \Phi^{(n)}(x) + (-1)^{n+1}(n+1)! x^{-n-2}. \tag{3.5}$$

Let H be any non–zero differential polynomial over $\mathbb{R}$. As in the original proof that the gamma function satisfies no algebraic differential equation, [43], one can use (3.5) to show that $H(\Phi'(x)) \neq 0$; see [79] for further details. Hence for some $i \in \mathbb{Z}$, $H(\Phi'(x))x^i$ tends to a finite, non–zero real number. This implies that $\mathbb{R}\langle\Phi'\rangle$ is a Hardy field, and $\mathbb{R}\langle\Phi\rangle$ is obtained from it by adjoining an integral; see Corollary 1 to Theorem 10. Thus $\mathcal{G}_0$ is a Hardy field.

However the argument using the asymptotic series for the derivatives of Φ apply equally well to $H(\Phi(x) + \lambda e^{-x} \sin x)$. Thus for each λ, the field $\mathcal{G}_\lambda$ is a Hardy field. However there can be no Hardy field which contains both $\mathcal{G}_0$ and $\mathcal{G}_1$, since it would have to contain $\Psi_1(x) - \Psi_0(x) = e^{-x} \sin x$.

Obviously the above construction, depends heavily on the fact that Φ does not satisfy any algebraic differential equation. One might have thought that the union of two Hardy fields whose elements all satisfy such equations would be contained in a single Hardy field. Nonetheless [14] contains a counter example.

Although we cannot assert that the union of two Hardy fields is contained in a Hardy field, we do have the following result.

Proposition 1 *If $\mathcal{F}_n$ is a sequence of Hardy fields with $\mathcal{F}_n \subset \mathcal{F}_{n+1}$ for all n, then $\cup\mathcal{F}_n$is a Hardy field.*

For the proof it is only necessary to observe that any non–zero element of $\mathcal{F}$ is contained in one of the $\mathcal{F}_n$s and so cannot have arbitrarily large zeros.

Since we want to use Hardy fields to compare the asymptotic behaviour of functions, it is obviously in our interests to have large Hardy fields at our disposal and to be able to extend existing Hardy fields in a variety of ways. In view of the previous paragraph, this is a non–trivial matter. What we *are* able to do is to add elements of a specific type - for example exponentials or integrals of existing elements.

From time to time, we shall need to consider roots of algebraic equations over some Hardy field, $\mathcal{F}$. However we certainly do not want to become involved with complex roots since our theory is essentially about real-valued functions. Of course it would be possible to treat real and imaginary parts separately, but functional composition would bring into play the limiting behaviour of functions of a complex variable, which is very different from that of Hardy-field elements, at least in the neighbourhood of an essential singularity, [42].

Thus the right object of study in this context is not the algebraic closure of $\mathcal{F}$ but its *real closure* . Real fields are treated in many texts on algebra, such as [120], and the reader is referred to them for details. Here we just give a brief sketch of the properties we need.

DEFINITION **4** *A field,* **P**, *is* formally real *if* -1 *cannot be expressed as a sum of squares in it.*

P *is* real closed *if it is formally real, but no algebraic extension of it is formally real.*

Clearly any Hardy field is formally real.

The adjunction of a square root of -1 turns an a real closed field into an algebraically closed one. Furthermore, any polynomial over a real closed field can be written as a product of linear and quadratic factors.

The following result was first given in [77].

THEOREM **9** *Let $\mathcal{F}$ be a Hardy field. Then its real closure, $\mathcal{F}_{cl}$, is also a Hardy field.*

Proof of Theorem 9

Let f be a continuous real–valued function defined on some interval (a, ∞) and suppose that P be a non–zero polynomial of minimal degree over $\mathcal{F}$, such that $P(f) \equiv 0$. We show that $\mathcal{F}(f)$ is a Hardy field.

On differentiating the relation $P(f) = 0$, we obtain

$$f' \,\frac{\partial P}{\partial y}(f) + \frac{\partial P}{\partial x}(f) = 0. \tag{3.6}$$

By the minimality of the degree of P, we have $(\partial P/\partial y)(f) \neq 0$, and so $f' = R_1(f)$ where R_1 is a rational function over $\mathcal{F}$. By repeatedly differentiating we obtain $f^{(j)} = R_j(f)$ with R_j rational over $\mathcal{F}$. Now let $S \in \mathcal{F}\langle y\rangle$, and suppose that $S\langle f\rangle$ vanishes on a sequence $\{x_n\}$ tending to infinity. On substituting for the derivatives of f we obtain

$$S\langle f\rangle = \frac{S_1(f)}{S_2(f)},$$

where S_1 and S_2 are polynomials over $\mathcal{F}$. Then $S_1(f)$ vanishes on $\{x_n\}$. Moreover by a standard property of fields, $S_1(f)$ satisfies a polynomial equation over $\mathcal{F}$, [125](Ch. II). Let $Q(y) = q_0 + \cdots q_m y^m$ be the minimal polynomial of $S_1(f)$. Then $Q(S_1(f))$ is identically zero, and so $q_0 = -q_1 S_1(f) - \cdots - q_m S_1^m(f)$ vanishes on $\{x_n\}$. However $q_0 \in \mathcal{F}$ which is a Hardy field, and hence $q_0 = 0$. Since Q was chosen to be minimal, this implies that $S_1(f) = 0$, as required.

Thus any simple algebraic extension of a Hardy field by a real function is again a Hardy field, and our conclusion follows from Proposition 1.

Our next result has a long history. In particular, special cases were established by Hardy, [40], and Marić, [59]. The present version is from [78], and is due to Michael Singer.

THEOREM **10** *Let $\mathcal{F}$ be a Hardy field, and let $A(y)$ and $B(y)$ be polynomials over $\mathcal{F}$. Let f be a differentiable real-valued function on an interval (a, ∞) of the real line on which $B(f)$ has no zeros and f satisfies the differential equation*

$$y' = A(y)/B(y). \tag{3.7}$$

Then the ring of germs $\mathcal{F}[f]$ is an integral domain, and its field of quotients is a Hardy field.

Proof of Theorem 10
Let $P \in \mathcal{F}[y]$ and suppose that $P(f)$ vanishes on a sequence, $\{x_n\}$, which tends to infinity. We show that this implies that $P(f)$ is the zero function. The conclusion of the theorem will then follow.

After the previous result, we may assume that $\mathcal{F}$ is real–closed. This implies that P may be factored into a product of linear and quadratic factors over $\mathcal{F}$. The discriminants of the quadratic factors are elements of $\mathcal{F}$ and so must be either zero, positive for all sufficiently large x or negative for sufficiently large x. Any quadratic factors with zero or positive discriminant can be broken down into linear factors, and those with negative discriminant cannot have real zeros for x large. Hence there is a linear factor, $y - \zeta$, of P such that $f - \zeta$ vanishes on a subsequence, $\{x_{n_k}\}$. We make a change of variable, $y \longrightarrow y + \zeta$. The form of equation (3.7) is unchanged, and so by a slight abuse of notation, we may continue to use the same symbols, A and B. Similarly we may take $\{x_{n_k}\}$ to be the whole sequence $\{x_n\}$. Then f vanishes on the sequence $\{x_n\}$.

If f is a constant, it must be zero, which is what we wished to prove. Otherwise $A(y) \not\equiv 0$ and we may write (3.7) in the form

$$y' = A(y)/B(y) = y^r \bar{A}(y)/\bar{B}(y); \tag{3.8}$$

here r is a non–negative integer, and $\bar{A}$ and $\bar{B}$ are polynomials over $\mathcal{F}$ such that $\bar{A}(0) \neq 0$ and $\bar{B}(f)$ has no zeros on some interval (a, ∞). If $r > 0$, then both f and the function 0 satisfy the differential equation (3.8) and the initial condition $y(x_1) = 0$. The uniqueness theorem for differential equations then gives $f \equiv 0$ as required. So there is only the case $r = 0$ left to consider.

We may assume that we have chosen a sufficiently large to ensure that $\bar{A}(0)$ and $\bar{B}(0)$ have constant sign on (a, ∞). Let x_n and x_{n+1} be consecutive zeros of f in the interval (a, ∞). Then f has constant sign on the interval (x_n, x_{n+1}) and $f(x_n) = f(x_{n+1}) = 0$. Moreover

$$f'(x_n) = \frac{\bar{A}(x_n)(f(x_n))}{\bar{B}(x_n)(f(x_n))} = \frac{\bar{A}(x_n)(0)}{\bar{B}(x_n)(0)}$$

and

$$f'(x_{n+1}) = \frac{\bar{A}(x_{n+1})(0)}{\bar{B}(x_{n+1})(0)}$$

have the same sign. This is impossible and so this case cannot occur. Thus we do have $P(f) \equiv 0$, which is what we needed to prove.

So any non-zero element of $\mathcal{F}[f]$ is either eventually positive or eventually negative, and hence has an inverse in $\mathcal{X}$. Thus $\mathcal{F}(f)$ is a Hardy field, and this completes the proof of Theorem 10.

COROLLARY
If $\mathcal{F}$ is a Hardy field and $f \in \mathcal{F}$ then the following are Hardy fields: $\mathcal{F}(\exp f)$, $\mathcal{F}(\int f)$, $\mathcal{F}(\log f)$, $\mathcal{F}(x)$.

The last of these follows from the fact that $1 \in \mathcal{F}$ and $x = \int 1$. A consequence is that if f is any Hardy-field element, comparisons like $f < \exp x$ or $f > \log(\log x)$ make sense asymptotically, using the ordering in a Hardy field containing f and the element $\exp x$ (respectively $\log(\log x)$). Thus we can compare f with the elements of a *scale of growth* ; see [39, 85].

The following consequence of Theorem 10 was noted by Rosenlicht.

COROLLARY
If we close a Hardy field under the taking of real powers of positive elements the result is another Hardy field.

To see this, note that it follows from Theorem 10 that a real power of any particular positive element can always be added to a Hardy field. Moreover the union of an increasing chain of Hardy fields is a Hardy field, by Proposition 1. Hardy fields of finite rank (see below) which contain the real constants and are closed under taking real powers were called *Rosenlicht fields* in [99]. Rosenlicht fields will play an important role in our study of the asymptotics of differential equations in Chapter 6.

Partial results, along the lines of Theorem 10, are obtainable for solutions of differential equations of order one and degree greater than one, [78], and for some second order equations, [78, 82]. However the fact that the sine function satisfies both $(y')^2 + y^2 = 1$ and $y'' + y = 0$ limits the possibilities here.

We have seen that the union of two Hardy fields is not always contained in a Hardy field, but that there are some elements of $\mathcal{X}$, such as $\exp x$, which can be added to any Hardy field to generate a new Hardy field. It is natural to ask which elements of $\mathcal{X}$ have this good property. The union of an increasing chain of Hardy fields is a Hardy field, and hence by Zorn's Lemma any Hardy field is contained in a maximal Hardy field. The following is from [9].

DEFINITION 5 *The class* E *is defined to be the intersection of all maximal Hardy fields.*

From Theorem 10, it can be seen that E includes the set of exp–log functions, and from the corollary to Theorem 9, it is real closed; hence E contains Hardy's class L. However it is also apparent from Theorem 10 that it contains many other functions. The question as to what E contains is of considerable interest but is not completely resolved at the time of writing. For further details, the reader is referred to the papers of Michael Boshernitzan on this subject, in particular [9, 10].

3.3 Rough Comparisons

We have already seen that the order on a Hardy field plays a fundamental role. However we often want to make less precise comparisons which might express the idea that f is "not much bigger than" g, in the spirit of the "big–O" notation. Let $\mathcal{F}$ be a Hardy field, and let f and g be two non–zero elements of $\mathcal{F}$. We write $f \cong g$ if the limit of f/g is finite and non–zero. It is easy to see that $\cong$ is an equivalence relation. We set Υ_0 to be the set of equivalence classes, and use γ_0 to denote the map from $\mathcal{F}^*$ to Υ_0 which sends an element to its equivalence class; here we have used the notation $\mathcal{F}^*$ for $\mathcal{F} \setminus \{0\}$. If u and v belong to Υ_0, we choose $f, g \in \mathcal{F}$ such that $u = \gamma_0(f)$ and $v = \gamma_0(g)$, and we define $u + v = \gamma_0(fg)$. It is easy to see that this definition is independent of the choices of f and g and makes Υ_0 into an abelian group with identity element $0 = \gamma_0(1)$. We set $f \prec_0 g$ to mean that $f/g \to 0$. Since this clearly depends only on $\gamma_0(f)$ and $\gamma_0(g)$, we may alternatively write $\gamma_0(f) < \gamma_0(g)$. Then Υ_0 becomes an ordered group. The map γ_0 is essentially the *non–Archimedean valuation*, ν on $\mathcal{F}$ considered by Robinson, Rosenlicht and others; see [50, 78]. However to fit in with our other notation we have reversed the usual ordering. In an attempt to avoid confusing readers who are already familiar with the valuation, we reiterate

$$\gamma_0(f) < \gamma_0(g) \iff \nu(f) > \nu(g).$$

Henceforth we shall not use the symbol 'ν' in this way, nor the term 'valuation'.

We can also define a second operation on $\Upsilon_0(\mathcal{F})$ by setting

$$\gamma_0(f) \tilde{+} \gamma_0(g) = \gamma_0(f+g), \text{ where the representatives } f \text{ and } g \text{ are chosen so that } f \not\sim -g$$

When supplied with both its operations, $\Upsilon_0(\mathcal{F})$ becomes a semi-ring. We call the set

$$\{(f,g) \in \mathcal{F}^* \times \mathcal{F}^*; f \sim -g\}$$

the *singular locus* of the operation $\tilde{+}$. Off the singular locus, the map γ_0 acts as a ring homomorphism from $\mathcal{F}^*$ to $\Upsilon_0(\mathcal{F})$.

The singular locus is one manifestation of a phenomenon that causes many of the difficulties in symbolic asymptotics, namely *cancellation.* The point is immediately apparent; if (f,g) belongs to the singular locus, we cannot calculate $\gamma_0(f+g)$ from a knowledge of $\gamma_0(f)$ and $\gamma_0(g)$.

We summarize the basic properties of $\Upsilon_0(\mathcal{F})$ in the following proposition.

PROPOSITION **2** *Let $\mathcal{F}$ be a Hardy field. There exists a semi-ring $\Upsilon_0(\mathcal{F})$ and a map $\gamma_0 : \mathcal{F}^* \to \Upsilon_0(\mathcal{F})$ with the following properties.*

The map γ_0 is a group homomorphism with respect to the multiplication. It also acts as an additive homomorphism off the singular locus, which is the set

$$\{(f,g) \in \mathcal{F}^* \times \mathcal{F}^*; f \sim -g\}.$$

If (f,g) belongs to the singular locus, either $f+g=0$ in which case $\gamma_0(f+g)$ is undefined, or else $\gamma_0(f+g) < \gamma_0(f)$ $(= \gamma_0(g))$.

The Hardy field $\mathbb{R}(x, \log x)$ has the property that any non-zero element, f is asymptotic to $Kx^n(\log x)^m$ for some $K \in \mathbb{R}^*$ and $n, m \in \mathbb{Z}$. So in this case, the group Υ_0 is generated by the elements $\gamma_0(x)$ and $\gamma_0(\log x)$, and if f is given by (3.1), we have $\gamma_0(f) = n\gamma_0(x) + m\gamma_0(\log x)$.

The following result from [78] lists some further properties of γ_0.

LEMMA **6** *Let $\mathcal{F}$ be a Hardy field and let $\gamma_0 : \mathcal{F}^* \to \Upsilon_0$ be as above. Then if $f, g \in \mathcal{F}^*$ we have the following:*

(1) $\gamma_0(fg) = \gamma_0(f) + \gamma_0(g)$.
(2) $\gamma_0(f) \leq 0$ *if and only if $f(x)$ has a finite limit as $x \to \infty$.*
(3) *If $f+g \in \mathcal{F}^*$ then $\gamma_0(f+g) \leq \max\{\gamma_0(f), \gamma_0(g)\}$ with equality if $\gamma_0(f) \neq \gamma_0(g)$.*
(4) *If both $\gamma_0(f)$ and $\gamma_0(g)$ are non-zero then $\gamma_0(f) \leq \gamma_0(g)$ if and only if $\gamma_0(f') \leq \gamma_0(g')$.*
(5) *If $\gamma_0(f) < \gamma_0(g) \neq 0$ then $\gamma_0(f') < \gamma_0(g')$.*
(6) *If $f \sim g$ and $\gamma_0(f)$ $(= \gamma_0(g)) \neq 0$ then $f' \sim g'$.*

Proof of Proposition 2
The first two statements merely reflect the respective definitions of addition and the order on Υ_0. To prove (3) we may assume without loss of generality that $\gamma_0(f) \leq \gamma_0(g)$. Then $|f| \leq K|g|$ for some constant K, and $|f+g| \leq (K+1)|g|$. Hence $\gamma_0(f+g) \leq \gamma_0(g) = \max\{\gamma_0(f), \gamma_0(g)\}$. If $\gamma_0(f) < \gamma_0(g)$ then $|f+g| > |g|/2$, and hence $\gamma_0(f+g) = \gamma_0(g)$.

We next establish (5). Suppose firstly that f and g are positive and that $\gamma_0(f) \leq 0 < \gamma_0(g)$. Then f has a finite limit and $g \to \infty$. If $\gamma_0(f') \geq \gamma_0(g')$

then $f' > Kg'$ for some constant K. On integrating, we get $f > Kg+C$ where C is another constant. This is obviously impossible, and so $\gamma_0(f) \leq 0 < \gamma_0(g)$ implies $\gamma_0(f') < \gamma_0(g')$. If $\gamma_0(f) < \gamma_0(g) < 0$ or $0 < \gamma_0(f) < \gamma_0(g)$, the same conclusion follows from L'Hôpital's Rule. This gives (5) for the case when f and g are positive and the other cases are easily reduced to this one.

Now to prove (4) suppose that $\gamma_0(f)$ and $\gamma_0(g)$ are both non–zero and $\gamma_0(f') \leq \gamma_0(g')$. We may immediately infer from (5) that $\gamma_0(f) \leq \gamma_0(g)$. For the reverse implication, we need only consider the case when $\gamma_0(f) = \gamma_0(g)$, and here L'Hôpital's Rule again gives that $\gamma_0(f') = \gamma_0(g')$. This gives (4), and a further use of L'Hôpital's Rule gives (6). Thus the proof of Proposition 2 is complete.

The idea of different orders of growth is familiar from elementary analysis. For example, a fixed positive power of x grows more rapidly than any fixed power of $\log x$ but not as rapidly as any fixed positive power of $\exp x$. The first systematic study of these matters seems to have been made by du Bois–Reymond in [31]. Hardy devoted his tract [39] to them. Hardy fields provide a good vehicle for studying growth orders via the notion of a comparability class, [79], which we now give.

Definition 6 *Let f_1 and f_2 be two elements of a Hardy field $\mathcal{F}$ with $f_1, f_2 \to \infty$. We say that f_1 and f_2 are* comparable *if there exists a positive integer n such that $f_1 < f_2^n$ and $f_2 < f_1^n$.*

We also specify firstly that for any f whose modulus tend to zero or infinity, $\pm f$ and $\pm f^{-1}$ are all comparable to each other, and secondly that any two elements which both tend to a non–zero real number are comparable.

Comparability is then an equivalence relation on $\mathcal{F}^*$. We refer to the equivalence classes as *comparability classes* and use $\gamma_1(f)$ to denote the comparability class of f. We write $\Upsilon_1(\mathcal{F})$ for the set of comparability classes of $\mathcal{F}^*$. Now with f_1 and f_2 tending to infinity, we write $f \prec_1 g$ if $f_2^n < f_1$ for all $n \in \mathbb{N}$; i.e. if $\log f_2 / \log f_1 \to 0$. It is easily seen that the relation $f \prec_1 g$ depends only on the comparability classes, and so may be written $\gamma_1(f_2) < \gamma_1(f_1)$. If we also specify that $\gamma_1(1)$ be the smallest class, we obtain a total order on the set of comparability classes. Thus

$$\begin{aligned}\gamma_1(-2+3/x) = \gamma_1(1) < \gamma_1(\log\log x) &= \gamma_1(\log\log x/\{(\log\log x)^2-3\}) \\ < \gamma_1(\log x) < \gamma_1(x) &< \gamma_1(\exp\{(\log x)^2\}) < \gamma_1(\exp x),\end{aligned}$$

etc. The fact that $\gamma_1(\log x) < \gamma_1(1/x) < \gamma_1(\exp x)$ may take a little bit of getting used to, but the notion of comparability class is a very useful one. The following proposition results directly from the application of Proposition 2 to the logarithms of the appropriate functions.

Proposition 3 *Let $\mathcal{F}$ be a Hardy field. The map $\mathcal{F}^* \longrightarrow \Upsilon_1(\mathcal{F})$ satisfies*

$$\gamma_1(fg) = \max\{\gamma_1(f), \gamma_1(g)\},$$

provided that $\log|f| \not\sim -\log|g|$.

The set where $\log|f| \sim -\log|g|$ is the *multiplicative singular locus* of $\mathcal{F}$. In fact we can obtain a complete analogue of Proposition 2 for γ_1 by using the star products introduced in the next chapter.

The number of comparability classes excluding $\gamma_1(1)$ (for historical reasons) is called the *rank* of $\mathcal{F}$. An important point, observed in [79], is that if $f \in \mathcal{X}$ belongs to a Hardy field and satisfies an algebraic differential equation of order n over $\mathbb{R}$ then the rank of the Hardy field $\mathbb{R}\langle f\rangle$ is at most n. For if $\mathbb{R}\langle f\rangle$ contains $m+1$ different comparability classes (counting $\gamma_1(1)$) then the transcendence degree of the extension $\mathbb{R}\langle f\rangle : \mathbb{R}$ must be at least m, and this ensures that $m \le n$. We shall see in Chapter 6 that under these circumstances, it is possible to say quite a lot about the possibilities for $\gamma_1(f)$.

Examples
Let $f = e^{x^2}$. Then $x = f'/2f$ and hence

$$\frac{f''}{2f} - \frac{(f')^2}{2f^2} = \left(\frac{f'}{2f}\right)' = 1.$$

It follows that any derivative of f may be expressed as a rational function of f and f', and so $\mathbb{R}\langle f\rangle = \mathbb{R}(x, e^{x^2})$. Thus from Theorem 10, the field $\mathbb{R}\langle e^{x^2}\rangle$ is a Hardy field, and we see that its rank is 2.

We now examine the Hardy field $\mathbb{R}(x, e^{x^2}, \int e^{x^2})$. Induction and integration by parts gives

$$\int e^{x^2} = e^{x^2}\left\{\frac{1}{2x} + \frac{1}{4}\sum_{k=1}^{n}\frac{(k-1)!}{(2k-1)!x^{2k+1}}\right\} + \frac{n!}{4(2n+1)!}\int\frac{e^{x^2}}{x^{2n+2}}.$$

Hence

$$\int e^{x^2} = e^{x^2}\left\{\frac{1}{2x} + \frac{1}{4}\sum_{k=1}^{\infty}\frac{(k-1)!}{(2k-1)!x^{2k+1}}\right\}, \tag{3.9}$$

the series converging uniformly on compact subsets of $(0,\infty)$. Now consider a polynomial $P \in \mathbb{R}[x, e^{x^2}, \int e^{x^2}]$ with terms collected according to the total degree in e^{x^2} and $\int e^{x^2}$. The leading term will be of the form $e^{Mx^2}Q(x, S(x))$, where $M \in \mathbb{N}$ and $S(x) = e^{-x^2}\int e^{x^2}$. Since $S(x)$ is given by the series on the right of (3.9), we see that $P \sim x^N e^{Mx^2}$ for some $N \in \mathbb{Z}$. Thus the comparability classes of $\mathbb{R}(x, e^{x^2}, \int e^{x^2})$ are $\gamma_1(1)$, $\gamma_1(x)$ and $\gamma_1(e^{x^2})$, and the rank is again 2.

We next give two technical lemmas which will be needed in later chapters. Let f and g be positive elements of $\mathcal{F}$. From Theorem 10, we may assume that $\log f$ and $\log g$ are in some Hardy field containing $\mathcal{F}$. Then by applying Lemma 6 we may obtain the following result. Its first part is taken from [79] and its second from [103].

LEMMA 7 *Let $\mathcal{F}$ be a Hardy field and let a and b belong to $\mathcal{F}^*$, with $\gamma_0(a)$ and $\gamma_0(b)$ both non-zero. Then*

(1) $\gamma_0(a'/a) = \gamma_0(b'/b)$ *if and only if* $\gamma_1(a) = \gamma_1(b)$*;*
(2) $\gamma_0(a'/a) < \gamma_0(b'/b)$ *if and only if* $\gamma_1(a) < \gamma_1(b)$*.*

Proof of Lemma 7
Directly from the definition, we have that $\gamma_1(a) = \gamma_1(b)$ (respectively $\gamma_1(a) < \gamma_1(b)$) if and only if $\gamma_0(\log a) = \gamma_0(\log b)$ (respectively $\gamma_0(\log a) < \gamma_0(\log b)$). The conclusion now follows from Lemma 6(5).

The second lemma, which appeared in [103], is concerned with comparability classes of derivatives.

LEMMA 8 *Let h be an element of a Hardy field.*
(1) *If* $\gamma_1(h) > \gamma_1(x)$ *then* $\gamma_1(h') = \gamma_1(h)$*.*
(2) *If* $\gamma_1(h) < \gamma_1(x)$ *and* $\gamma_0(h) \neq \gamma_0(1)$*, then* $\log h' \sim -\log x$ *(and in particular* $\gamma_1(h') = \gamma_1(x)$*).*
(3) *If* $\gamma_1(h) = \gamma_1(x)$ *and* $\log|h| \not\sim \log x$ *then* $\gamma_1(h') = \gamma_1(x) = \gamma_1(h)$*.*

Proof of Lemma 8.
To prove (1) we may suppose that $|h| \to \infty$; for otherwise we may replace h by h^{-1}. Then $\log|h| > K \log x$ for every $K \in \mathbb{R}$. On differentiating, we obtain that $h'/h > K/x$, and it follows that $\gamma_1(h') \geq \gamma_1(h)$. On the other hand, $h^{-1} \to 0$ and hence $h'/h^2 \to 0$. Since the hypotheses ensure that $|h'| \to \infty$, this implies that $\gamma_1(h') \leq \gamma_1(h)$. Thus $\gamma_1(h') = \gamma_1(h)$.

For (2), we may assume that $h \to \infty$. Then for every $\delta \in \mathbb{R}^+$, we have $-x^{-\delta} < h < x^{\delta}$. Therefore $\delta x^{-\delta-1} < h' < \delta x^{\delta-1}$, and hence $\log h' \sim -\log x$ as asserted.

To establish (3) we note that its hypotheses imply that $\log|h| \sim k \log x$ for some real constant $k \neq 1$. Let ε be any positive real number which is sufficiently small to ensure that $k-1-\varepsilon$ and $k-1+\varepsilon$ have the same sign. Then $x^{k-\varepsilon} < |h| < x^{k+\varepsilon}$, and hence $|k-\varepsilon|x^{k-1-\varepsilon} < |h'| < |k+\varepsilon|x^{k-1+\varepsilon}$. So $\gamma_1(h) = \gamma_1(x)$ as required. This completes the proof of Lemma 8.

3.4 The Map γ_2

We have seen that γ_1 can be obtained by applying γ_0 to the logarithms of the desired functions. We can use the same idea again with $\log_2$ replacing log. Here we used the notation $\log_n(x)$ for the n-times-iterated logarithm of x. Similarly we use $\exp_n(x)$ for the iterated exponential. A recursive definition is obtained by setting

$$\log_0(x) = x = \exp_0(x),$$

and for $n \geq 1$,

$$\log_n(x) = \log(\log_{n-1}(x)) \qquad \& \qquad \exp_n(x) = \exp(\exp_{n-1}(x)).$$

It will be convenient to use the convention that if n is a negative integer then $\exp_n$ means $\log_{-n}$ and similarly $\log_n = \exp_{-n}$.

DEFINITION **7** *Let $\mathcal{F}$ be a Hardy field and let $\mathcal{F}^\infty$ denote the set of elements of $\mathcal{F}$ which tend to infinity. For $f, g \in \mathcal{F}^\infty$, we define $f \approx_2 g$ to mean that $\log_2 f / \log_2 g$ tends to a non-zero constant, and $f \prec_2 g$ to mean that $\log_2 f / \log_2 g \to 0$.*

We stipulate that $f \approx_2 g$ for $g = -f$ and $g = \pm f^{-1}$, and moreover that this relation holds whenever both f and g tend to non-zero finite limits.

Of course, one can similarly define $\approx_n$ and $\prec_n$ for any $n \in \mathbb{N}$, and this is done is a slightly more general context in Chapter 9. Here we are only concerned with the case $n = 2$.

The relation $\approx_2$ is an equivalence relation, and $f \prec_2 g$ depends only on the equivalence classes of f and g. We write $\Upsilon_2(\mathcal{F})$ for the set of equivalence classes of the relation $\approx_2$, and denote the map which takes an element of $\mathcal{F}^*$ to its equivalence class under $\approx_2$ by γ_2. We write $\gamma_2(f) < \gamma_2(g)$ whenever $f \prec_2 g$. As an example, note that $\gamma_1(x) < \gamma_1(\exp(\log^2(x)))$ but by contrast $\gamma_2(x) = \gamma_2(\exp(\log^2(x)))$.

One question we might ask is what sort of growths are possible for elements of a Hardy field of a given rank. Lemma 8 opens up the possibility of getting some answers. The essential ingredient in this is to show that if a Hardy field possesses an element of a certain growth, there must be other elements with different growths.

If $f \in \mathcal{F}$ with $f \to \infty$ and $i \in \mathbb{N}$, then $\gamma_i(\log f) < \gamma_i(f)$, but we cannot expect $\log f$ to be an element of our existing Hardy field, $\mathcal{F}$ (the fact that $\mathcal{F}(\log f)$ is also a Hardy field does not help us here). However the logarithmic derivative $f^\triangle = f'/f$ is an element of $\mathcal{F}$, and in some cases that suffices. In other cases we have to do something a little more complicated.

We write $L_p = x \log(x) \log_2(x) \cdots \log_{p-1}(x) = 1/(\log_p x)'$ for $p \in \mathbb{N}^+$, and for $p = 0$ we put $L_0 = 1$.

PROPOSITION **4** *Let f be a non-zero element of a Hardy field, $\mathcal{F}$, with $\gamma_0(f) \neq 0$. Then for $p \in \mathbb{N}$ we have the following:*
(1) If $\gamma_2(f) > \gamma_2(\log_{p-1}(x))$, then $\gamma_1(L_p f^\triangle) = \gamma_1(\log f) < \gamma_1(f)$.
(2) If $\gamma_2(f) < \gamma_2(\log_{p-1}(x))$, then $\log(L_p f^\triangle) \sim -\log_{p+1}(x)$; in particular $\gamma_1(L_p f^\triangle) = \gamma_1(\log_p(x))$.
(3) If $\gamma_2(f) = \gamma_2(\log_{p-1}(x))$ and $\log_2(f) \not\sim \log_{p+1}(x)$, then $\gamma_1(L_p f^\triangle) = \gamma_1(\log_p(x))$.
(4) If $\log_2(f) \sim \log_{p+1}(x)$ then $\gamma_1(L_p f^\triangle) < \gamma_1(\log_p(x)) = \gamma_1(\log f)$.

This result is somewhat technical, but its use will soon become apparent.

Proof of Proposition 4

We first consider the case $p = 0$. Then the hypotheses of (1) imply that $\log x / \log_2 f \to 0$, which yields $\gamma_1(\log f) > \gamma_1(x)$. Then Lemma 8 gives $\gamma_1((\log f)') = \gamma_1(\log f)$ and (1) follows. Similarly $\gamma_2(f) < \gamma_2(e^x)$ implies that $\gamma_1(\log f) < \gamma_1(x)$, which by Lemma 8 gives (2) for this case, and (3) is proved in a similar fashion. As for (4), we have $\log_2(f) \sim \log(x) \Rightarrow (\log_2(f))' \sim (\log(x))'$ and hence $f^\triangle/\log f \sim x^{-1}$. Therefore $\log f^\triangle \sim \log_2(f) - \log(x) = o(\log(x))$, which implies that $\gamma_1(f^\triangle) < \gamma_1(x)$.

For the case when $p > 0$, we make the change of variable $t = \log_p(x)$ and write $G(t) = f(\exp_p(t))$. The function–of–a–function rule then yields $G^\triangle(t) = (L_p f^\triangle)(\exp_p(t))$. Now the proposition for the case $p = 0$ applied to G gives the conclusion for f. For example, if $\gamma_2(f) > \gamma_2(\log_{p-1}(x))$ then $\gamma_2(G) > \gamma_2(e^t)$. Hence $\gamma_1(G^\triangle) = \gamma_1(\log G)$, by (1) in the case $p = 0$. This gives $\gamma_1(L_p f^\triangle) = \gamma_1(\log f)$, which is (1) for the general case. Similar arguments give (2), (3) and (4), and so establish Proposition 4.

Next for any C^∞ function, f, and any Hardy field $\mathcal{F}$, we write $\mathcal{F}\langle\langle f\rangle\rangle$ for the field of rational functions of real powers of f and its derivatives with coefficients in $\mathcal{F}$. Here we take $\mathcal{F}$ to be a field of constants, $\mathbb{K}$ and assume that $\mathbb{K}(f)$ is a Hardy field. Then by the second corollary to Theorem 10 $\mathcal{F}\langle\langle f\rangle\rangle$ is also a Hardy field.

On the assumption that $f \to \infty$, we define a sequence $\{f_i\}$, of elements of $\mathbb{K}\langle\langle f\rangle\rangle$ such that $\gamma_1(f_i)$ strictly decreases. The sequence $\{f_i\}$ may turn out to be finite or infinite.

We take $f_0 = f$, and define the other f_i recursively as follows:

(1) If $\gamma_2(f_i) > \gamma_2(e^x)$, we take $f_{i+1} = f_i^\triangle$. Then Proposition 4 (1), with $p = 0$, implies that $\gamma_1(f_{i+1}) = \gamma_1(\log f_i) < \gamma_1(f_i)$ as required.

(2) If $\gamma_2(f_i) = \gamma_2(e^x)$, let $r = \lim\{\log_2(f_i)/\log(x)\}$. Then $r \in \mathbb{R}^+$, and we have an asymptotic expression for f_i, namely $f_i = \exp(x^r\phi)$ where $\gamma_1(\phi) < \gamma_1(x)$. We refer to such an expression as a *partial nested form* (or *pnf*) for f_i. We take $f_{i+1} = f_i^\triangle$. Logarithmic differentiation of the pnf gives $f_{i+1} \sim rx^{r-1}\phi$. So if $r \neq 1$, $\gamma_1(f_{i+1}) = \gamma_1(x)$, and since then $f_{i+1}^\triangle \sim (r-1)x^{-1}$, we have

$$\frac{f_{i+1}}{r}\left(\frac{f_{i+1}^\triangle}{r-1}\right)^{r-1} \sim \phi.$$

Similarly, if $r = 1$, we obtain $f_{i+1} \sim \phi$ and so of course $\gamma_1(f_{i+1}) = \gamma_1(\phi)$. In both cases, we will have $\gamma_2(f_{i+1}) \le \gamma_2(x) < \gamma_2(e^x)$. Since we will have defined $f_0, \ldots, f_i$ in such a way that $\gamma_1(f_j)$ decreases, we must have $\gamma_2(f_j) > \gamma_2(e^x)$ for $j < i$. Hence $\gamma_1(f_i) = \gamma_1(\log_i(f))$ by case (1), and we have a pnf for f, namely

$$f = \exp_{i+1}(x^r\phi).$$

Moreover we have an element of $\mathbb{K}\langle\langle f\rangle\rangle$ which is asymptotic to ϕ.

(3) Now suppose that $\gamma_2(\log_p x) < \gamma_2(f_i) < \gamma_2(\log_{p-1} x)$, $p \ge 0$. We observe that if there is an element, g, of $\mathbb{K}\langle\langle f\rangle\rangle$ which is asymptotic to L_p, then Proposition 4(2) implies that

$$\log(gf_i^\triangle) \sim \log(L_p f_i^\triangle) \sim -\log_{p+1}(x).$$

Differentiation yields $-1/(gf_i^{\triangle})^{\triangle} \sim L_{p+1}$. Now we certainly have an element asymptotic to $L_0 = 1$, and since our use of Proposition 4(2) only requires that $\gamma_2(f_i) < \gamma_2(\log_{p-1}(x))$, we can apply the above argument with p replaced by $0, 1, \ldots, p-1$. Hence $\mathbb{K}\langle\langle f\rangle\rangle$ will indeed contain an element g asymptotic to L_p. We then set $f_{i+1} = gf_i^{\triangle}$ giving $\gamma_1(f_{i+1}) = \gamma_1(\log_p x)$; of course $f_{i+1}^{\triangle} \sim -(\log_{p+1} x)' = -L_{p+1}^{-1}$. Now we define $f_{i+2} = -f_i^{\triangle}/f_{i+1}^{\triangle}$ so that $f_{i+2} \sim L_{p+1}f_i^{\triangle}$. Then Proposition 4(1) with p replaced by $p+1$ implies that $\gamma_1(f_{i+2}) = \gamma_1(\log f_i)$. The hypotheses of this case then give directly that $\gamma_1(\log_{p+1} x) < \gamma_1(f_{i+2}) < \gamma_1(\log_p x)$.

(4) If $\gamma_2(f_i) = \gamma_2(\log_p x)$ with $p \geq 0$, we have the pnf $f_i = \exp(\log_{p+1}^r(x)\phi)$, where $r = \lim\{\log_2(f_i)/\log_{p+2}(x)\}$ and $\gamma_1(\phi) < \gamma_1(\log_{p+1}(x))$. If $r > 1$ or if $r = 1$ and $\phi \to \infty$, then $\gamma_1(f_i) > \gamma_1(\log_p(x))$, while if $r < 1$ or if $r = 1$ and $\phi \to 0$, then $\gamma_1(f_i) < \gamma_1(\log_p(x))$. We consider the case when $r = 1$ and $\phi \to k \in \mathbb{K}^*$ later.
As in (3), we may assume that there is an element g of $\mathbb{K}\langle\langle f\rangle\rangle$ which is asymptotic to L_p. Now if a, b are Hardy-field elements with $\gamma_1(b) < \gamma_1(a)$ then
$$(ab)' = ab\left(\frac{a'}{a} + \frac{b'}{b}\right) \sim a'b,$$
by Lemma 7. Hence
$$f_i^{\triangle} \sim r\phi \log_{p+1}^{r-1}(x)L_{p+1}^{-1} = r\phi\log_{p+1}^{r-1}(x)(\log_p x)^{-1}L_p^{-1}.$$
We take $f_{i+1} = gf_i^{\triangle}$, so that $f_{i+1} \sim r\phi\log_{p+1}^{r-1}(x)\log_p^{-1}(x)$. Then $\gamma_1(f_{i+1}) = \gamma_1(\log_p x)$ and $f_{i+1}^{\triangle} \sim -(\log_{p+1} x)' = -L_{p+1}^{-1}$. Next we set $f_{i+2} = -f_i^{\triangle}/f_{i+1}^{\triangle}$, and observe that
$$f_{i+2} \sim r\phi\log_{p+1}^{r-1} x. \tag{3.10}$$
If $r = 1$, then $\gamma_1(f_{i+2}) = \gamma_1(\phi)$. Otherwise $\gamma_1(f_{i+2}) = \gamma_1(\log_{p+1} x) = \gamma_1(\log f_i)$. Here too we can obtain an element of $\mathbb{K}\langle\langle f\rangle\rangle$ asymptotic to ϕ. We note that $f_{i+2}^{\triangle}/f_{i+1}^{\triangle} \sim (r-1)/\log_{p+1} x$. On putting
$$f_{i+3} = \frac{f_{i+2}}{r(r-1)^{r-1}}\left(\frac{f_{i+2}^{\triangle}}{f_{i+1}^{\triangle}}\right)^{r-1},$$
we see that $f_{i+3} \sim \phi$.
If $r = 1$ and $\phi \to k$ where $k \in \mathbb{K}^*$, then $f_i = \exp(\log_{p+1} x(k + o(1))) = \log^k(x)\psi$, where $\gamma_1(\psi) < \gamma_1(\log(x))$. Now $f_i^{\triangle} \sim kL_{p+1}$, and taking g as above, we then have $f_i^{\triangle}/(kg) \sim \log_p x$. Hence we may take
$$f_{i+1} = f_i\left(\frac{kg}{f_i^{\triangle}}\right)^k \sim \psi.$$

We introduced above the notion of a pnf. We now give a definition of this.

DEFINITION **8** *Let f be a positive $\mathcal{C}^\infty$ function which tends to zero or infinity as the variable, x, does. A* partial nested form *(or* pnf*) for f is a finite sequence $\{(\epsilon_i, s_i, m_i, d_i, \phi_i), i = 1, 2, \ldots, k\}$, such that*

1. *$\epsilon_i = \pm 1$, s_i and m_i are non-negative integers and d_i is a positive real number.*
2. *$f(x) = \exp_{s_1}^{\epsilon_1}(\log_{m_1}^{d_1}(x)\phi_1(x))$ and for $i = 1, \ldots, k$, we have $\phi_i(x) = \exp_{s_i}^{\epsilon_i}(\log_{m_i}^{d_i}(x)\phi_{i+1}(x))$.*
3. *For $i = 1, \ldots, k$, $\gamma_1(\phi_i(x)) < \gamma_1(\log_{m_i}(x))$.*

We refer to the expression $f(x) = \exp_{s_1}^{\epsilon_i}(\log_{m_1}^{d_1}(x)\phi_1(x))$ as the first pnf for f.

For a negative function tending to zero or to $-\infty$, a pnf is as above with a minus sign prefixed to the expression; i.e. $-\exp_{s_1}^{\epsilon_1}(\log_{m_1}^{d_1}(x)\phi_1(x))$.

If $f \to A \in \mathbb{K}^*$ the first pnf for f is the expression $f = A + \rho$, where it is understood that $\rho = f - A \to 0$. If $f \equiv A$ the only pnf is A. Otherwise the second pnf for f is $A +$ 1st-pnf(ρ), and so on.

We note that in cases (2) and (4) above we obtain a first pnf for f_i, while in cases (1) and (3) we obtain an element of $\mathbb{K}\langle\langle f\rangle\rangle$ asymptotic to $\log f_i$. Also if for some $p \in \mathbb{N}$, $\log_p f$ has a first pnf, then f has a first pnf. We therefore have the following conclusions.

THEOREM **11** *Suppose that f is an element of a Hardy field with $f \to \infty$. Then either*

[1] *f has a first pnf, $f = \exp_s(\log_m^d(x)\phi_1(x))$, or*
[2] *$\gamma_2(f) > \gamma_2(\exp_s(x))$ for all $s \geq 0$, or*
[3] *$\gamma_2(f) < \gamma_2(\log_m(x))$ for all $m \geq 0$, or*
[4] *There is a $q \in \mathbb{Z}$ such that for every $p \in \mathbb{N}$, $\gamma_2(\log_{q+p} x) < \gamma_2(\log_p f) < \gamma_2(\log_{q+p-1} x)$.*

If [2], [3] or [4] occur then $\mathbb{K}\langle\langle f\rangle\rangle$ contains elements asymptotic to each $\log_p(f)$, $p \geq 0$. Furthermore in case [4], $\mathbb{K}\langle\langle f\rangle\rangle$ contains elements asymptotic to $\log_m x$ for every $m \geq p$.

The statement regarding case [3] follows immediately from Proposition 4(2). Cases [2], [3] and [4] can certainly occur, as we shall see in Chapter 9. However for applications case [1] is usually the most important. In particular this is the only case which can occur when f satisfies an algebraic differential equation over $\mathbb{R}(x)$. We note that in case [1], $\mathbb{K}\langle\langle f\rangle\rangle$ contains an element asymptotic to ϕ_1. We examine this further in Chapter 6 and show that $\Upsilon_1(\mathbb{K}\langle\langle f\rangle\rangle)$ contains $s+m+\delta+1$ elements (including $\gamma_1(1)$) plus any additional classes generated by the element asymptotic to ϕ_1; here δ is 1 unless $d = 1$ and $m = 0$, in which case $\delta = 0$.

Some other properties of Hardy fields will be given in Chapter 9. Moreover Hardy fields will be used extensively in other chapters. Further properties of these interesting objects may be found in the literature cited at the beginning of the present chapter.

4

Output Data Structures

4.1 Asymptotic Power Series

Like his contemporaries, Euler was happy to use divergent series for the purposes of calculation. He was able to avoid errors firstly because his instincts were sure, but also because the functions he considered were relatively simple. For the mathematicians of Euler's day a function was something given explicitly by a formula, as indeed it is for us. Later, Abel and Cauchy began to regard a function as a process acting on numbers, and thereby came to consider much more general types of function. When the functions were no longer given so explicitly much more care was needed with the logic. Whether or not this lay behind their reasoning, Abel and Cauchy sought to impose a ban on the use of divergent series. From the point of view of pure mathematics at that time this was undoubtedly justified, but divergent series continued to give surprisingly accurate results, and continued to have their adherents. The controversy raged long until it was finally resolved in brilliant fashion by Poincaré, and independently by Stieltjes.

DEFINITION **9** *Let $\{a_n\}$ be any sequence of real numbers. If for each value of n,*

$$f(x) - \left(a_0 + \frac{a_1}{x} + \cdots + \frac{a_{n-1}}{x^{n-1}}\right) = \mathrm{O}(x^{-n})$$

as $x \to \infty$, we say that $\sum a_r x^{-r}$ is an asymptotic series *for f, and we write*

$$f(x) \simeq a_0 + \frac{a_1}{x} + \frac{a_2}{x^2} + \cdots. \tag{4.1}$$

Note that the partial sums may be used as approximations to the function $f(x)$ for sufficiently large values of x even though the series may diverge. The definition may be generalized in a number of ways. For example the a_ns may be complex and the exponents of x may be any sequence tending to $-\infty$. Asymptotic series may be added, multiplied and otherwise combined, very much as Taylor series can. However special care is needed when differentiating them; see [30](Chapter 1.6) for example.

One of the most important uses of asymptotic series lies in their application to the solution of differential equations. The approach is as follows, [19]. One first looks for a formal series solution and then tries to prove by independent means that there does indeed exist a solution which has such an asymptotic series. There are powerful techniques of summability available for the latter, and a considerable amount is known concerning the analytic continuation of solutions and the description of their singularities. It is not part of our purpose to survey the huge volume of work on divergent series that has taken place during the last hundred years and more. For details of classical work, we refer the reader to [19], [41], [30], [121] and [62]. An introduction to some more recent results may be found in [64]. Work on formal series solutions of differential equations is surveyed in [108].

One limitation of asymptotic series is that many functions do not have the sort of asymptotic behaviour that can be described by a power series; $\exp x$ and $\log x$ are obvious examples. A natural idea for overcoming this defect is to widen the definition of asymptotic series to allow series in other base functions, such as $\exp x$ or $\log x$. So for example one could consider series of the form

$$\sum_{\mathbf{n}} c_{\mathbf{n}} \mathbf{t}_1^{n_1} \mathbf{t}_2^{n_2} \cdots \mathbf{t}_k^{n_k}, \tag{4.2}$$

where $\mathbf{n} = (n_1, n_2, \ldots, n_k)$, the $c_{\mathbf{n}}$ are constants and $\mathbf{t}_1, \mathbf{t}_2, \ldots, \mathbf{t}_k$ are base functions. It is reasonable to stipulate that $\mathbf{t}_1, \mathbf{t}_2, \ldots, \mathbf{t}_k$ all have different comparability class, say $\gamma_1(\mathbf{t}_1) < \gamma_1(\mathbf{t}_2) < \cdots < \gamma_1(\mathbf{t}_k)$, and we may adopt the convention that each $\mathbf{t}_i$ tends to zero. However there are problems inherent in this approach.

One quite simple point that tends to get overlooked is the following. Suppose that we make a choice of base functions, and search for solutions, say to a differential equation, in terms of these. Then clearly we are only going to find solutions which have that sort of growth. This is obvious enough, but it means that the whole basis of the method requires us to prejudge the asymptotic growth of the solutions to be found.

Secondly, many quite ordinary functions do not have asymptotic series in terms of simpler base functions. One example is

$$\exp\left\{\frac{xe^x}{x-1}\right\}. \tag{4.3}$$

Here one can get an expansion as a product, but if exp is replaced by $\exp_2$ even that is no longer possible. Functions of this sort would seem to be quite common among solutions of non-linear differential equations.

The problem is particularly acute with regard to inverse functions. As conjectured by Hardy in [40], even the entire set of exp-log functions is an insufficient set of base functions in order to give first terms of asymptotic series for them; see the beginning of Chapter 7.

Finally, there is the problem of the meaning to be attached to (4.2). The definition of an asymptotic power series allows for different functions to have the same asymptotic series. So suppose that f and g are such. Then it makes sense to ask whether $f - g$ has an expansion in terms of other base functions, for example e^{-x}.

If one function, f, is chosen as standard from among all the functions having a particular expansion, E_0, and if f_1 also has expansion E_0, we can write

$$f_1(x) \sim \mathsf{E}_0(x) + \mathsf{E}_1(x), \tag{4.4}$$

where E_1 is the expansion of $f_1 - f$. We stress that (4.4) only makes sense with respect to a particular choice of f. There will be many functions with expansion $\mathsf{E}_0 + \mathsf{E}_1$ in the sense indicated above. Again, if f_1 is chosen as standard one can look at the expansion of the difference between f_1 and another such function, f_2. We would thus arrive at a formula

$$f_2(x) \sim \mathsf{E}_0(x) + \mathsf{E}_1(x) + \mathsf{E}_2(x),$$

say. This process can be continued, and one can regard (4.2) as a realisation of this idea. However if (4.2) is to be an asymptotic expansion of a function, one has to have standard representatives for various truncations of (4.2), for example

$$\sum_{n_1=0}^{\infty} c_{n_1,0,\ldots,0} \mathbf{t}_1^{n_1},$$

just as above one needs a standard function f having expansion E_0.

It is generally too restrictive to require all the expansions considered to converge, but in a particular context there may be a canonical way of associating a function with an expansion, such as a summation method. For transseries, Ecalle has given a number of powerful summation methods, which are apparently sufficient for almost all of the natural cases that arise. This is rightly hailed as an advantage of transseries, and some interesting work has been done concerning the physical interpretation of the *terms beyond all orders*, that is to say those following the first asymptotic expansion (E_0 in the above discussion); see [6] for example.

However for the purposes of symbolic computation, it is not enough to know that a series can be summed. One would need effective algorithms to find the sum. Since these seem unlikely to be forthcoming in the near future, it is necessary to take a somewhat different approach. We look at some possibilities in the following sections of this chapter.

4.2 Multiseries

Example (4.3) indicates where the difficulties with choosing simple base functions lie. Expansions may occur inside iterated exponentials, and any method which aspires to full generality must accommodate this fact. It turns out that once this has been done, all the generality one could reasonably ask for is obtained, [32].

Our algorithmic approach demands some further properties. We begin by formalizing our requirements for the base functions. The following definition is mutually recursive with that of a multiseries (Definitions 11 and 12).

DEFINITION **10** *Let $\mathcal{F}$ be a Hardy field and let $\mathbf{t}_1, \ldots, \mathbf{t}_n$ be elements of $\mathcal{F}$ which tend to zero and satisfy $\gamma(\mathbf{t}_1) < \gamma(\mathbf{t}_2) < \cdots < \gamma(\mathbf{t}_n)$. We say that $\{\mathbf{t}_1, \ldots, \mathbf{t}_n\}$ is an* asymptotic scale *if the following properties hold:*

1. $x^{-1} \in \{\mathbf{t}_1, \ldots, \mathbf{t}_n\}$.
2. *Each $\mathbf{t}_i$ is either of the form $\log_k^{-1} x$ or else $\log \mathbf{t}_i$ has a multiseries expansion in the scale $\{\mathbf{t}_1, \ldots, \mathbf{t}_{i-1}\}$ with every top-level summand tending to plus or minus infinity.*
3. *If $\log_k^{-1} x$ belongs to $\{\mathbf{t}_1, \ldots, \mathbf{t}_n\}$ for some $k > 0$ then so do $\log^{-1} x, \ldots, \log_{k-1}^{-1} x$.*

The reason for the last part of condition 2 is that summands which tend to finite limits inside an exponential can cause technical difficulties (see Chapter 5.2). As we shall see in Section 4.3.3, it is both natural and easy to remove such terms from scale elements.

For the definition of a multiseries, we consider first the case when $n = 1$.

DEFINITION **11** *We say that an element g of $\mathcal{F}$ has an* asymptotic $\mathbf{t}_1$-expansion $\sum c_m \mathbf{t}_1^{r_m}$ *if $\{c_m\}$ and $\{r_m\}$ are sequences of real numbers, with the following properties: r_m is strictly increasing and tends to infinity, and for each $N \geq 0$ there is a strictly positive real number, δ_N, such that*

$$g - \sum_{m=0}^{N} c_m \mathbf{t}_1^{r_m} = \mathrm{O}(\mathbf{t}_1^{r_N + \delta_N}). \tag{4.5}$$

Now for the general case.

DEFINITION **12** *Let $f \in \mathcal{F}$, and suppose that there exists a strictly increasing sequence of real numbers, $\{r_m\}$ with $r_m \uparrow \infty$, and a sequence of elements $\{g_m\} \subset \mathcal{F}$ such that for each $N \geq 0$ there is a $\delta_N \in \mathbb{R}^+$ with*

$$g - \sum_{m=0}^{N} g_m \mathbf{t}_n^{r_m} = \mathrm{O}(\mathbf{t}_n^{r_N + \delta_N}). \tag{4.6}$$

Then we say that g has $\{\mathbf{t}_1, \ldots, \mathbf{t}_n\}$ multiseries expansion $\sum g_m \mathbf{t}_n^m$ *provided that each g_m has a $\{\mathbf{t}_1, \ldots, \mathbf{t}_{n-1}\}$ multiseries expansion.*

We shall refer to (4.6) as the $\mathbf{t}_n$*-expansion* of g. It is not hard to see that both scales and multiseries are well defined despite the recursion.

Note that in our expansions we allow real powers of the base elements, not just integer powers. In the same way, when we use the term *monomial*, we shall mean a constant times a product of real powers of the base elements in question.

As a simple example we consider the function $\exp(x^{-1}+\log^{-1}x)$. We take as scale elements $\mathbf{t}_1(x)=\log^{-1}x$ and $\mathbf{t}_2(x)=x^{-1}$. Then

$$\begin{aligned}\exp(x^{-1}+\log^{-1}x) &= \sum_{m=0}^{\infty}\exp(\log^{-1}x)\frac{\mathbf{t}_2^m(x)}{m!}\\ &= \sum_{m=0}^{\infty}\left(\sum_{n=0}^{\infty}\frac{\mathbf{t}_1^n(x)}{n!}\right)\frac{\mathbf{t}_2^m(x)}{m!}.\end{aligned}$$

Of course the coefficient of $\mathbf{t}_k^r$ will generally depend on r. Moreover as we shall see later, care needs to be exercised when using standard power series expansions to produce multiseries.

To say that the asymptotic $\mathbf{t}_1$-expansion (4.5) is *computable* means that we have an algorithm to effectively compute c_m and r_m for any given $m\geq 0$. To say that the asymptotic $\{\mathbf{t}_1,\ldots,\mathbf{t}_n\}$-expansion (4.6) is computable will mean that we have an algorithm which, for any m, outputs r_m, a closed form for g_m and an algorithm to compute its asymptotic $\{\mathbf{t}_1,\ldots,\mathbf{t}_{n-1}\}$-expansion. In practice the closed form for g_m will need to be one to which zero-equivalence tests can be applied. In [114], van der Hoeven takes a different line, in keeping with his stance of regarding multiseries more as formal objects in the spirit of Ecalle. Zero-equivalence problems are thereby circumvented, but at the price of weakening the connection with functions. Our present treatment follows [103] and [73].

It will be useful to have a standard notation for the coefficients of the various series of a $\{\mathbf{t}_1,\ldots,\mathbf{t}_n\}$ expansion. Suppose then that the element g has a computable $\{\mathbf{t}_1,\ldots,\mathbf{t}_n\}$-expansion. We write the $\mathbf{t}_n$-expansion of g as $g\sim\sum_{m_n}g_{\{m_n\}}\mathbf{t}_n^{r\{m_n\}}$. We denote the coefficients in the $\mathbf{t}_{n-1}$ expansion of $g_{\{m_n\}}$ by $g_{\{m_n,m_{n-1}\}}$, so that $g_{\{m_n\}}=\sum g_{\{m_n,m_{n-1}\}}\mathbf{t}_{n-1}^{r\{m_n,m_{n-1}\}}$. Similarly the coefficients in the $\mathbf{t}_{n-2}$-expansion of $g_{\{m_n,m_{n-1}\}}$ are denoted by $g_{\{m_n,m_{n-1},m_{n-2}\}}$, and so on.

4.3 Operations on Multiseries

Arithmetic operations involving two multiseries expansions with the same scale are derived from the corresponding operations for power and puiseux series. When the scales are different, we have to start by amalgamating them; see §4.3.5.

Let $\{\mathbf{t}_1,\ldots,\mathbf{t}_n\}$ be an asymptotic scale. Suppose that f and g are functions having multiseries expansions in this scale

$$f=\sum_{m=0}^{\infty}f_m\mathbf{t}_n^{r_m}\quad\&\quad g=\sum_{m=0}^{\infty}g_m\mathbf{t}_n^{s_m}$$

respectively. We first check that $f+g$ is not functionally equivalent to zero, since this would not otherwise be recognized by our algorithm; in fact the algorithm would perpetually generate zero terms. In this situation some algebra systems just return an answer like "$0+O(x^{-25})$", which is less than satisfactory, especially if the answer is going to be fed into another calculation.

Having discovered that $f+g \not\equiv 0$, we let $\{v_m\}$ be the sequence obtained by merging $\{r_m\}$ and $\{s_m\}$ and for $k \in \mathbb{N}$ define

$$F_k = \begin{cases} f_i, & v_k = r_i \ \& \ v_k \notin \{s_m\} \\ g_j, & v_k = s_j \ \& \ v_k \notin \{r_m\} \\ f_i + g_j, & v_k = r_i = s_j. \end{cases}$$

Then the $\mathbf{t}_n$ expansion of $f+g$ is

$$f+g = \sum_{k=0}^{\infty} F_k \mathbf{t}_n^{v_k}.$$

Since $f+g \not\equiv 0$ we must eventually find a non-zero F_k. The expansions of the F_k can be obtained in the same way from the expansions of the appropriate f_i and g_j, and so on recursively.

Subtraction is almost identical, but multiplication is more complicated. The main problem is to ensure the correct ordering of terms $f_i g_j \mathbf{t}^{r_i+s_j}$. We define a partial order $\prec$ on $\mathbb{N}\times\mathbb{N}$ by setting $(i_1,j_1) \prec (i_2,j_2)$ if $i_1 \le i_2$ and $j_1 < j_2$ or if $j_1 = j_2$ and $i_1 < i_2$. Then we maintain a '*waiting list*', W, of pairs that might give the index of the next term. The waiting list will satisfy two rules. The first is that if $(i_1,j_1) \in W$ and $(i_1,j_1) \prec (i_2,j_2)$, then $(i_2,j_2) \notin W$. The second is that for any $(i,j) \in (\mathbb{N}\times\mathbb{N})\setminus W$ which has not already been used in the expansion of the product there exists a $(\alpha,\beta) \in W$ with $(\alpha,\beta) \prec (i,j)$. Thus W will be the largest set of pairs that are minimal among those that have not already been used in the product expansion.

We start with $W = \{(0,0)\}$. The pair $(0,0)$ is used to generate the first product term, $f_0 g_0 \mathbf{t}_n^{r_0+s_0}$, after which W becomes empty. Now suppose that we have the first $N-1$ terms of the $\mathbf{t}_n$–expansion of fg, with $N>1$. We let V be the subset of W consisting of those pairs, $(i,j) \in W$, for which $r_i + s_j$ is minimal. The next terms of the $\mathbf{t}_n$–expansion of fg are the summands of $\sum_{(i,j)\in V} f_i g_j \cdot \mathbf{t}_n^{r_i+s_j}$. Then W is changed by replacing each $(i,j) \in V$ by the two elements $(i+1,j)$ and $(i,j+1)$ *provided that the first rule is not broken.* If $(i+1,j)$ is larger under $\prec$ than an existing element of W then (i,j) is still removed but $(i+1,j)$ is not added, and similarly for $(i,j+1)$. Note that both rules will be satisfied by the new waiting list. To see that this is the case for the second rule, observe that any (i',j') which is greater than (i,j) is either greater than $(i+1,j)$ or greater than $(i,j+1)$ or is equal to one of these pairs. Of course the $\mathbf{t}_{n-1}$–expansions of the various $f_i g_j$ are obtained in the same way.

To see that the algorithm orders the terms of the product expansion correctly, we note that if $(i,j) \prec (i',j')$ then $r_i + s_j < r_{i'} + s_{j'}$. So at each stage we should use pairs from W.

As with addition and subtraction, we have to consider the possibility that an expansion might terminate finitely. So if we want an expansion up to a certain number of non-zero terms, N, then when a partial expansion of $M < N$ terms of fg has been obtained, say $\sum_{m=0}^{M-1} h_m \mathbf{t}^{v_m}$, we must check whether $fg - \sum_{m=0}^{M-1} h_m \mathbf{t}^{v_m} \equiv 0$. In practice one would probably only do this after a certain number of cancellations had been observed.

As an example of the working of the algorithm, we consider the case when

$$f \sim \sum_{m=0}^{\infty} f_m x^{-r_m} = \sum_{m=0}^{\infty} (m+2) x^{-\log(m+1)}$$

and

$$g \sim \sum_{m=0}^{\infty} g_m x^{-s_m} = \sum_{m=0}^{\infty} (m+1)^2 x^{\sqrt{m}}.$$

Initially $W = \{(0,0)\}$ and the first term of the product is $2 \cdot 1^2 x^{0+0} = 2$. Now $(0,0)$ is removed from W and $(1,0)$ and $(0,1)$ are added; so $W = \{(1,0),(0,1)\}$. Since $r_1 + s_0 = \log 2 < 1 = r_0 + s_1$, the set V is just $\{(1,0)\}$ and the second term of the product is $3x^{-\log 2}$. The pairs $(2,0)$ and $(1,1)$ are candidates for adding to W but $(1,1)$ is larger than the existing element $(0,1)$. So $(1,0)$ is replaced by $(2,0)$ and W becomes $\{(0,1),(2,0)\}$. Since $\log 3 > 1$, the new V is $\{(0,1)\}$ and the next term is $8x^{-1}$. After a few more steps, the same process yields the expansion

$$fg \sim 2 + 3x^{-\log 2} + 8x^{-1} + 4x^{-\log 3} + 5x^{-\log 4} + 9x^{-\sqrt{2}} + 6x^{-\log 5} \\ +12x^{-1-\log 2} + \cdots .$$

Very often the expansions that arise have integral exponents, or at least sets of exponents finitely generated over the integers, and then faster methods can be used; see [25, 26, 86, 118].

4.3.1 Substituting into a Power Series

Next we look at substituting a multiseries into a power series. The results will be applied to the problems of taking reciprocals, logarithms and exponentials of multiseries. So let f have $\mathbf{t}_n$–expansion $\sum_{m=0}^{\infty} f_m \mathbf{t}_n^{r_m}$ as above, and let $\{a_n\}$ be a sequence of real numbers. We consider the effect of substituting $z \mapsto f$ into $g(z) = \sum a_n z^n$. It is fairly quickly apparent that there is an algorithmic problem if $r_0 \le 0$. For example if $r_0 = 0$ *every* term $a_n f^n$ will contribute to the coefficient of $\mathbf{t}_n^0$ in the final expansion, so how can we know whether that coefficient is functionally equivalent to zero or not?

If g is analytic at the limit of f_0 is finite we can use the following device: Let $\eta = \sum_{m=1}^{\infty} f_m \mathbf{t}_n^{r_m}$. Then if $r_0 = 0$, we have

$$g(f) = g(f_0 + \eta) = g(f_0) + \eta g'(f_0) + \cdots + \eta^k g^{(k)}(f_0) + o(\eta^k) \tag{4.7}$$

for every $k \geq 1$. So the coefficient of $\mathbf{t}_n^c$ may be obtained from expanding η in $\eta g'(f_0) + \cdots + \eta^k g^{(k)}(f_0)$ where k is the greatest integer such that $kr_1 \leq c$. In some important special cases, for example when g is the exponential function, more efficient methods are available.

If $r_0 < 0$ then substitution into $\sum a_n z^n$ will generate something that is not an asymptotic series. We are not able to overcome such a difficulty at this level of generality, although again in some important special cases this can be done.

For now, we shall assume that $r_0 > 0$. Essentially we have to multiply out the powers of $\sum_{m=0}^{\infty} f_m \mathbf{t}_n^{r_m}$ and collect together terms. So it is a matter of giving an implementable algorithm for this. Let us refer to the series we are generating as S. Our algorithm maintains a list of lists, L. The p–th element of L, call it L_p, is a list of non–decreasing p–tuples of natural numbers. Each p–tuple is a set of indices into the sequence $\{r_m\}$, and represents the possible terms from $(\sum_{m=0}^{\infty} f_m \mathbf{t}_n^{r_m})^p$ that might be used next. We call $(\sum_{m=0}^{\infty} f_m \mathbf{t}_n^{r_m})^p$ the *p-power*. Clearly the first term of S is a_0. To compute the next term, we put $L = \{\{(0)\}\}$ to indicate that the $\mathbf{t}^{r_0}$ term from the 1–power is the next candidate. Since it is the only one, the next term is $a_1 f_0 \mathbf{t}_n^{r_0}$. Candidates for the next term come from the $\mathbf{t}_n^{2r_0}$ term in the 2–power and the $\mathbf{t}_n^{r_1}$ term in the 1–power. So $L = \{\{(1)\}, \{(0,0)\}\}$, where (1) indicates the $\mathbf{t}_n^{r_1}$ term and $(0,0)$ the $\mathbf{t}_n^{2r_0}$ term. If $2r_0 < r_1$ the third term of S is $a_2 f_0^2 \mathbf{t}_n^{2r_0}$ and L becomes $\{L_1, L_2, L_3\} = \{\{(1)\}, \{(0,1)\}, \{(0,0,0)\}\}$. Similarly if $2r_0 > r_1$ the third term of S is $a_1 f_1 \mathbf{t}^{r_1}$ and L becomes $\{\{(2)\}, \{(0,0)\}\}$. Finally if $2r_0 = r_1$, the third term of S is $(a_2 f_0^2 + a_1 f_1)\mathbf{t}^{r_1}$ and L becomes $\{\{(2)\}, \{(0,1)\}, \{(0,0,0)\}\}$. So far we have only one p–tuple in each L_p ($p = 1, \ldots, 3$), but that will change as the algorithm progresses.

To describe the way that L changes, we introduce a partial order, $\prec_p$, on non–decreasing tuples of natural numbers. We set

$$(m_1, m_2, \ldots, m_p) \prec_p (n_1, n_2, \ldots, n_p)$$

if $m_i \leq n_i$ for all $i = 1, \ldots, p$ and for at least one value of i, $m_i < n_i$. The rules for L_p are similar to those for W in the multiplication algorithm. L_p must not contain two comparable elements, and it contains a minimal set of unused p–tuples,. The last element, L_p, of L is always a list of one tuple consisting of p zeros.

To each p–tuple, $\pi = (n_1, n_2, \ldots, n_p)$, in L_p we attach the number s_π given by

$$s_\pi = \sum_{i=1}^{p} r_{n_i}.$$

When we need to choose the next term in the result expansion, S, we take a set, V, consisting of the elements, π, of $\cup L_p$ for which the appropriate s_π is minimal. The exponent of $\mathbf{t}_n$ in the next term of S is then this common value of s_π. However more work is needed to calculate the coefficient.

Suppose that $(n_1, n_2, \ldots, n_p)$ is an element of V consisting of k_1 repeated N_1s, followed by k_2 repeated N_2s, and so on, ending with k_j repeated N_js. Then the coefficient of $\mathbf{t}_n^{s_\pi}$ in S has a factor which is the number of combinations, namely $|\pi|!/(k_1!k_2!\cdots k_j!)$. Here $|\pi|$ is the length of π, that is to say the number of entries in the tuple. Thus the next term of S is

$$\left(\sum_{\pi\in V} a_{|\pi|}\frac{|\pi|!}{k_1!k_2!\cdots k_j!} f_{N_1}^{k_1} f_{N_2}^{k_2}\cdots f_{N_j}^{k_j}\right)\mathbf{t}_n^{s_\pi}.$$

Now we have to change L. Each element of V is removed from the L_p in which it appears and is replaced by all the p–tuples which are minimally larger than it under $\prec_p$ provided that the rules are not violated. So if $(n_1, n_2, \ldots, n_p)$ as above is removed, we replace the last k_1 by k_1+1, and provided that the resulting tuple is not larger than any now in L_p this tuple is added. Then the last k_2 in the original removed tuple is replaced by k_2+1 and the result added provided it is not larger than an existing element of L_p. This process is continued, so that finally we replace the last k_j by k_j+1 and check as before. If the single element of the last L_p, consisting of repeated zeros, is replaced, a new L_{p+1} with a single tuple consisting of $p+1$ zeros is also added.

As with the arithmetic operations, it will generally be necessary to use zero–equivalence methods at various stages to check whether an expansion has terminated. Indeed this is something that always needs to be done whenever we want a particular number of non–zero terms in an expansion.

The proof of correctness of the algorithm is similar to that used for the multiplication algorithm. Because L contains all the minimal unused tuples and $\{r_m\}$ increases, the smallest value of s_π will be obtained only when $\pi\in L$. The procedure for changing L ensures that this key property of L is maintained.

Example
As a relatively simple example, we compute the asymptotic series of $f(x) = \exp\{\sqrt{x}/(x+1)\}$. We take $t = x^{-1}$. Then

$$\begin{aligned} f(x) &= \exp\left\{\frac{t^{-\frac{1}{2}}}{t^{-1}+1}\right\} = \exp\left\{\frac{t^{\frac{1}{2}}}{1+t}\right\} \\ &= \exp\left\{\sum_{m=0}^{\infty}(-1)^m t^{m+\frac{1}{2}}\right\} \\ &= \sum_{n=0}^{\infty}\frac{1}{n!}\left(\sum_{m=0}^{\infty}(-1)^m t^{m+\frac{1}{2}}\right)^n . \end{aligned} \tag{4.8}$$

So $a_n = 1/n!$, $f_m = (-1)^m$ and $r_m = m+\frac{1}{2}$. Clearly the expansion of f begins $f(x)\sim 1+t^{\frac{1}{2}}$. At this stage, L becomes $\{\{(1)\},\{(0,0)\}\}$, having previously been $\{\{(0)\}\}$. Now $s_{(0,0)} = 2r_0 = 1$ and $s_{(1)} = r_1 = 1\frac{1}{2}$. Therefore

$V = \{(0,0)\}$ and the next term in the expansion is $t/2$. Now L changes to $\{\{(1)\},\{(0,1)\},\{(0,0,0)\}\}$. We have $s_{(0,1)} = r_0 + r_1 = 2$, while $s_{(0,0,0)} = 3r_0 = 1\frac{1}{2} = s_{(1)}$. Thus $V = \{(1),(0,0,0)\}$ and the next term is $(-1+\frac{1}{3!})t^{\frac{3}{2}}$. The expansion thus far is

$$f \sim 1 + t^{\frac{1}{2}} + \frac{1}{2}t - (1 - \frac{1}{3!})t^{\frac{3}{2}} + \cdots .$$

Then (1) is removed from L_1 and replaced by (2), $(0,0,0)$ is replaced in L_3 by $(0,0,1)$ and a new $L_4 = \{(0,0,0,0)\}$ is added to L, which therefore becomes

$$\{\{(2)\},\{(0,1)\},\{(0,0,1)\},\{(0,0,0,0)\}\}.$$

A simple check gives $V = \{(0,1),(0,0,0,0)\}$. The new term to be added is thus $(-2\cdot 1/2! + 1/4!)t^2$. Then L becomes

$$\{\{(2)\},\{(1,1),(0,2)\},\{(0,0,1)\},\{(0,0,0,1)\},\{(0,0,0,0,0)\}\},$$

and the expansion is

$$f \sim 1 + t^{\frac{1}{2}} + \frac{1}{2}t - (1 - \frac{1}{3!})t^{\frac{3}{2}} - (1 - \frac{1}{4!})t^2 + \cdots .$$

4.3.2 The Logarithm of a Multiseries

Let f have $\mathbf{t}_n$–expansion $\sum_{m=0}^{\infty} f_m \mathbf{t}_n^{r_m}$ as above. By considering the first term in the $\mathbf{t}_{n-1}$–expansion of f_0, and the first term in its expansion, and so on, we can compute an expression for f of the form

$$f = A\mathbf{t}_n^{r_0^n}\mathbf{t}_{n-1}^{r_0^{n-1}}, \ldots, \mathbf{t}_1^{r_0^1}(1+\epsilon), \tag{4.9}$$

where A is a non–zero constant and ϵ tends to zero and has a computable $\{\mathbf{t}_1 \ldots, \mathbf{t}_n\}$ multiseries expansion. If $r_0^1 \neq 0$, then we begin by adding $-\log^{-1}\mathbf{t}_1$ as a new scale element; it will be of the form $\log_j^{-1} x$. Then

$$\log f = r_0^n \log \mathbf{t}_n + r_0^{n-1}\log \mathbf{t}_{n-1} + \cdots + r_0^1 \log \mathbf{t}_1 + \log A - \log(1+\epsilon). \tag{4.10}$$

Suppose that the $\mathbf{t}_n$–expansion of ϵ is $\sum_{i=0}^{\infty} g_i \mathbf{t}_n^{s_i}$. If $s_0 > 0$ we may expand $\log(1+\epsilon)$ immediately using the method of Section 4.3.1. On the other hand, if $s_0 = 0$, we write

$$\begin{aligned}\log(1+\epsilon) &= \log\left(1 + g_0 + \sum_{i=1}^{\infty} g_i \mathbf{t}_n^{s_i}\right)\\ &= \log(1+g_0) + \log\left(1 + \sum_{i=1}^{\infty}\frac{g_i}{1+g_0}\mathbf{t}_n^{s_i}\right),\end{aligned} \tag{4.11}$$

and obtain the $\mathbf{t}_n$–expansion of $\log f$ by substituting from (4.11) into (4.10) and using the method of Section 4.3.1.

4.3.3 The Exponential of a Multiseries

Let f have $\mathbf{t}_n$–expansion $\sum_{m=0}^{\infty} f_m \mathbf{t}_n^{r_m}$ as before. By inspecting the signs of successive r_m (and if some $r_{m_0} = 0$ considering signs of exponents in the multiseries of f_{m_0}, and so on), we can write the multiseries for f as the sum $F + c + g$ where $F \to \infty$ (or else $F = 0$), c is a constant and $g \to 0$. Clearly

$$\exp(f) = \exp(F)e^c \exp(g) = \exp(F)e^c \sum_{i=0}^{\infty} g^i/i!. \tag{4.12}$$

The exponential series here can be handled in a similar way to the logarithmic series in the previous section; further details are given below. e^c may be a new constant, in which case our constant field needs to be appropriately extended.

If F is not zero and is not given as a logarithm, $\exp F$ is a candidate for a new scale element. We compare F with $\log \mathbf{t}_j$ for all $j = 1, \ldots, n$, adding an extra logarithm to the scale if necessary. By condition 2 of Definition 10, we have a $\{\mathbf{t}_1, \ldots, \mathbf{t}_n\}$ multiseries expansion of each $\log \mathbf{t}_j$, and by hypothesis we can compute a multiseries for F in the same scale. If we discover that $F \sim C \log \mathbf{t}_j$ for some j with C constant, then we write $\psi = \exp(F)\mathbf{t}_j^{-C}$ and apply the same consideration to ψ, iterating the process as many times as is needed. Of course this can only be a finite number of times because we must have $\log \psi = o(\log \mathbf{t}_j)$ and so the value of j must decrease with each iteration.

At some stage we may discover that $F/\log \mathbf{t}_j$ does not tend to a non–zero finite limit for any $j = 1, \ldots, n$. Then $\gamma_1(\exp F)$ is different from each $\gamma_1(\mathbf{t}_j)$, $j = 1, \ldots, k$, and we do indeed have to add a new element to the scale. Since we will already have calculated the limit of each $F/\log \mathbf{t}_j$, $j = 1, \ldots, n$, we know the order relation between $\gamma_1(\exp(F))$ and the various $\gamma_1(\mathbf{t}_j)$. Hence we can insert $\exp(\pm F)$ in its proper position in the new scale.

When it comes to dealing with the infinite series in (4.12) it is possible, as in the previous section, that the $\mathbf{t}_n$–expansion of g begins with a term in $\mathbf{t}_n^0$. Then, rather than employing (4.7), we use the functional equation of the exponential function. Thus if g has $\mathbf{t}_n$–expansion $g(x) = \sum_{m=0}^{\infty} g_m \mathbf{t}_n^{r_m}$ with $r_0 = 0$,

$$\exp(g) = \exp(g_0) \exp\left(\sum_{m=1}^{\infty} g_m \mathbf{t}_n^{r_m} \right) = \exp(g_0) \sum_{i=0}^{\infty} \sum_{m=1}^{\infty} \frac{(g_m \mathbf{t}_n^{r_m})^i}{i!}.$$

Now the methods of Section 4.3.1 can be used to deal with the double series, and the required condition on the lead exponent is satisfied.

4.3.4 Powers of a Multiseries

Let f be as above. To calculate a multiseries for $f^c, c \in \mathbb{R}$ we first write f in the form (4.9). If it turns out that the $\mathbf{t}_n$–expansion of ϵ begins with a zero power, we next apply the equation

$$(1+\epsilon_0+\eta)^c = (1+\epsilon_0)^c \left(1+\frac{\eta}{1+\epsilon_0}\right)^c,$$

where η is the tail of the $\mathbf{t}_n$–expansion of ϵ. Otherwise,

$$f^c = A^c \mathbf{t}_n^{cr_0^n} \mathbf{t}_{n-1}^{cr_0^{n-1}}, \dots, \mathbf{t}_1^{cr_0^1} \sum_{i=0}^{\infty} \frac{\Gamma(c+1)}{\Gamma(c+1-i)} \epsilon^i.$$

Of course we apply the method of Section 4.3.1 to the sum.

It is by now apparent that, modulo zero–equivalence problems with constants, we can calculate a multiseries expansion for any given exp–log function by representing the function as a tree and building the multiseries from the leaves upwards. The algorithm thus obtained first appeared in [73] and was derived from an earlier (and more clumsy) method in [97].

4.3.5 Amalgamating Scales

It may often be necessary to combine in some way two multiseries whose associated scales are different, and we then need to amalgamate these scales.

There is clearly no difficulty with reciprocals of iterated logarithms, and so it is not hard to see that we can reduce to the situation where we have an existing scale $\{\mathbf{t}_1, \dots, \mathbf{t}_n\}$ and a potential new scale element $\exp F$with F having a multiseries in $\{\mathbf{t}_1, \dots, \mathbf{t}_n\}$. However we have already covered this in §4.3.3.

4.3.6 Substituting One Multiseries Into Another

Let f have multiseries as above and let g be a function having a multiseries in a scale $\{\mathbf{s}_1, \dots, \mathbf{s}_p\}$ with $\mathbf{s}_p$–expansion $g = \sum_{j=0}^{\infty} g_j \mathbf{s}_p^{q_j}$. We wish to obtain a multiseries for $g \circ f$. We begin by substituting into the scale elements $\{\mathbf{s}_1, \dots, \mathbf{s}_p\}$. For scale elements which are reciprocals of iterated logarithms, we can use §4.3.2 and §4.3.4. On the other hand if $\mathbf{s}_j = \exp h$, for some j, we may assume by induction that we can calculate a multiseries for $h \circ f$. Then the multiseries for $\mathbf{s}_j \circ f$ is obtained using §4.3.3.

Now we have to substitute into the $\mathbf{s}_p$–expansion of g. By induction we may assume that we can do this for the coefficients, i.e. we can calculate expansions for $g_m \circ f$. We now need to adapt the algorithm of §4.3.1 to our present circumstances. There are two additional complications. The coefficients are not in general constant but have multiseries themselves, and the powers q_j are not in general integral.

The first problem is easily overcome, since the algorithm of §4.3.1 produces coefficients which are finite combinations of the coefficients, a_n, of the power series and the coefficients of the $\mathbf{t}_n$–expansion of f. When the a_ns are replaced by coefficients from the $\mathbf{s}_p$–expansion of g, we still only have to handle finite combinations, which we can do by amalgamating scales and using our methods for addition and multiplication.

As regards the second problem, suppose that $\mathbf{s}_p \circ f$ has top–level multiseries expansion $\sum_{k=0}^{\infty} h_k W^{c_k}$. Then we may write

$$g \circ f = \sum_{j=0}^{\infty} \left\{ g_j(f) h_0^{q_j} W^{c_0 q_j} \left(1 + \sum_{i=1}^{\infty} h_i h_0^{-1} W^{c_i - c_0}\right)^{q_j} \right\}.$$

The q_j–th power of the series may be expanded as in §4.3.4, and hence the exponents of W occurring in the expansion of $g \circ f$ are of the form $c_0 q_j + N(c_i - c_0)$ with $N \in \mathbb{N}$. It is thus a matter of getting these in the correct order and attaching the required coefficients. The algorithm of §4.3.1 may be suitably adapted. This may be regarded as an exercise for readers who like such things!

4.4 Nested Expansions

The idea behind nested expansions is to use iterated logarithms of the variable, x, as base elements and to regard exponentiation as an operator on a par with addition. Once an exponential 'exp($\cdots$' is opened the matching ')' only appears if and when the expansion terminates. So expansion typically takes place inside exponentials, and the exponentials nest inside one another; hence the name. On the one hand, this theory can be seen as attempted generalisation of the theory of asymptotic series, or perhaps more accurately the theory of continued fractions. On the other, it may be regarded as a development of the work on orders of infinity initiated by du Bois Reymond, [31], and continued by Hardy, [39].

We recall the notation $\log_n$ and $\exp_n$ for the n-times-iterated logarithm and exponential respectively.

An example of a nested expansion is $\phi(x)$ given by

$$\exp_2\{x^2 \exp(\log_2^{\pi/2} x(3 + \exp^{-1}(\log_3^3 x\{\sqrt{7} - \exp_2^{-1}\{\log_6^3 x - 3\log_6^2 x + \cdots. \tag{4.13}$$

However an expression such as $x^2 + e^{-x} - e^{-2x}$ is not a nested expansion because the exponentials fail to nest. It can be made into one by rewriting

$$x^2 + e^{-x} - e^{-2x} = x^2 \left(1 - \frac{e^{-x}}{x^2}(1 - e^{-x})\right).$$

For convenience, we give again the definition of a pnf from Chapter 3.

DEFINITION **(8)** *Let f be a positive C^∞ function which tends to zero or infinity as the variable, x, does. A* partial nested form *or* pnf *for f is a finite sequence $\{(\epsilon_i, s_i, m_i, d_i, \phi_i),\ i = 1, 2, \ldots, k\}$, such that*

1. *$\epsilon_i = \pm 1$, s_i and m_i are non-negative integers and d_i is a positive real number.*
2. *$f(x) = \exp_{s_1}^{\epsilon_1}(\log_{m_1}^{d_1}(x)\phi_1(x))$ and for $i = 1,\ldots,k$, we have $\phi_i(x) = \exp_{s_i}^{\epsilon_1}(\log_{m_i}^{d_i}(x)\phi_{i+1}(x))$.*
3. *For $i = 1,\ldots,k$, $\gamma_1(\phi_i(x)) < \gamma_1(\log_{m_i}(x))$.*

We refer to the expression $f(x) = \exp_{s_1}^{\epsilon_i}(\log_{m_1}^{d_1}(x)\phi_1(x))$ as the first pnf for f.

For a negative function tending to zero or to $-\infty$, a pnf is as above with a minus sign prefixed to the expression; e.g. $-\exp_{s_1}^{\epsilon_1}(\log_{m_1}^{d_1}(x)\phi_1(x))$.

If $f \to A \in \mathbb{K}^*$ the first pnf for f is the expression $f = A + \rho$, where it is understood that $\rho = f - A \to 0$. If $f \equiv A$ the only pnf is A. Otherwise the second pnf for f is $A +$ 1st-pnf(ρ), and so on.

Condition 3 implies that

$$m_{i+1} \geq m_i + \max\{1, s_{i+1}\}. \tag{4.14}$$

Note also that negative powers of the $\log_i$ are obtained by taking the appropriate $\epsilon_i = -1$, since d_i is required to be positive; thus x^{-2} is given by the sequence $\{(-1, 0, 0, 2, 1)\}$.

If in the expression $f = \exp_{s_1}(\log_{m_1}^{d_1}(x)\phi_1(x))$, we have $m_1 > 0$, $s_1 > 0$, $d_1 = 1$ and $\phi_1 = A + o(1)$, then it would make more sense to express our information by $f = \exp_{s_1-1}(\log_{m_1-1}^{A}(x)\psi(x))$, where $\psi(x) = \log_{m_1-1}^{\phi_1-A}(x)$. If $A > 0$ this is a 'simpler' first pnf than $f = \exp_{s_1}(\log_{m_1}^{d_1}(x)\phi_1(x))$. If $A < 0$ and $s_1 - 1 > 0$, then the exponentials in $\exp_{s_1-1}(\log_{m_1-1}^{A}(x)\psi(x))$ need to be expanded as in Section 4.4.2. If $A = 1$, $m_1 - 1 > 0$, $s_1 - 1 > 0$ and ψ tends to a non-zero constant, a further reduction can be performed. This multiplicity of representation is clearly a drawback of pnfs, which we rectify in the next definition.

DEFINITION **13** *A* nested form *for a function f tending to zero or to plus or minus infinity is a pnf, $\pm\{(\epsilon_i, s_i, m_i, d_i, \phi_i), i = 1, 2, \ldots, k\}$, such that*

1. *ϕ_k tends to a non-zero finite limit.*
2. *If $s_k \geq 1$ and $m_k \geq 1$ then $d_k \neq 1$.*

If $f \to A \in \mathbb{K}^*$ the nested form for f is $A +$ nested-form$(f - A)$.

In example (4.13), the nested form for ϕ is the sequence $\{(+1, 2, 0, 2, \phi_1)$, $(+1, 1, 2, \pi/2, \phi_2)\}$, where

$$\phi_2 = 3 + \exp^{-1}(\log_3^3 x\{\sqrt{7} - \exp_2^{-1}\{\log_6^3 x - 3\log_6^2 x + \cdots.$$

The nested form of ϕ_2 is $3 + \psi$ where $\psi = \{(-1, 1, 3, 3, \psi_1)\}$, with

$$\psi_1 = \sqrt{7} - \exp_2^{-1}\{\log_6^3 x - 3\log_6^2 x + \cdots$$

and ψ_1 having nested form $\sqrt{7} - \eta$ with $\eta = \{(-1, 2, 6, 3, \lambda)\}$. Here we have a succession of nested forms, ϕ, ϕ_2, ψ, ψ_1, η, λ, ... which build an expansion for $\phi(x)$ somewhat analogous to an asymptotic series. This is a nested expansion. The formal definition is as follows.

DEFINITION **14** *A* nested expansion *for a function f is a sequence, finite or infinite, of nested forms $\{\mathbf{n}_j\}$ such that $\mathbf{n}_1$ is a nested form for f and if*

$$\mathbf{n}_j = \{(\epsilon_{j,i}, s_{j,i}, m_{j,i}, d_{j,i}, f_{j,i}),\ i = 1, \ldots, k_j\},$$

$j \geq 1$, then $\mathbf{n}_{j+1}$ is a nested form for f_{j,k_j}.

It is possible that some f_{J,k_J} is actually equal to a non-zero constant, in which case both the nested form $\mathbf{n}_J$ and the nested expansion $\{\mathbf{n}_j, j = 1, \ldots, J\}$ are called *precise* .

Usually we consider nested forms and expansions of functions in some Hardy field, $\mathcal{F}$. If f belongs to $\mathcal{F}$ and has nested form $\pm\{(\epsilon_i, s_i, m_i, d_i, \phi_i),$ $i = 1, \ldots, k\}$, it follows from Theorem 10 that $\mathcal{F}$ can be extended to a Hardy field containing ϕ_k. Similarly if f is an exp-log function, so is ϕ_k. Thus we can often arrange that the nested forms making up the nested expansion of a given function belong to the class of functions being considered. Moreover if we have an exact expression for f above, we can obtain one for ϕ_k and hence for any other nested form appearing in the nested expansion of f. In this situation it may be convenient to work with finite nested expansions, retaining, at least in principle, exact expressions for the last nested form in the expansion; i.e. for the 'error term'.

PROPOSITION **5** *If a function f has a nested form it is uniquely determined.*

Proof of Proposition 5
Clearly if f has nested form $A + \rho$, then A is determined as the limit of f. Similarly, if the nested form of f is $\pm\{(\epsilon_i, s_i, m_i, d_i, \phi_i), i = 1, \ldots, k\}$, the initial sign and the value of ϵ_1 is clearly determined by the limiting behaviour of f.

So suppose $f = \exp_s(\log_m^d(x)\phi(x)) = \exp_t(\log_n^c(x)\psi(x))$ where the conditions of Definition 13 apply. If $t > s$ then

$$\log_t(f(x)) = \log_n^c(x)\psi(x) = \begin{cases} \log_{m+1}(x)(d + o(1)) & \text{if } t = s+1 \\ \log_{m+t-s}(x)(1 + o(1)) & \text{if } t > s+1. \end{cases}$$

It follows that $c = 1$ and ψ is asymptotic to a non-zero constant. This contravenes Definition 13.2. A similar contradiction is obtained if $t < s$, and hence $t = s$. Then

$$\log_s(f(x)) = \log_m^d(x)\phi(x) = \log_n^c(x)\psi(x),$$

and hence $m = n$, $d = c$ and $\phi = \psi$, which gives the conclusion of Proposition 5.

The following is now evident.

COROLLARY *If f has a nested expansion, the values of the parameters in it are uniquely determined.*

In the converse direction, the function is **not** uniquely determined by the parameters in the nested expansion. As remarked in Section 4.1, this is a familiar fact for asymptotic series expansions. For example, both $(1-x)^{-1}$ and $(1-x)^{-1}+e^{-x}$ have nested expansion

$$1+x(1+x(1+x(\cdots.$$

4.4.1 Comparison of Nested Forms

One often wants to compare the growths of two nested forms or two nested expansions. The following lemma allows us to compare their γ_0 values.

LEMMA **9**
Suppose f and g have nested forms $\exp_s(\log_m^d(x)\phi(x))$ and $\exp_t(\log_n^c(x)\psi(x))$ respectively. Then $\gamma_0(f)=\gamma_0(g)$ if and only if either

1.1 $s=t=0$, $m=n$, $c=d$ *and* $\gamma_0(\phi)=\gamma_0(\psi)$, *or*
1.2 $s=t\geq 1$, $m=n$, $c=d$ *and*

$$J_s=(\phi(x)-\psi(x))\log_n^d(x)\exp_1(\log_n^d(x)\psi(x))\exp_2(\log_n^d(x)\psi(x))\cdots\exp_{s-1}(\log_n^d(x)\psi(x))$$

has a finite limit.

Secondly, $\gamma_0(f)>\gamma_0(g)$ if and only if either

2.1 $s-m>t-n$, *or*
2.2 $s-m=t-n$, $s>t$ *and either* $d>1$ *or* $d=1$ *and* $\phi\to\infty$, *or*
2.3 $s-m=t-n$, $s<t$ *and either* $c<1$ *or* $c=1$ *and* $\psi\to 0$, *or*
2.4 $s=t$, $m=n$ *and either* $d>c$ *or* $d=c$ *and* $\gamma_0(\phi)>\gamma_0(\psi)$, *or*
2.5 $s=t\geq 1$, $m=n$ $d=c$ *and* $J_s\to\infty$, *where J_s is as above.*

Proof of Lemma 9
Condition 1.1 is obviously sufficient for $\gamma_0(f)=\gamma_0(g)$. To show that 1.2 is too, we use the following lemma from [99].

LEMMA **10** *Suppose that $s\geq 1$ and let J_s be as in the previous lemma. Write $K_s=\exp_{s-1}(\log_m^d(x)\phi(x))-\exp_{s-1}(\log_m^d(x)\psi(x))$, and suppose that $\phi(x)>\psi(x)$ for all sufficiently large x. Then*

1. *$K_s\to\infty$ if and only if $J_s\to\infty$.*
2. *$K_s\to 0$ if and only if $J_s\to 0$, and in this case $K_s\sim J_s$.*
3. *K_s tends to a positive limit if and only if J_s does, and then also $K_s\sim J_s$.*

Proof of Lemma 10
We use induction. If $s=1$, then $K_s=J_s$ and the lemma is trivial. Suppose therefore that the lemma holds for a particular value of s. Then

$$
\begin{aligned}
K_{s+1} &= \exp_s(\log_m^d(x)\phi(x)) - \exp_s(\log_m^d(x)\psi(x)) \\
&= \{\exp[\exp_{s-1}(\log_m^d(x)\phi(x)) - \exp_{s-1}(\log_m^d(x)\psi(x))] - 1\} \\
&\qquad \times \exp_s(\log_m^d(x)\psi(x)) \\
&= \{\exp(K_s) - 1\}\exp_s(\log_m^d(x)\psi(x)). \qquad (4.15)
\end{aligned}
$$

There are just two possibilities. If $K_s \to 0$, then $\exp(K_s) - 1 \sim K_s \sim J_s$, by the induction hypothesis. Hence $K_{s+1} \sim J_s \exp_s(\log_m^d(x)\psi(x)) = J_{s+1}$.

On the other hand, if $K_s \not\to 0$, then by (4.15), $K_{s+1} \to \infty$. Also the induction hypothesis tells us that $J_s \not\to 0$, and it follows directly from the definition that $J_{s+1} \to \infty$. This completes the proof of Lemma 10.

Continuation of the Proof of Lemma 9

If 1.2 holds, then K_s tends to a finite limit and hence so does $f/g = \exp K_s$.

We now turn our attention to the second set of conditions. If $s-m > t-n$ then

$$
\gamma_0\left(\frac{\log_{s+t}(f)}{\log_{s+t}(g)}\right) = \gamma_0\left(\frac{\log_{m+t}(x)}{\log_{n+s}(x)}\right),
$$

and so

$$
\frac{\log_{s+t}(f)}{\log_{s+t}(g)} \to \infty.
$$

Hence $f/g \to \infty$ and so 2.1 is sufficient.

If $s - m = t - n$ and $s > t$, then

$$
\frac{\log_s(f)}{\log_s(g)} \sim \frac{\log_m^d(x)\phi(x)}{C\log_n(x)},
$$

where C is equal to c or 1 according to whether $s = t+1$ or $s > t+1$. It is then clear that 2.2 implies that $\log_s(f)/\log_s(g) \to \infty$, which is sufficient to ensure that $\gamma_0(f) > \gamma_0(g)$.

The sufficiency of 2.3 is proved in a similar fashion, and that of 2.4 follows from the fact that if $t = s$ and $n = m$ then

$$
\frac{\log_s(f)}{\log_s(g)} = \frac{\log_m^d(x)\phi(x)}{\log_m^c(x)\psi(x)}.
$$

The sufficiency of 2.5 follows from Lemma 10.

If we interchange, f and g, we obtain conditions 3.1-3.5, analogous to 2.1-2.5, which are sufficient for $\gamma_0(f) < \gamma_0(g)$. However, bearing in mind that $K_s \to \infty$ if and only if $J_s \to \infty$, we see that conditions 1.1, 1.2, 2.1-2.5 and 3.1-3.5 constitute a breakdown into cases. That is to say they are mutually exclusive and cover all the possibilities. It follows that they are necessary as well as sufficient, and the proof of Lemma 9 is complete.

By using the fact that for functions f and g tending to infinity, $\gamma_1(f) = \gamma_1(g)$ if and only if $\gamma_0(\log f) = \gamma_0(\log g)$, and similarly $\gamma_1(f) > \gamma_1(g)$ if and only if $\gamma_0(\log f) > \gamma_0(\log g)$, we can use Lemma 9 to obtain the following result from [99].

LEMMA **11** *Suppose f and g have respective nested forms* $\exp_s(\log_m^d(x)\phi(x))$ *and* $\exp_t(\log_n^c(x)\psi(x))$. *Then* $\gamma_1(f) = \gamma_1(g)$ *if and only if either*

1.1 $s = t = 0$ *and* $m = n$ *or*
1.2 $s = t = 1$, $m = n$, $c = d$ *and* $\gamma_0(\phi) = \gamma_0(\psi)$, *or*
1.3 $s = t \geq 2$, $m = n$, $c = d$ *and* J_{s-1} *has a finite limit.*

Secondly, $\gamma(f) > \gamma(g)$ *if and only if either*

2.1 $s - m > t - n$, *or*
2.2 $s - m = t - n$, $s > t$ *and either* $d > 1$ *or* $d = 1$ *and* $\phi \to \infty$, *or*
2.3 $s - m = t - n$, $s < t$ *and either* $c < 1$ *or* $c = 1$ *and* $\psi \to 0$, *or*
2.4 $s = t = 1$, $m = n$ *and either* $d > c$ *or* $d = c$ *and* $\gamma_0(\phi) > \gamma_0(\psi)$, *or*
2.5 $s = t \geq 2$, $m = n$ $d = c$ *and* $J_{s-1} \to \infty$.

4.4.2 Operations on Nested Forms and Expansions

In this section, we consider arithmetic operations on nested forms and expansions, and also the actions of the exponential and logarithm functions. We use $\oplus$, $\otimes$, $//$, Log and Exp for operations on forms and expansions to distinguish them from the corresponding operations on functions.

In certain cases cancellation will arise and extra nested forms in the expansions of the operands will be needed to compute the nested form of the result. In extreme cases, the entire nested expansions of the operands may be insufficient.

The Action of Log

Firstly, for C a non-zero constant and ϕ a nested form tending to zero, we have

$$\mathrm{Log}|C + \phi| = \log|C| + \frac{\phi}{C} \oplus \frac{-\phi^2}{2C^2} \oplus \frac{\phi^3}{3C^3} \cdots \tag{4.16}$$

The terms in the sum are handled successively as given below. Next, for a nested form $\exp_s^{\pm 1}(\log_m^d(x)\phi(x))$, we have

$$\mathrm{Log}|\exp_s^{\epsilon}(\log_m^d(x)\phi(x))| = \begin{cases} \epsilon \exp_{s-1}(\log_m^d(x)\phi(x)), & \text{if } s > 0 \\ \log_{m+1}(x)\left(\epsilon d + \epsilon \frac{\log \phi(x)}{\log_{m+1}(x)}\right), & \text{if } s = 0. \end{cases}$$

We note that by the definition of a nested form, $\log(\phi(x))/\log_{m+1}(x) \to 0$.

The Action of Exp

Let $\exp_s^{\epsilon}(\log_m^d(x)\phi(x))$ be a nested form. Then

$$\mathrm{Exp}(\exp_s^{\epsilon}(\log_m^d(x)\phi(x))) =$$

$$\begin{cases} 1 \oplus \exp_s^{-1}(\log_m^d(x)\phi^n(x)) \oplus \frac{1}{2!}\exp_s^{-2}(\log_m^d(x)\phi^n(x)) \oplus \cdots, & \text{if } \epsilon = -1, \\ \log_{m-1}^A(x) \otimes \mathrm{Exp}[\log_m(x) \otimes (\phi(x) - A)], & \text{if } \epsilon = 1, s = 0,\ m > 0,\ d = 1 \\ & \ \& \ A = \lim \phi \in \mathbb{R} \setminus \{0\}, \\ \exp_{s+1}(\log_m^d(x)\phi(x)), & \text{otherwise.} \end{cases}$$

Note that the difficult second case arises because $\exp_1(\log_m(x)\phi(x))$ is not a legitimate nested form if ϕ tends to a finite, non-zero limit. Again, the terms of the sum will have to be handled as given below.

Addition and Subtraction

Suppose that $f_1(x) = \exp_s(\log_m^d(x)\phi(x))$, $f_2(x) = \exp_t(\log_n^c(x)\psi(x))$ and $f_1 \geq f_2$. We have

$$f_1 \oplus f_2 = \begin{cases} \mathrm{Exp}(\mathrm{Log} f_1 \oplus \mathrm{Log}(1 + f_2/f_1)), & \text{if } s > 0 \\ f_1(1 + (f_2/f_1)) \quad \text{if } s = 0 \ \& \ f_2 = o(f_1) \\ \log_m^d(x)(\phi \oplus \psi), & \text{if } s = 0,\ m = n \ \& \ c = d \end{cases}$$

If $f_1 = A_1 + \rho_1$ and $f_2 = A_2 + \rho_2$, then

$$f_1 \oplus f_2 = (A_1 + A_2) + (\rho_1 \oplus \rho_2).$$

The definitions for subtraction are entirely analogous. However it should be said that in certain cases, the above recursive scheme may fail to terminate because the corresponding terms of the two nested expansions annihilate each other indefinitely. As we have already pointed out, this can happen even when the functions f_1 and f_2 are different; see example (1.2) of Chapter 1. In order to have even a chance of handling such cases, we must keep hold of the underlying functions when working with the nested expansions. It is for this reason that we have taken nested forms and finite nested expansions to be exact representations of the functions. Ways of handling the more difficult cancellation cases for particular classes of functions will appear in later chapters.

Multiplication and Division

With f_1 and f_2 as above, we define

$$f_1 \otimes f_2 = \begin{cases} \mathrm{Exp}(\mathrm{Log} f_1 \oplus \mathrm{Log} f_2), & \text{if } s > 0 \text{ or } t > 0 \\ \log_m^d(x) \cdot \log_n^c(x)(\phi \otimes \psi), & \text{if } s = t = 0, \end{cases}$$

where it is understood that $\log_m^d(x) \cdot \log_n^c(x)$ is written as $\log_m^{c+d}(x)$ if $m = n$. The formula for division is once again analogous, and again cancellation problems may arise since $\oplus$ occurs in the definition.

4.5 The Algebra of Star Products

Star products can be considered as a generalization of both nested expansions and multiseries, and so might offer the advantages of both. Moreover the algebraic structure may be of independent interest. The basic idea is this: To multiply two numbers one can add their logarithms and exponentiate. Indeed many of us who were taught arithmetic before computers became commonplace learned to do just this. Multiplication can then be generalized by using iterated logarithms and exponentials. These ideas first appeared in [104]

We recall that for $k < 0$, $\exp_k$ means $\log_{-k}$ and $\log_k$ means $\exp_{-k}$.

4.5.1 Definitions and Elementary Properties

Let $\mathcal{A}$ be a field with an operation $\exp : \mathcal{A} \to \mathcal{A}$, and a partial operation $\log : \mathcal{A} \to \mathcal{A}$. The usual properties of exp and log are assumed to hold, and if we write $\log(a)$ for some element $a \in \mathcal{A}$, it will be implicitly assumed that log is defined at a.

Definition **15** *Let a and b be elements of $\mathcal{A}$ and let $k \in \mathbb{Z}$. We define*

$$a \star_k b = \exp_{k-1}\{\log_{k-1}(a) \cdot \log_{k-1}(b)\}.$$

Note that $a \star_k b$ will not be defined unless both $\log_{k-1}(a)$ and $\log_{k-1}(b)$ are. The following proposition gives some elementary properties of the star product, $\star$. The proofs are straightforward.

Proposition **6**

1. $a \star_0 b = a + b.$
2. $a \star_1 b = a \cdot b.$
3. *Provided* $\log_k(a)$ *and* $\log_k(b)$ *are defined,*
$$a \star_k b = \exp_k\{\log_k(a) + \log_k(b)\}.$$
4. *For each $k \in \mathbb{Z}$, $\star_k$ is commutative and associative.*
5. *$\star_k$ is distributive over $\star_{k-1}$, for all $k \in \mathbb{Z}$; i.e.*
$$a \star_k (b \star_{k-1} c) = (a \star_k b) \star_{k-1} (a \star_k c).$$
6. $\exp_{k-1}(0)$ *is the zero element for the operation $\star_k$, and* $\exp_k(0)$ *is the identity. Moreover the $\star_k$-inverse of a is* $\exp_{k-1}(\log_{k-1}^{-1}(a))$ *provided that* $a \neq \exp_{k-1}(0)$.
7. *We have*
$$\log(a \star_k b) = (\log a) \star_{k-1} (\log b)$$
and
$$\exp(a \star_k b) = (\exp a) \star_{k+1} (\exp b).$$

The reason why we have used Definition 15 for the star product rather than property 3 above is that $\log_{k-1} a$ may be defined when $\log_k a$ is not.

As a generalization of the rule for multiplication and addition, we regard $\star_n$ as taking precedence over $\star_m$ for $n > m$. So if $n > m$, $a \star_n b \star_m c$ means $(a \star_n b) \star_m c$. Note that this is generally different from $a \star_n (b \star_m c)$ just as it is in the case when $n = 1$ and $m = 0$.

We think of $\star_k$ as a generalized multiplication, and from that point of view it is natural to introduce notation for generalized quotients and powers.

DEFINITION **16** *For $a, b \in \mathcal{A}$, we define*

$$a /_k b = \exp_{k-1}\{\log_{k-1} a / \log_{k-1} b\},$$

and for $b \in \mathcal{A}$ for example, we set

$$a^{\star_k^b} = \exp_k\{b \log_k a\} = \exp_{k-1}(\log_{k-1}^b a) = a \star_{k+1} \exp_k b.$$

Note that subtraction is obtained as $a - b = a /_0 b$.

The standard expanded form of a polynomial is as a sum of products. We may want to rewrite expressions involving star products in an analogous fashion. For this we shall require the following result.

PROPOSITION **7 (The Generalized Distributive Law)**
For $j \geq 1$, we have

$$a \star_{k+j} (b \star_k c) = (a \star_{k+j} b) \star_{k+j-1} (a \star_{k+j} \exp_{k+j-2}(1 + d_{j-2})), \quad (4.17)$$

where $d_0 = \log_k c / \log_k b$, and $d_i = \log(1 + d_{i-1}) / \log_{k+i} b$, $i = 1, \ldots, j-2$. The right-hand side may also be written as

$$(a \star_{k+j} b) \star_{k+j-1} (a \star_{k+j} \lambda_j),$$

where $\lambda_1 = c$ and for $i \geq 2$,

$$\lambda_i = (\lambda_{i-1} /_{k+i-1} b)^{\star_{k+i-3}^e}.$$

This result is complicated in comparison with the usual distributive law, but that merely reflects the complexity of asymptotics when iterated exponentiation is involved.

Proof of Proposition 7
$a \star_{k+j} (b \star_k c)$ is equal to

$$\begin{aligned}
&\exp_{k+j-1}\{\log_{k+j-1} a \cdot \log_{k+j-1}[\exp_{k-1}(\log_{k-1} b \cdot \log_{k-1} c)]\} \\
&\quad = \exp_{k+j-1}\{\log_{k+j-1} a \cdot \log_j[\log_{k-1} b \cdot \log_{k-1} c]\} \\
&\quad = \exp_{k+j-1}\{\log_{k+j-1} a \cdot \log_{j-1}[\log_k b(1 + d_0)]\} \\
&\quad = \exp_{k+j-1}\{\log_{k+j-1} a \cdot \log_{j-2}[\log_{k+1} b + \log(1 + d_0)]\},
\end{aligned}$$

where $d_0 = \log_k c / \log_k b$. We show by induction on j that

$$\log_{j-2}\left(\log_{k+1} b + \log(1+d_0)\right) = \log_{k+j-1} b + \log(1+d_{j-2}),$$

with $d_i = \log(1+d_{i-1})/\log_{k+i} b$, for $i = 1,\ldots,j-2$.

For $j = 2$, this is a triviality, so suppose it holds for $j \geq 2$. Then

$$\begin{aligned}\log_{j-1}(\log_{k+1} b + \log(1+d_0)) &= \log(\log_{j-2}(\log_{k+1} b + \log(1+d_0)))\\ &= \log(\log_{k+j-1} b + \log(1+d_{j-2}))\\ &= \log_{k+j} b + \log(1+d_{j-1}),\end{aligned}$$

as required. But then

$$\begin{aligned}a \star_{k+j} (b \star_k c) &= \exp_{k+j-1}\{\log_{k+j-1} a \log_{k+j-1} b + \log_{k+j-1} a \log(1+d_{j-2})\}\\ &= \exp_{k+j-1}[\log_{k+j-1}(a \star_{k+j} b) + \log_{k+j-1}(a \star_{k+j} \exp_{k+j-1}(1+d_{j-2}))]\\ &= (a \star_{k+j} b) \star_{k+j-1} (a \star_{k+j} \exp_{k+j-2}(1+d_{j-2})).\end{aligned}$$

Now we show by induction on j that

$$\lambda_j = \exp_{k+j-2}(1+d_{j-2}).$$

For the case $j = 2$, we have

$$\begin{aligned}\lambda_2 &= (c \,/_{k+1}\, b)^{\star_{k-1}^{e}} = \exp_{k-1}(e \cdot \log_{k-1}(c \,/_{k+1}\, b))\\ &= \exp_{k-1}(\exp_1(1 + \log_k c/\log_k b)) = \exp_k(1+d_0).\end{aligned}$$

For the induction step, suppose the formula holds when j is replaced by i. Then

$$\begin{aligned}\lambda_{i+1} &= (\lambda_i \,/_{k+i}\, b)^{\star_{k+i-2}^{e}} = \exp_{k+i-2}\{e \cdot \log_{k+i-2}(\lambda_i \,/_{k+i}\, b)\}\\ &= \exp_{k+i-1}\{1 + \log_{k+i-1} \lambda_i/\log_{k+i-1} b\}\\ &= \exp_{k+i-1}\{1 + \log(1+d_{i-2})/\log_{k+i-1} b\}, \quad \text{by induction}\\ &= \exp_{k+i-1}\{1 + d_{i-1}\}.\end{aligned}$$

Proposition 7 now follows.

The general distributive law plays an important role in star–product expansions.

Definition 17 *A* star polynomial *is a finite expression formed from the elements of $\mathcal{A}$ using the binary operations $\star_k$ for various $k \in \mathbb{Z}$. A* star rational expression *is a finite expression formed using the operations $\star_k$ and $/_k$ for various $k \in \mathbb{Z}$.*

We have the following result.

Theorem 12 *The exp-log functions are precisely the star-rational functions, that is to say the functions obtained as finite combinations of the elements $\{\exp_n x;\ n \in \mathbb{Z}\}$ using $\star_k$ and $/_k$ for $k \in \mathbb{Z}$.*

The proof of this theorem is a straightforward application of Proposition 6.

4.5.2 Differentiation and Star Products

We first note that if $f(x)$ and $g(x)$ are differentiable functions,

$$(f \star_k g)' = (f \star_k g) \star_1 (\log f \star_{k-1} \log g)'. \tag{4.18}$$

For $k > 1$, this gives a recursive formula for the derivative of $f \star_k g$. The proof is obtained by noting that

$$f \star_k g = \exp\{\log f \star_{k-1} \log g\},$$

by Proposition 6.7. For negative values of k, we can use (4.18) with f, g, k respectively replaced by $\exp f$, $\exp g$, $k+1$, and re–arrange. This gives

$$(f \star_k g)' = \frac{(\exp f \star_{k+1} \exp g)'}{\exp f \star_{k+1} \exp g}. \tag{4.19}$$

Written in non–recursive form, these formulae are a little complicated, but that just reflects the complexity of differentiating $\exp_k x$ and $\log_k x$ for large values of k.

We return to star products in Chapter 9.

5

Algorithms for Function Towers

Our main approach to calculating limits and asymptotic expressions is to regard functions as elements in a tower of Hardy fields,

$$\mathcal{F}_0 \subset \mathcal{F}_1 \subset \cdots \subset \mathcal{F}_N, \tag{5.1}$$

with $\mathcal{F}_i = \mathcal{F}_{i-1}(f_i)$, for some function f_i, $i = 1, \ldots, N$. The field $\mathcal{F}_0$ will usually be a field of constants, $\mathbb{K}$, with $\mathbb{Q} \subseteq \mathbb{K}$, and often $\mathcal{F}_1 = \mathbb{K}(x)$. On the inductive assumption that we know how to calculate the asymptotics in $\mathcal{F}_{i-1}$, $1 \le i \le N$, we would like to be able to do the same in $\mathcal{F}_i$. We ask what conditions are required on f_i in order to be able to do this.

Firstly we need to be able to carry out zero-equivalence tests in $\mathcal{F}_i$. This is obviously necessary since we must be able to recognize the zero function if it is presented to us. Also zero-equivalence testing forms an integral part of many algorithms, as we shall see.

Secondly we need to be able to expand f_i in a scale containing the scale for $\mathcal{F}_{i-1}$. If f_i is of different comparability class from the elements of the existing scale, $|f_i|$ (or $|f_i|^{-1}$) is a candidate for a new scale element. However there may be reasons for making other choices. For example if the existing scale includes e^{-x} and $f_i = \exp(x - x^2)$ it would obviously make more sense from the user's point of view to take $\exp(-x^2)$ as the new scale element rather than f_i. There may also be technical reasons for not choosing f_i. One of these was highlighted by Dominik Gruntz in [37]. Suppose that we are computing the asymptotics of

$$\frac{1}{\exp(-x + e^{-x})} - e^x.$$

If we take $f_1 = \exp(-x+e^{-x})$, then we soon arrive at the problem of expressing e^x in terms of f_1. Since $\log(e^x)/f_1 \to -1$ we would naturally write $e^x = f_1^{-1} \cdot h$ where $h = \exp(e^{-x})$. But now e^x occurs as a subexpression of h and we are led into an infinite loop. For this reason, when we introduce an exponential e^u as a new scale element we remove any parts of the expansion of u which tend to finite limits.

Our third requirement is that we must be able to handle cancellation problems in $\mathcal{F}_i$. This is the reason why we need different scale elements to be of different comparability class. However we also have to be able to detect when a coefficient function is equivalent to zero. For if our scale is $\{\mathbf{t}_1, \ldots, \mathbf{t}_n\}$, and the $\mathbf{t}_n$–expansion of a function f is

$$f = \sum_{m=1}^{\infty} f_m \mathbf{t}_n^{r_m},$$

then $f_1 \mathbf{t}_n^{r_1}$ dominates the expansion provided that f_1 is not the zero function. This is obvious enough but it needs attention in the algorithm. A slightly more subtle point, already made in Chapter 1, is that we must avoid getting into a situation where we test all the coefficient functions in some infinite expansion and successively discover each one to be zero. Then the algorithm would fail to terminate. Without due care this can happen even if the function being expanded is non–zero, see (1.2). We proceed as follows:

We check that our function is not itself zero. Then having obtained our scale, we expand in the scale element of largest comparability class, i.e. in the notation above we compute the $\mathbf{t}_n$–expansion. We then check the coefficient functions for zero equivalence. We must succeed in finding a non-zero coefficient since the function is non–zero *and there can be no terms beyond the* $\mathbf{t}_n$*-expansion.* Then we expand the first non–zero coefficient in terms of $\mathbf{t}_{n-1}$, and so on. We can use zero-equivalence tests to determine if there are other non-zero coefficients in the $\mathbf{t}_n$–expansion, and similarly expand any of these if desired.

5.1 The Exp-Log Algorithm

We now take a detailed look at how these ideas work out for exp–log functions. Some of the groundwork has already been done in Section 4.3, but here we add some more detail and present the algorithm in a form that is suitable for generalization.

We recall that exp–log functions are given by finite expressions in the constants, the variable x, arithmetic operations and the function symbols exp and log. It is clear that for any such function, f, we can find a tower of function fields

$$\mathbb{K} \subset \mathbb{K}(x) \subset \mathcal{F}_2 \subset \cdots \subset \mathcal{F}_N$$

with $\mathcal{F}_i = \mathcal{F}_{i-1}(f_i)$, $2 \leq i \leq N$, and each f_i either a logarithm or an exponential of an element in $\mathcal{F}_{i-1}$. It is convenient for us to take the constant field, $\mathbb{K}$, to be $\mathbb{R}_E$, the field of real exp–log constants, although in implementation one would probably want to take $\mathbb{K} = \mathbb{Q}$ and add new constants as they arise.

It is well known that any element of $\mathbb{K}(x)$ can be expanded in an asymptotic Laurent series, and fast algorithms exist for carrying this out, [25, 26, 86]. Hence our prime concern is the adding of the logarithms and exponentials.

We suppose that we have a scale $\{\mathbf{t}_1, \ldots, \mathbf{t}_k\}$ and an algorithm for deriving multiseries in this scale for elements of the field $\mathcal{F}_{i-1}$. We wish to arrive at a similar position with respect to $\mathcal{F}_i = \mathcal{F}_{i-1}(f_i)$. Since we can divide multiseries in the same scale, it is sufficient to treat $\mathcal{F}_{i-1}[f_i]$, though there is one point which needs care; see Section 5.1.3.

5.1.1 Case 1: $f_i = \log h$, $h \in \mathcal{F}_{i-1}$

Since $h \in \mathcal{F}_{i-1}$, we can compute a $\mathbf{t}_k$–expansion for h,

$$h \sim \sum_{m_k} h_{\{m_k\}} \mathbf{t}_k^{r\{m_k\}},$$

and then a $\mathbf{t}_{k-1}$–expansion for $h_{\{1\}}$ and so on. As in Section 4.3.2, we arrive at an expression for h of the form

$$f = A\mathbf{t}_k^{r_0^k} \mathbf{t}_{k-1}^{r_0^{k-1}}, \ldots, \mathbf{t}_1^{r_0^1}(1+\epsilon),$$

where A is a non–zero constant and ϵ tends to zero and has a computable $\{\mathbf{t}_1 \ldots, \mathbf{t}_k\}$ multiseries expansion. If $j = 1$ and $r \neq 0$, then we need to insert $-\log^{-1} \mathbf{t}_1$ as new first scale element. Otherwise $\log \mathbf{t}_j$ will have a computable $\{\mathbf{t}_1 \ldots, \mathbf{t}_k\}$ expansion. In either case the expansion of f_i can be obtained from

$$f_i = r_0^k \log \mathbf{t}_k + r_0^{k-1} \log \mathbf{t}_{k-1} + \cdots + r_0^1 \log \mathbf{t}_1 + \log A + \log(1+\epsilon)$$

by using the power–series expansion of $\log(1+\epsilon)$, as in Chapter 4.3. However there is a potential problem here. For example, suppose that $\mathbf{t}_1 = \log^{-1} x$, $\mathbf{t}_2 = x^{-1}$ and $\epsilon = -\mathbf{t}_1 - \mathbf{t}_2$. The logarithmic expansion gives

$$\log(1 - \mathbf{t}_1 - \mathbf{t}_2) = \sum_{m=1}^{\infty} \frac{(\mathbf{t}_1 + \mathbf{t}_2)^m}{m}. \tag{5.2}$$

However expanding out the m–th powers in (5.2) does not give a multiseries according to our requirements since the coefficients of the various $\mathbf{t}_2^j$ are not in closed form. Thus for example the coefficients of $\mathbf{t}_2$ and $\mathbf{t}_2^2$ are given respectively as $\sum \mathbf{t}_1^{m-1}$ and $\sum (m-1)\mathbf{t}_1^{m-2}/2$. In these particular cases, it is easy to find sums of the series, but $\log(1 - \mathbf{t}_1 - \mathbf{t}_2)$ might merely be part of a much larger and more complicated expression. Thus we should not apply the log series to expressions which contain sums of terms of different comparability class. Similar remarks apply to exponential series and infinite binomial expansions. Notice that the above problems do not arise with $\log(1 - \mathbf{t}_1\mathbf{t}_2)$.

The way out of the trap is to use the functional equation of log. We write

$$\begin{aligned} \log(1 - \mathbf{t}_1 - \mathbf{t}_2) &= \log(1 - \mathbf{t}_1) + \log\left(1 - \frac{\mathbf{t}_2}{1 - \mathbf{t}_1}\right) \\ &= \log(1 - \mathbf{t}_1) + \sum_{m=1}^{\infty} \frac{\mathbf{t}_2^m}{m(1 - \mathbf{t}_1)^m}, \end{aligned} \tag{5.3}$$

and now the coefficient of $\mathbf{t}_2^m$, $m \geq 1$, is given in finite form by $(1-\mathbf{t}_1)^{-m}/m$. We show that this works in general.

Suppose that ϵ tends to 0 and has a multiseries in the scale $\{\mathbf{t}_1, \ldots, \mathbf{t}_k\}$. Let the $\mathbf{t}_k$–expansion be

$$\epsilon = \sum_{m=0}^{\infty} \epsilon_m \mathbf{t}_k^{r_m}, \tag{5.4}$$

with $r_0 = 0$ (which is the difficult case). Then $\epsilon = \epsilon_0 + \sigma$ where σ has $\mathbf{t}_k$–expansion $\sum_{m=1}^{\infty} \epsilon_m \mathbf{t}_k^{r_m}$. We have

$$\begin{aligned} \log(1+\epsilon) &= \log(1+\epsilon_0) + \log\left(1 + \frac{\sigma}{1+\epsilon_0}\right) \\ &= \log(1+\epsilon_0) + \sum_{s=1}^{\infty} \frac{(-1)^{s-1}\sigma^s}{s(1+\epsilon_0)^s}. \end{aligned} \tag{5.5}$$

Since $r_1 > 0$, when we collect terms in some fixed power of $\mathbf{t}_k$ these will now only come from a finite number of terms of the series on the right of (5.5), and so we will have finite expressions for them. Of course if $r_0 > 0$ we can just expand $\log(1+\epsilon)$ directly. We may apply similar consideration to the $\mathbf{t}_{k-1}$–expansion of $\log(1+\epsilon_0)$, and so on.

5.1.2 Case 2: $f_i = \exp h$, $h \in \mathcal{F}_{i-1}$

Our first concern is whether we need a new element of the scale. If h is a logarithm, we have nothing more to do since $f_i \in \mathcal{F}_{i-1}$. Otherwise we compare h with $\log \mathbf{t}_j$ for all $j = 1, \ldots, k$ for which $\mathbf{t}_j$ is an exponential. By Condition 2 of Definition 10, we have a $\{\mathbf{t}_1, \ldots, \mathbf{t}_k\}$ multiseries expansion of $\log \mathbf{t}_j$ for such values of j, and by hypothesis we can compute a multiseries for h in the same scale.

If we discover that $h \sim C \log \mathbf{t}_j$ for some j, then we write $\psi_i = f_i \mathbf{t}_j^{-C}$ and apply the same consideration to ψ_i as given to f_i. Once this has been concluded, we can compute in $\mathcal{F}_i$ since

$$f_i = \psi_i \mathbf{t}_j^C.$$

Since $\log(f_i/\mathbf{t}_j^C) = o(\log f_i)$, the comparability class of ψ_i is strictly smaller than that of f_i. So while we may need to rewrite ψ_i in terms of the existing scale elements, this process can only happen at most k times since there are only k scale elements. Note however that it is not generally true that the comparability classes of a Hardy field are well ordered. For example

$$\gamma_1(e^x) > \gamma_1(e^{x/\log x}) > \cdots > \gamma_1(e^{x/\log^n x}) > \cdots.$$

Another possibility is that h tends to a finite limit, l, so that $h = l + \epsilon$ where ϵ tends to zero and has a $\{\mathbf{t}_1, \ldots, \mathbf{t}_k\}$–multiseries expansion. Then

$f_i = e^l \exp \epsilon$. Notice that if e^l is not already in our field of constants, it will be necessary to add it. Also the same point arises with the expansion of $\exp \epsilon$ as arose with the log series above. So if the expansion of ϵ is given by (5.4) with $r_0 = 0$, we write

$$f_i = e^l \exp \epsilon_0 \sum_{s=0}^{\infty} \frac{\sigma^s}{s!}, \tag{5.6}$$

where $\sigma = \sum_{m=1}^{\infty} \epsilon_m \mathbf{t}_k^m = \exp(\epsilon - \epsilon_0)$. Here again we use the algorithm of Chapter 4.3.

The remaining case is when $|h| \to \infty$ and $h/\log \mathbf{t}_j$ does not tend to a non-zero finite limit for any $j = 1, \ldots, k$. Now $\gamma_1(f_i)$ is different from each $\gamma_1(\mathbf{t}_j)$, $j = 1, \ldots, k$, and we need to add a new element to the scale. The only point of any substance concerns the removal of any terms from h which do not tend to $\pm\infty$. These terms can be identified from the multiseries for h. First we examine the $\mathbf{t}_k$-expansion of h and remove all the terms with a positive power of $\mathbf{t}_k$. Then if there is a term with a zero-th power of $\mathbf{t}_k$, we examine its $\mathbf{t}_{k-1}$-expansion, remove any positive-power terms and examine the $\mathbf{t}_{n-2}$-expansion of the zero-power term. Continuing in this way we remove all multiseries terms tending to zero from h. Finally, if there is a constant term remaining, that is removed. So if $h = h' + h''$ with h' containing the terms whose modulus tends to infinity, we introduce a new scale element $\exp(-|h'|)$. Since we will already have calculated the limit of each $h/\log \mathbf{t}_j$, $j = 1, \ldots, k$, we know the order relation between $\gamma_1(\exp(-|h'|)) = \gamma_1(f_i)$ and the various $\gamma_1(\mathbf{t}_j)$. Thus we can insert $\exp(-|h'|)$ in its proper position in the new scale. Then $f_i = \exp(h')\exp(h'')$, and we can expand $\exp(h'')$ as we expanded $\exp \epsilon$ above.

5.1.3 Handling Denominators and Other Powers

The above-mentioned difficulty with the logarithmic and exponential series may also occur with powers, when the infinite binomial expansion needs to be used. Again the solution is along the same lines. If ϵ has $\mathbf{t}_k$-expansion (5.4), $r_0 = 0$ and c is a non-zero real number, then we can expand $(1+\epsilon)^c$ as

$$\begin{aligned}(1+\epsilon)^c &= (1+\epsilon_0)^c \left(1 + \frac{\sigma}{1+\epsilon_0}\right)^c \\ &= (1+\epsilon_0)^c \sum_{s=0}^{\infty} \frac{\Gamma(c+1)}{\Gamma(c-s)} \left(\frac{\sigma}{1+\epsilon_0}\right)^s,\end{aligned}$$

and Chapter 4.3 applies.

As we saw in 4.3.1, we can play a similar trick in a fairly general setting. Thus if f is analytic at the origin and we can compute its derivatives, we can use the power series for f,

$$f(\epsilon_0 + \sigma) = \sum_{s=0}^{\infty} \frac{f^{(s)}(\epsilon_0)}{s!} \sigma^s.$$

Of course at this point in the book we do not have any candidates for f other than logarithms, exponentials and powers.

5.1.4 Summary of the Algorithm

1. Check that f is not functionally equivalent to zero.
2. From the expression given for the function, f, construct a tower of function fields (5.1) with $\mathcal{F}_0$ a suitable field of constants, $\mathbb{K}$, $\mathcal{F}_1 = \mathbb{K}(x)$ and each $\mathcal{F}_i = \mathcal{F}_{i-1}(f_i)$, $i = 2, \ldots, N$ with f_i either a logarithm or an exponential.
3. Start from the scale $\mathcal{S}_1 = \{x^{-1}\}$ for $\mathcal{F}_1$ and build successively a scale for each of $\mathcal{F}_2, \ldots, \mathcal{F}_N$ together with a multiseries expansion for f_i in that scale. The procedure for the two cases, when f_i is a logarithm and when it is an exponential, is as given above.
4. Our function f is now given by a rational expression R in the 'variables' $x, f_1(x), \ldots, f_N(x)$, and we can obtain the multiseries for f by substituting the multiseries for the f_is into R. However we must be careful to avoid indefinite cancellation. If the scale for $\mathcal{F}_N$ is $\mathcal{S}_N = \{\mathbf{t}_1, \ldots, \mathbf{t}_k\}$, we first form the $\mathbf{t}_k$–expansion of f keeping the coefficients in closed form as indicated in Section 5.1.3, successively testing the coefficients for zero equivalence and discarding any that are zero. Then we can compute the $\mathbf{t}_{k-1}$–expansion of any coefficients we want to expand. This process can be continued right down to $\mathbf{t}_1$–expansions, whose coefficients will be constants.

We note that at each stage we must ultimately find a non–zero coefficient, since we know that the function being expanded is not zero and we always expand in terms of the largest comparability class.

5.1.5 Examples

We use an example from [73]. The function is given by

$$f(x) = \log\log(xe^{xe^x} + 1) - \exp\exp(\log\log x + x^{-1}), \tag{5.7}$$

which is sufficiently complicated to provide a realistic test of the method but small enough to be followed step by step.

There are a number of possibilities for the function tower. We will use the following: $\mathcal{F}_0 = \mathbb{R}_E$, $\mathcal{F}_1 = \mathbb{R}_E(x)$, $\mathcal{F}_2 = \mathcal{F}_1(e^x)$, $\mathcal{F}_3 = \mathcal{F}_2(\exp(xe^x))$, $\mathcal{F}_4 = \mathcal{F}_3(\log(x\exp(xe^x) + 1))$, $\mathcal{F}_5 = \mathcal{F}_4(\log(\log(x\exp(xe^x) + 1)))$, $\mathcal{F}_6 = \mathcal{F}_5(\log x)$, $\mathcal{F}_7 = \mathcal{F}_6(\log\log x)$, $\mathcal{F}_8 = \mathcal{F}_7(\exp(\log\log x + x^{-1}))$ and $\mathcal{F}_9 = \mathcal{F}_8(\exp(\exp(\log\log x + x^{-1})))$.

Starting with $\mathcal{F}_1$, we have the scale $\mathcal{S}_1 = \{x^{-1}\}$. We have to add the exponential e^x. We know that $x/\log x \to \infty$, and so e^x is of a different comparability class from any existing scale element. (Of course in this particular case we can see this more easily.) We therefore form a new scale $\mathcal{S}_2 = \{x^{-1}, e^{-x}\}$.

Next we have to add $\exp(xe^x)$. Comparing xe^x with the log of our top scale element, e^x, we have

$$\frac{\log(\exp(xe^x)}{\log(e^x)} = \frac{xe^x}{x} = \frac{1}{e^{-x}},$$

which tends to infinity since e^{-x} is a scale element. Hence we add a new scale element to form $\mathcal{S}_3 = \{x^{-1}, e^{-x}, \exp(-xe^x)\}$, which is the scale for $\mathcal{F}_3$.

Now we consider $\log(\exp(xe^x) + 1)$. We have

$$\log(x \exp(xe^x) + 1) = \log x + xe^x + \log(1 + x^{-1} \exp(-xe^x)).$$

However $\log x$ is of smaller comparability class than any element of our existing scale. We therefore add $\log^{-1} x$ at the beginning to form a new scale $\mathcal{S}_4 = \{\mathbf{t}'_1 = \log^{-1} x, \mathbf{t}'_2 = x^{-1}, \mathbf{t}'_3 = e^{-x}, \mathbf{t}'_4 = \exp(-xe^x)\}$. In terms of the new scale we have

$$\log(x \exp(xe^x) + 1) = \mathbf{t}'^{-1}_3 \mathbf{t}'^{-1}_2 + \mathbf{t}'^{-1}_1 + \log(1 + \mathbf{t}'_2 \mathbf{t}'_4).$$

We can expand the last logarithm to give a multiseries, but we defer doing so in order to keep closed forms.

Similarly for $\mathcal{F}_5$ we need to add a new scale element $\log^{-1}(\log x)$. To keep the notation tidier, we will use $\mathbf{t}_i$, $i = 1, \ldots, 5$ to denote the elements of $\mathcal{S}_5$, so that

$$\mathcal{S}_5 = \{\mathbf{t}_1 = \log^{-1}(\log x), \mathbf{t}_2 = \log^{-1} x, \mathbf{t}_3 = x^{-1}, \mathbf{t}_4 = e^{-x}, \mathbf{t}_5 = \exp(-xe^x)\}.$$

Then

$$\begin{aligned} \log\log(x \exp(xe^x) + 1) &= \log(\mathbf{t}'^{-1}_3 \mathbf{t}'^{-1}_2 + \mathbf{t}'^{-1}_1 + \log(1 + \mathbf{t}'_2 \mathbf{t}'_4)) \\ &= \log\{\mathbf{t}_4^{-1} \mathbf{t}_3^{-1} + \mathbf{t}_2^{-1} + \log(1 + \mathbf{t}_3 \mathbf{t}_5)\} \\ &= \mathbf{t}_3^{-1} + \mathbf{t}_2^{-1} + \log\{1 + \mathbf{t}_4 \mathbf{t}_3 (\mathbf{t}_2^{-1} + \log(1 + \mathbf{t}_3 \mathbf{t}_5))\}. \end{aligned}$$

We now turn our attention to the second term on the right of (5.7). Clearly $\log\log x + x^{-1}$ is expandable in our scale, so we examine $\exp(\log\log x + x^{-1})$. Here

$$\exp(\log\log x + x^{-1}) = \log x \exp(x^{-1}) = \mathbf{t}_2^{-1} \exp(\mathbf{t}_3),$$

and we see that no additions to $\mathcal{S}_5$ are needed. Finally

$$\begin{aligned} \exp\exp(\log\log x + x^{-1}) &= \exp(\mathbf{t}_2^{-1} \exp \mathbf{t}_3) \\ &= \mathbf{t}_3^{-1} \exp\{\mathbf{t}_2^{-1} (\exp \mathbf{t}_3 - 1)\}. \end{aligned}$$

The point about the second expression is that $\mathbf{t}_2^{-1}(\exp \mathbf{t}_3 - 1) \to 0$ and so the exponentials may be expanded out to give a multiseries. Once again no additions to the scale are needed. We can therefore give our function f in terms of the scale $\mathcal{S}_5$ as

$$\begin{aligned} f(x) &= \log\log(xe^{xe^x}+1) - \exp\exp(\log\log x + x^{-1}) \\ &= \mathbf{t}_3^{-1} + \mathbf{t}_2^{-1} + \log\{1 + \mathbf{t}_4\mathbf{t}_3(\mathbf{t}_2^{-1} + \log(1+\mathbf{t}_3\mathbf{t}_5))\} \\ &\quad -\mathbf{t}_3^{-1}\exp\{\mathbf{t}_2^{-1}(\exp\mathbf{t}_3 - 1)\}. \end{aligned}$$

We now give the computation of an asymptotic equivalent of f. The first step consists in computing the dominant term with respect to $\mathbf{t}_5$. We have

$$\mathbf{t}_2^{-1} + \log(1+\mathbf{t}_3\mathbf{t}_5) \sim \mathbf{t}_2^{-1} + O(\mathbf{t}_3\mathbf{t}_5).$$

Hence our first asymptotic expression for f is the coefficient of $\mathbf{t}_5^0$, namely

$$\mathbf{t}_3^{-1} + \mathbf{t}_2^{-1} + \log[1 + \mathbf{t}_2^{-1}\mathbf{t}_3\mathbf{t}_4] - \mathbf{t}_3^{-1}\exp\{\mathbf{t}_2^{-1}(\exp\mathbf{t}_3 - 1)\}.$$

We then proceed with the computation of the dominant term of this expression with respect to $\mathbf{t}_4$. The same type of computation leads to

$$\mathbf{t}_3^{-1} + \mathbf{t}_2^{-1} - \mathbf{t}_3^{-1}\exp\{\mathbf{t}_2^{-1}\exp\mathbf{t}_3 - \mathbf{t}_2^{-1}\}.$$

Next, we compute the dominant term of this expression with respect to $\mathbf{t}_3$. Let $h = \mathbf{t}_2^{-1}\exp\mathbf{t}_3 - \mathbf{t}_2^{-1}$, so that h is the argument of the outermost exponential. The computation of the dominant term of h leads to the cancellation $\mathbf{t}_2^{-1} - \mathbf{t}_2^{-1}$ which is recognized by the zero-equivalence test, whereas the function h itself is not zero. By computing the next term of the expansion, we obtain the dominant term $\mathbf{t}_2^{-1}\mathbf{t}_3$ of h. The dominant term of $\mathbf{t}_3^{-1}\exp(h)$ is $\mathbf{t}_3^{-1}$, whence a new cancellation $\mathbf{t}_3^{-1} - \mathbf{t}_3^{-1}$. Computing the next term leads to another cancellation $\mathbf{t}_2^{-1} - \mathbf{t}_2^{-1}$. One more term is necessary before arriving at the conclusion that the dominant term with respect to $\mathbf{t}_3$ is

$$-\frac{(\mathbf{t}_2^{-2} + \mathbf{t}_2^{-1})\mathbf{t}_3}{2}.$$

Thus we obtain the desired equivalence

$$f \sim -\frac{\mathbf{t}_2^{-2}\mathbf{t}_3}{2} = -\frac{\log^2 x}{2x}.$$

The k first coefficients with respect to any element of the scale can be computed in a similar way.

Our second example illustrates the dangers of indefinite cancellation. Let

$$G(x) = \log\left(1 + \frac{1}{\log(x+e^{-x})}\right) - \log\left(1 + \frac{1}{\log x}\right). \tag{5.8}$$

By a process similar to that used in the previous example, we arrive at a suitable scale, $\mathcal{S} = \{\log^{-1} x, x^{-1}, e^{-x}\}$. If we just expand the outermost logarithms, we obtain

$$G(x) = \sum_{n=1}^{\infty} \frac{1}{n \log^n (x + e^{-x})} - \sum_{n=1}^{\infty} \frac{1}{n \log^n x}. \tag{5.9}$$

The trouble comes when we try to use (5.9) to find the leading term of the multiseries for G. We can write

$$\log(x + e^{-x}) = \log x + \log(1 + x^{-1}e^{-x}) = \log x + \sum_{m=1}^{\infty} \frac{x^{-m}e^{-mx}}{m}$$

and substitute into the first sum of (5.9). However when we collect terms in $\log^{-p} x$ for any $p \in \mathbb{N}$ we will obtain two cancelling infinite series, one from each of the sums in (5.9). In fact if we look at the coefficients of the terms of largest comparability class, namely the powers of e^{-x}, we see that the coefficient of the zero–th power is equivalent to zero. The algorithm as given has no difficulty here since it expands first with respect to e^{-x}. This gives

$$\begin{aligned}\log\left(1 + \frac{1}{\log(x + e^{-x})}\right) &= \log\left(1 + \frac{1}{\log x}\right) + e^{-x}[D_y^1]_{y=0} + \frac{e^{-2x}}{2!}[D_y^2]_{y=0} \\ &\quad + \cdots \\ &= \log\left(1 + \frac{1}{\log x}\right) - \frac{x^{-1}e^{-x}}{\log x + \log^2 x} + \cdots,\end{aligned}$$

where we wrote D_y^1 and D_y^2 for the first and second partial derivatives of $\log\{1 + \log^{-1}(x + y)\}$ with respect to y. The first term on the right cancels with the term $\log(1 + \log^{-1} x)$ on the right of (5.8), which accounts for the indefinite cancellation in (5.9). We see that G is asymptotic to the second term of the expansion, i.e.

$$G \sim -\frac{x^{-1}e^{-x}}{\log x + \log^2 x}.$$

Further terms may be calculated as required. Note the use of the Taylor expansion for the terms of $\log(1 + \log^{-1}(x + e^{-x}))$.

The class of exp-log functions represents something of a *pons asinorum* for algorithms to compute limits. The work of Hardy, [40] established that every function in the class tends to a limit, perhaps infinite. It was not until [28] that an algorithm for finding the limit existed. An independent algorithm, more in tune with the needs of computer algebra was given in [97]. Some refinements were added by Dominik Gruntz in [37]. Gruntz's implemention formed part of the MAPLE library. It only calculates the limit, but it could be very easily adapted to give the multiseries.

An algorithm for calculating the multiseries of an exp-log function, based on [97], appeared in [73], and has recently been implemented by Bruno Salvy and the present author as part of a project funded by the Anglo–French Alliance initiative.

At the time of writing, Bruno Salvy and his team at INRIA are in the process of updating MAPLE's series handling to incorporate recent developments in multiseries.

5.2 Asymptotic Fields

We have seen in the previous section how to calculate multiseries for exp–log functions. From our perspective, it was important to retain closed forms for the coefficients so that zero–equivalence tests could be carried out on rational expressions in them. We were able to do this with exp–log functions essentially because we could write any given exp–log function in terms of scale elements using rational expressions and some transcendental functions with well known power series expansions, namely $\exp \mathbf{t}$, $\log \mathbf{t}$ and $(1+\mathbf{t})^c$, where $\mathbf{t} \to 0$ and $c \in \mathbb{R} \setminus \{0\}$.

One way of viewing this is as follows. Suppose we have an exp–log expression $R(\mathbf{t}_1, \ldots, \mathbf{t}_{n-1}, \mathbf{t}_n)$. If this has a $\mathbf{t}_n$–expansion

$$R(\mathbf{t}_1, \ldots, \mathbf{t}_{n-1}, \mathbf{t}_n) = \sum_{m=0}^{\infty} R_m \mathbf{t}^m,$$

then (obviously) $R_0 = R(\mathbf{t}_1, \ldots, \mathbf{t}_{n-1}, 0)$. The other coefficients can be found by differentiating partially before replacing $\mathbf{t}_n$ by zero. So in essence we get the coefficients by setting $\mathbf{t}_n = 0$.

Our present interest is to look beyond exp–log functions. We shall still keep the idea of building towers of function fields, extending each $\mathcal{F}_{i-1}$ by a function f_i to give $\mathcal{F}_i$, but now we may want to use f_i which are not exponentials or logarithms of elements in $\mathcal{F}_{i-1}$. For example we might want some f_i to be given as an integral. It is possible to compute an asymptotic expansion for an integral in terms of exp–log functions, and hence to determine whether any new comparability classes need to be introduced (see Section 5.2.2). However it is not clear how we are going to compute multiseries coefficients. The direct analogue of 'putting $\mathbf{t}_n$ to zero' does not always work.

Another problem reappears now. We want to be sure that arithmetic operations on multiseries remain valid, so that for example the multiseries of the sum of two functions can be obtained by adding the multiseries. For exp–log functions this was guaranteed by the fact that each series generated converged to the corresponding function. However asymptotic series for integrals can diverge. We therefore have to guard against the possibility that, for example, we might calculate the same multiseries for two different functions. Then we would not obtain the multiseries for the difference by subtracting the multiseries! In theory the problem of divergent asymptotic series might be handled via resummation, [32], but our algorithmic viewpoint requires something different. What we need is a system for making consistent choices in such a way that we can compute closed forms. This means that the way that we expand an integral may depend on the choices we have made in fields lower in the tower. Also we must not adversely prejudice later choices. This section examines one way of handling the above problems without resorting to resummation, using the concept of an *asymptotic field* from [103].

Obviously we must not 'put $\mathbf{t}_n$ to zero' in an expression which contains negative powers of $\mathbf{t}_n$. To encapsulate this principle in a more general context we start with the following definition.

DEFINITION **18** *Let $\mathcal{F}$ be a Hardy field and let $\mathbf{t}$ be an element of $\mathcal{F}$ which tends to zero. Then we write*

$$\mathcal{R}_{\mathbf{t}}(\mathcal{F}) = \{f \in \mathcal{F} : \forall \varepsilon \in \mathbb{R}^+, \ |f| < \mathbf{t}^{-\varepsilon}\}.$$

We shall generally take $\mathbf{t}$ to be a scale element. Suppose we have a scale $\{\mathbf{t}_1, \ldots, \mathbf{t}_n\}$ for $\mathcal{F}$. For a general $f \in \mathcal{F}$, we would hope to be able to find a constant c such that $f = \mathbf{t}_n^{c+o(1)}$. Then $f\mathbf{t}_n^{-c} \in \mathcal{R}_{\mathbf{t}_n}(\mathcal{F})$, and so it would be sufficient to be able to compute multiseries for elements of $\mathcal{R}_{\mathbf{t}_n}(\mathcal{F})$.

LEMMA **12** *$\mathcal{R}_{\mathbf{t}}(\mathcal{F})$ is a differential subring of $\mathcal{F}$.*

Proof of Lemma 12
Suppose that $\varepsilon \in (0,1) \subset \mathbb{R}^+$, and let $f, g \in \mathcal{R}_{\mathbf{t}}(\mathcal{F})$. Then $|f|, |g| < \mathbf{t}^{-\varepsilon/2}$ and so

$$|f \pm g| < 2\mathbf{t}^{-\varepsilon/2} < \mathbf{t}^{-\varepsilon} \quad \text{and} \quad |fg| < \mathbf{t}^{-\varepsilon}.$$

Hence $\mathcal{R}_{\mathbf{t}}(\mathcal{F})$ is a subring of $\mathcal{F}$, and we only need to check that it is closed under differentiation.

Let $f \in \mathcal{R}_{\mathbf{t}}(\mathcal{F})$. Then for any $\varepsilon \in \mathbb{R}^+$, we have $\gamma_0(f) < \gamma_0(\mathbf{t}^{-\varepsilon})$, and so by Lemma 6(5), $\gamma_0(f') < \gamma_0(\mathbf{t}^{-\varepsilon} \cdot \mathbf{t}'/\mathbf{t})$. However $\mathbf{t}^{\varepsilon} \to 0$ which implies that $\gamma_0(\mathbf{t}^{\varepsilon}) < \gamma_0(x)$. On applying Lemma 6(5) again, we obtain $\gamma_0(\mathbf{t}^{\varepsilon} \cdot \mathbf{t}'/\mathbf{t}) < \gamma_0(x') = 0$. Therefore $\gamma_0(\mathbf{t}'/\mathbf{t}) < \gamma_0(\mathbf{t}^{-\varepsilon})$. Hence

$$\gamma_0(f') < \gamma_0(\mathbf{t}^{-\varepsilon} \cdot \mathbf{t}'/\mathbf{t}) < \gamma_0(\mathbf{t}^{-2\varepsilon}).$$

Since this holds for every $\varepsilon \in \mathbb{R}^+$, we have $f' \in \mathcal{R}_{\mathbf{t}}(\mathcal{F})$, as required. This completes the proof of Lemma 12.

Next we want to look at the 'small' elements of $\mathcal{R}_{\mathbf{t}}(\mathcal{F})$. i.e. those that will disappear if we 'set $\mathbf{t}$ to zero'.

DEFINITION **19** *With $\mathcal{F}$ and $\mathbf{t}$ as above, we write*

$$\mathcal{I}_{\mathbf{t}}(\mathcal{F}) = \{f \in \mathcal{F} : \exists \delta \in \mathbb{R}^+, \ |f| < \mathbf{t}^{\delta}\}.$$

LEMMA **13** *$\mathcal{I}_{\mathbf{t}}(\mathcal{F})$ is a differential ideal of $\mathcal{R}_{\mathbf{t}}(\mathcal{F})$.*

Proof of Lemma 13
Let $f, g \in \mathcal{I}_{\mathbf{t}}(\mathcal{F})$ and let $h \in \mathcal{R}_{\mathbf{t}}(\mathcal{F})$. Then there exists a $\delta \in \mathbb{R}^+$ such that $|f|, |g| < \mathbf{t}^\delta$, and for all $\varepsilon \in \mathbb{R}^+$, $|h| < \mathbf{t}^{-\varepsilon}$. Then $|f \pm g| < 2\mathbf{t}^\delta < \mathbf{t}^{\delta/2}$ and $|fh| < \mathbf{t}^\delta \cdot \mathbf{t}^{-\delta/2} = \mathbf{t}^{\delta/2}$. So $\mathcal{I}_{\mathbf{t}}(\mathcal{F})$ is an ideal of $\mathcal{R}_{\mathbf{t}}(\mathcal{F})$, and again it is a matter of showing closure under differentiation.

Now $f \in \mathcal{I}_{\mathbf{t}}(\mathcal{F})$ implies that $\gamma_0(f) < \gamma_0(\mathbf{t}^\delta)$ for some $\delta \in \mathbb{R}^+$. Hence by Lemma 6(5), $\gamma_0(f') < \gamma_0(\mathbf{t}^\delta \cdot \mathbf{t}'/\mathbf{t})$, whence $\gamma_0(f') < \gamma_0(\mathbf{t}^{\delta/2})$ as in the proof of Lemma 12. So $\mathcal{I}_{\mathbf{t}}(\mathcal{F})$ is indeed a differential ideal, which completes the proof of Lemma 13.

DEFINITION **20** *Let $\mathcal{F}$ and $\mathbf{t}$ be as above, and let $\mathcal{S}$ be a subfield of $\mathcal{F}$. We say that $\mathcal{S}$ has the* shadow property *with respect to $\mathbf{t}$ if the following two conditions hold:*
SF(i) *$\mathcal{S}$ is closed under relative differentiation; that is to say if a and b belong to $\mathcal{S}$ and $b' \neq 0$ then $a'/b' \in \mathcal{S}$.*
SF(ii) $\mathcal{S} \cap \mathcal{I}_{\mathbf{t}}(\mathcal{F}) = \{0\}$.

The idea is that projection from $\mathcal{R}_{\mathbf{t}}(\mathcal{F})$ onto a suitable field $\mathcal{S}$ which has the shadow property with respect to $\mathbf{t}$ will provide a generalization of 'setting $\mathbf{t}$ to zero' inside an exp–log expression. Because $\mathcal{S}$ is a field, elements of $\mathcal{S}$ cannot combine arithmetically to produce an element of comparability class greater than or equal to $\gamma_1(\mathbf{t})$.

Note that to require $\mathcal{S}$ to be closed under differentiation would be too restrictive. For example if we are adding an integral and $\mathbf{t} = x^{-1}$, we are likely to need $\log x$ to be an element of $\mathcal{S}$, whereas x^{-1} cannot be if SF(ii) is to hold. Closure under relative differentiation does not suffer from the same problem and turns out to tie $\mathcal{S}$ sufficiently into the differential structure of $\mathcal{F}$. If $\mathcal{S}$ contains x, as in practice it will when $\gamma_1(\mathbf{t}) > \gamma_1(x)$, then SF(i) is of course equivalent to closure under differentiation.

Asymptotic fields will generally be built up in a tower of the form (5.1), and each asymptotic field in the tower will carry a set of shadow fields, $\mathcal{S}_i$, one for each $\mathbf{t}_i$ in the existing scale. When we add an asymptotic field on top of an existing tower, we shall want to be able to obtain the new shadow fields by extending the existing ones. The following lemma, from [103], shows that condition SF(i) will normally be satisfied in this situation.

LEMMA **14** *Let ϕ be an element of a differential field, $\mathcal{G}$, and let $\mathcal{S}$ be a subfield of $\mathcal{G}$ which is closed under relative differentiation. Suppose that ϕ satisfies a differential equation of the form*

$$\phi' = h_1 g_1' + h_2 g_2' + \cdots + h_r g_r', \tag{5.10}$$

where $h_1, \ldots, h_r \in \mathcal{S}(\phi)$ and $g_1, \ldots, g_r \in \mathcal{S}$. Then $\mathcal{S}(\phi)$ is closed under relative differentiation.

Proof of Lemma 14
We may suppose that $g_1' \neq 0$. Then

$$\frac{\phi'}{g_1'} = h_1 + h_2\frac{g_2'}{g_1'} + \cdots + h_r\frac{g_r'}{g_1'}.$$

Since $\mathcal{S}$ satisfies SF(i), $g_i'/g_1' \in \mathcal{S}$ for $i = 2\ldots, r$, and hence $\phi'/g_1' \in \mathcal{S}(\phi)$.

Let a be an element of $\mathcal{S}[\phi]$. Then

$$a = a_0 + a_1\phi + \cdots + a_j\phi^j,$$

for some j, with $a_0, \ldots, a_j \in \mathcal{S}$. Hence

$$\frac{a'}{g_1'} = \frac{a_0'}{g_1'} + \frac{a_1'}{g_1'}\phi + \cdots + \frac{a_j'}{g_1'}\phi^j + \frac{\phi'}{g_1'}(a_1 + 2a_2\phi + \cdots + ja_j\phi^{j-1}).$$

The right-hand side belongs to $\mathcal{S}(\phi)$ because $\phi'/g_1' \in \mathcal{S}(\phi)$ and similarly each $a_i'/g_1' \in \mathcal{S}$, since $a_i \in \mathcal{S}$. Thus $a'/g_1' \in \mathcal{S}(\phi)$.

Then if $a_1, a_2 \in \mathcal{S}[\phi] \setminus \{0\}$,

$$\frac{(a_1/a_2)'}{g_1'} = \frac{a_1'}{g_1'}\frac{1}{a_2} - \frac{a_1}{a_2^2}\frac{a_2'}{g_1'} \in \mathcal{S}(\phi).$$

Finally if $f_1, f_2 \in \mathcal{S}(\phi)$ with $f_1' \neq 0$ then

$$\frac{f_1'}{f_2'} = \frac{f_1'}{g_1'}\frac{g_1'}{f_2'} \in \mathcal{S}(\phi).$$

Hence $\mathcal{S}(\phi)$ is closed under relative differentiation, and Lemma 14 is established.

The following is of some interest in its own right, but its main purpose is as a tool in later proofs.

LEMMA **15** *If $\mathcal{S}$ has the shadow property with respect to* $\mathbf{t}$, *then so does $\mathcal{S}_{alg}$, the real closure of $\mathcal{S}$.*

Proof of Lemma 15
If $h \in \mathcal{S}_{alg}$ then there is a polynomial, P, over $\mathcal{S}$ such that $P(h) = 0$. Hence

$$h'\frac{\partial P}{\partial h}(h) + \frac{\partial P}{\partial x}(h) = 0,$$

where $\partial P/\partial x$ denotes the polynomial obtained by differentiating the coefficients of P. It then follows from Lemma 14 that $\mathcal{S}_{alg}$ is closed under relative differentiation.

To establish SF(ii), let $h \in \mathcal{S}_{alg} \cap \mathcal{I}_{\mathbf{t}}$, so that $|h| < \mathbf{t}^{\varepsilon}$ for some $\varepsilon \in \mathbb{R}^+$. Let $P(y) = p_0 + p_1 y + \cdots + p_m y^m$. If $h \neq 0$, we may take p_0 to be non-zero. Now for any $j = 1, \ldots, m$, if $p_j \neq 0$ then $p_j^{-1} \in \mathcal{S}$ and hence $|p_j| < \mathbf{t}^{-\varepsilon/3}$. Therefore

$$|p_0| = |h||p_1 + \cdots + p_m h^{m-1}| < \mathbf{t}^{\varepsilon/2}.$$

However since $\mathcal{S}$ satisfies SF(ii), this implies that $p_0 = 0$, which is contrary to our earlier assertion. Hence h must be zero, and $\mathcal{S}_{alg}$ satisfies SF(ii). We have therefore proved Lemma 15.

Lemma 15 is used in the proof of the following result .

Lemma **16** *Let $\mathcal{G}$ be a Hardy field with elements ϕ and $\mathbf{t}$, where $\mathbf{t} \to 0$. Let $\mathcal{S}$ be a subfield of $\mathcal{G}$ which has the shadow property with respect to $\mathbf{t}$. Suppose that $\phi \in \mathcal{R}_{\mathbf{t}}(\mathcal{G})$ and that for every $p \in \mathcal{S}_{alg}$ and every $\delta \in \mathbb{R}^+$,*

$$|\phi - p| < \mathbf{t}^\delta \quad \Rightarrow \quad \phi = p.$$

Then $\mathcal{S}(\phi)$ satisfies SF(ii) with respect to $\mathbf{t}$.

Proof of Lemma 16
Let $h \in \mathcal{S}(\phi)$ and suppose that $|h| < \mathbf{t}^\sigma$ for some $\sigma \in \mathbb{R}^+$. We must show that $h = 0$. Let $h = n/d$ where n and $d \neq 0$ are two polynomials in $\mathcal{S}[\phi]$. Now every element of $\mathcal{S}$ is smaller in modulus than $\mathbf{t}^{-\varepsilon}$ for every $\varepsilon \in \mathbb{R}^+$, and the same is true of ϕ. Hence $|d| < \mathbf{t}^{-\sigma/2}$. It follows that $|n| < \mathbf{t}^{\sigma/2}$, and so it suffices to consider the case when $h \in \mathcal{S}[\phi]$.

So let $h = P(\phi)$ and let $p_1, \ldots, p_k \in \mathcal{S}_{alg}$ be the real zeros of the polynomial P. We may write

$$P(y) = Q(y) \prod_{i=1}^{k} (y - p_i),$$

where Q is a polynomial over $\mathcal{S}_{alg}$ having no real zeros. Suppose that the minimum modulus of Q on the real line is attained at β. Then β is a zero of $\partial Q/\partial y$ and hence belongs to $\mathcal{S}_{alg}$. Since the coefficients of Q also belong to $\mathcal{S}_{alg}$, and by Lemma 15 $\mathcal{S}_{alg}$ has the shadow property, we have $|Q(\beta)| > \mathbf{t}^{\sigma/2}$. Therefore there is at least one $i \in \{1, \ldots, k\}$ such that $|\phi - p_i| < \mathbf{t}^{\sigma/2k}$. But the hypotheses of the Lemma now give that $\phi = p_i$, and so $h = 0$ as required. This completes the proof of Lemma 16.

We can use shadow fields to ensure that arithmetic operations on multiseries give the right results even when the series diverge. We arrange that the coefficients always lie within a predefined shadow field. Then SF(ii) implies that arithmetic operations on coefficients will not result in a function of too large a comparability class. A computable multiseries expansion with this property is called a *shadow expansion.*

Definition **21** *Let $\mathcal{F}$ be a Hardy field with constant field, $\mathbb{K}$. We say that $\mathcal{F}$ is an* asymptotic field *if the following conditions hold:*

1. *$\mathcal{F}$ contains a scale $\{\mathbf{t}_1, \ldots, \mathbf{t}_n\}$ satisfying Definition 10.*
2. *There is a sequence of fields*

$$\mathbb{K} = \mathcal{S}_1(\mathcal{F}) \subset \mathcal{S}_2(\mathcal{F}) \subset \cdots \subset \mathcal{S}_n(\mathcal{F})$$

   such that for $i = 1, \ldots, n$, $\mathcal{S}_i(\mathcal{F})$ has the shadow property with respect to $\mathbf{t}_i$. Also $\gamma_1(1), \gamma_1(\mathbf{t}_1), \ldots, \gamma_1(\mathbf{t}_{i-1})$ are precisely the comparability classes of $\mathcal{S}_i(\mathcal{F})$.
3. *Every element of $\mathcal{F}$ has a computable shadow expansion with respect to the shadow fields $\mathcal{S}_1(\mathcal{F}), \ldots, \mathcal{S}_n(\mathcal{F})$; this means in particular that the coefficients in every $\mathbf{t}_i$-expansion belong to $\mathcal{S}_i(\mathcal{F})$, for $i = 1, \ldots, n$.*

4. There exists a Hardy field, $\hat{\mathcal{F}}$ which contains $\mathcal{F}$ and each $\mathcal{S}_i(\mathcal{F})$.

Note that it is part of the requirement of (3) that we can determine zero equivalence in $\mathcal{F}$ and its attached shadow fields.

Condition (4) will normally be guaranteed by existing results in the theory of Hardy fields, in particular by Theorems 9 and 10. Moreover it will often be relatively easy to obtain condition (1). So the main task will be to establish (2) and (3). Suppose for the moment that we have established (2), so that we have a Hardy field $\mathcal{F}$ for which conditions (1), (2) and (4) hold. If we wish to compute the $\mathbf{t}_n$–expansion of an element f of $\mathcal{F}$, we may divide through by a $\{\mathbf{t}_1, \ldots, \mathbf{t}_n\}$–monomial in order to reduce to the case when $f \to 1$. If we can write f in the form $f = \phi + \psi$ where ϕ belongs to the shadow field $\mathcal{S}_n$ and $\psi \in \mathcal{I}_n(\mathcal{F})$, then ϕ will be first term of the $\mathbf{t}_n$–expansion of f. Provided that we have a suitable closed form for ψ, we may apply a similar process to it in order to obtain the second term in the $\mathbf{t}_n$–expansion of f, and so on. If we want to compute the $\mathbf{t}_{n-1}$–expansion of ϕ, we can again use a similar process. Thus the key is being able to split elements f as $f = \phi + \psi$. We use the following notation and terminology.

DEFINITION **22** *If $f = \phi + \psi$ as above with ϕ belonging to the shadow field with respect to $\mathbf{t}_i$ and $|\psi| < \mathbf{t}_i^\delta$ for some $\delta \in \mathbb{R}^+$, we write $\eta_i(f)$ for ϕ and $\xi_i(f)$ for ψ. We refer to $\eta_i(f)$ as the* shadow *of f with respect to $\mathbf{t}_i$ and $\xi_i(f)$ as the* ghost *of f with respect to $\mathbf{t}_i$.*

It should be pointed out that the maps η_i and ξ_i are only defined relative to the shadow field $\mathcal{S}_i$ in question.

For a fixed asymptotic field, $\mathcal{F}$ say, we shall generally write $\mathcal{S}_i$ for $\mathcal{S}_i(\mathcal{F}) = \mathcal{S}_{\mathbf{t}_i}(\mathcal{F})$, and similarly $\mathcal{R}_i$ for $\mathcal{R}_{\mathbf{t}_i}(\mathcal{F})$, etc. We write $\hat{\mathcal{F}}$ for the field generated by $\mathcal{F} \cup \mathcal{S}_1 \cup \cdots \cup \mathcal{S}_n$. The following Lemma further indicates the role of the maps η_i and ξ_i.

LEMMA **17** *Let $\mathcal{F}$ be an asymptotic field with scale $\{\mathbf{t}_1, \ldots, \mathbf{t}_n\}$ and shadow fields $\mathcal{S}_1, \ldots, \mathcal{S}_n$. Then for each $i = 1, \ldots, n$ there is a computable field homomorphism η_i from $\mathcal{R}_i$ to $\mathcal{S}_i$ with the property that for each $f \in \mathcal{R}_i$ $f - \eta_i(f) \in \mathcal{I}_i(\hat{\mathcal{F}})$; that is to say there is a $\delta \in \mathbb{R}^+$ such that $|f - \eta_i(f)| < \mathbf{t}_i^\delta$.*

Further for each i, j with $1 \leq i < j \leq n$ there is a computable homomorphism, $\eta_{i,j}$ from $\mathcal{R}_i(\mathcal{S}_j)$ to $\mathcal{S}_i$ such that $\eta_{i,j} \circ \eta_j = \eta_i$, and for any $g \in \mathcal{R}_i(\mathcal{S}_j)$, we have $g - \eta_{i,j}(g) \in \mathcal{I}_i(\hat{\mathcal{F}})$. Moreover if $1 \leq i < j < k \leq n$ then $\eta_{i,j} \circ \eta_{j,k} = \eta_{i,k}$.

Conversely let $\mathcal{F}$ be a Hardy field satisfying conditions (1), (2) and (4) of Definition 21. Suppose that computable functions η_i, $1 \leq i \leq n$ and $\eta_{i,j}$, $1 \leq i < j \leq n$ exist satisfying the above conditions, and that we can compute $\{\mathbf{t}_1, \ldots, \mathbf{t}_n\}$–monomials asymptotic to elements of $\mathcal{F}$ and their various shadows and ghosts (including ghosts of shadows etc.). Then $\mathcal{F}$ is an asymptotic field.

Usually the requirement to be able to compute monomials asymptotic to shadows and ghosts is met by presenting the shadows and ghosts in a form similar to that of f.

Proof of Lemma 17
Suppose that $\mathcal{F}$ is an asymptotic field, and let $g \in \mathcal{R}_i(\mathcal{F})$ where $1 \leq i \leq n$. The first term in the $\mathbf{t}_n$–expansion of g is $g_{\{1\}} t_n^{r_{\{1\}}}$, and we must have $r_{\{1\}} \geq 0$. If $r_{\{1\}} > 0$, we set $\eta_i(g) = 0$. In general, let $r_{\{1^\nu\}}$ denote $r_{\{1,1,\ldots,1\}}$, where ν ones appear in the list. If for some $\nu \leq n-i$, $r_{\{1\}} = r_{\{1,1\}} \cdots = r_{\{1^{\nu-1}\}} = 0$ but $r_{\{1^\nu\}} > 0$, then $\eta_i(g) = 0$. Otherwise, we must have $r_{\{1\}} = r_{\{1,1\}} \cdots = r_{\{1^{n-i}\}} = 0$, and we set $\eta_i(g) = g_{\{1^{n-i}\}}$. Then $|g - \eta_i(g)| < \mathbf{t}_i^\delta$ for some $\delta \in \mathbb{R}^+$.

Similarly if $g \in \mathcal{R}_i(\mathcal{S}_j(\mathcal{F}))$ with $1 \leq i < j \leq n$, then the $\{\mathbf{t}_1, \ldots, \mathbf{t}_{j-1}\}$–expansion of g can contain no negative powers of $\mathbf{t}_{j-1}$. We set $\eta_{i,j}(g) = 0$ if there is a $\nu \leq j-i-1$ such that in the expansion of g, $r_{\{1\}} = r_{\{1,1\}} \cdots = r_{\{1^{\nu-1}\}} = 0$ but $r_{\{1^\nu\}} > 0$. Otherwise, $r_{\{1\}} = r_{\{1,1\}} \cdots = r_{\{1^{j-i-1}\}} = 0$, and we set $\eta_{i,j}(g) = g_{\{1^{j-i-1}\}}$. It is easy to check that the required conditions hold.

Conversely suppose the conditions of the second paragraph of Lemma 17 are satisfied. Let g be any element of $\mathcal{F}$. To compute the $\mathbf{t}_n$–expansion of g we first check that $g \not\equiv 0$ and then calculate $\eta_n(g)$. Then we compute a monomial asymptotic to $\xi_n(g)$ (or at least the power of $\mathbf{t}_n$ in this), divide through by the appropriate $\mathbf{t}_n^{r_1}$ to ensure that $\gamma_1(\xi_n(g)\mathbf{t}_n^{-r_1}) < \gamma_1(\mathbf{t}_n)$ and calculate $\eta_n(\xi_n(g)\mathbf{t}_n^{-r_1})$. At this stage, unless one or both of the coefficients turns out to be zero, we have the first two terms of the $\mathbf{t}_n$–expansion; i.e.

$$g \sim \eta_n(g) + \eta_n(\xi_n(g)\mathbf{t}_n^{-r_1})\mathbf{t}_n^{r_1}.$$

We continue subtracting the terms we have from g, calculating the next power of $\mathbf{t}_n$ in the expansion and using η_n to get the coefficient. We must eventually find at least one non–zero term since $g \not\equiv 0$.

We can compute $\mathbf{t}_i$–expansions of coefficients in the same way. Thus condition (3) of Definition 21 holds and so we have proved Lemma 17.

The following result will be needed later in this chapter.

LEMMA **18** *If $\mathcal{F}$ is an asymptotic field with scale $\{\mathbf{t}_1, \ldots, \mathbf{t}_n\}$ and $\mathbf{t}_1 = \log_k^{-1} x$, then $\mathcal{F}(\log_{k+1}^{-1} x)$ is an asymptotic field.*

Proof of Lemma 18
First we add $\log_{k+1}^{-1} x$ as a new scale element at the beginning of the existing scale and re–label the scale as $\{\mathbf{s}_1, \mathbf{s}_2, \ldots, \mathbf{s}_{n+1}\}$; so $\mathbf{s}_1 = \log_{k+1}^{-1}$ and $\mathbf{s}_{i+1} = \mathbf{t}_i$ for $i = 1, \ldots, n$.

Next we define the shadow fields by setting

$$\mathcal{S}_1(\mathcal{F}(\log_{k+1}^{-1} x)) = \mathcal{S}_1(\mathcal{F}),$$

which is the field of constants, and

$$\mathcal{S}_i(\mathcal{F}(\log_{k+1}^{-1} x)) = \mathcal{S}_{i-1}(\mathcal{F})(\log_{k+1}^{-1} x)$$

for $i = 2, \ldots, n+1$. It is then easy to prove that the shadow property holds for the new shadow fields and the rest of the proof is straightforward.

Example
In order to illustrate the ideas above, we look at the case when $\mathcal{F} = \mathbb{R}(x, \log x)$; c.f. [103]. We take $\{\mathbf{t}_1, \mathbf{t}_2\}$ as our scale, where $\mathbf{t}_1 = \log^{-1} x$ and $\mathbf{t}_2 = x^{-1}$. Then

$$\begin{aligned}\mathcal{R}_2 &= \{f \in \mathbb{R}(x, \log x);\ \ \forall \varepsilon \in \mathbb{R}^+, |f| < x^{\varepsilon}\} \\ &= \{P/Q \in \mathbb{R}(x, \log x);\ \ \deg(P, x) \le \deg(Q, x)\}.\end{aligned}$$

We take $\mathcal{S}_2 = \mathbb{R}(\log x)$. We can write any $f \in \mathcal{R}_2$ uniquely in the form

$$f = \frac{p_0 + p_1 x^{-1} + \cdots + p_n x^{-n}}{q_0 + q_1 x^{-1} + \cdots + q_m x^{-m}},$$

where $n, m \in \mathbb{N}$ and $p_0, \ldots, p_n, q_0, \ldots, q_m \in \mathbb{R}(\log x)$ with $q_0 \neq 0$. We define $\eta_2(f) = p_0/q_0$. Similarly

$$\mathcal{R}_1 = \{f \in \mathbb{R}(x, \log x);\ \ \forall \varepsilon \in \mathbb{R}^+, |f| < \log^{\varepsilon} x\},$$

and we have $\mathcal{S}_1 = \mathbb{R}$. Any $f \in \mathcal{R}_1$ may be written uniquely in the form

$$f = \sum_{i,j=0}^{n} p_{i,j} x^{-i} \log^{-j} x \Big/ \sum_{i,j=0}^{m} q_{i,j} x^{-i} \log^{-j} x$$

with $p_{i,j}, q_{i,j} \in \mathbb{R}$ for $i, j = 1, \ldots, m$ and $q_{0,0} \neq 0$. Then $\eta_1(f) = p_{0,0}/q_{0,0}$.

Consider the element $g = (1 + x^{-1})/(1 - x^{-1} \log^{-1} x)$. This has an x-expansion

$$g = \sum (1 + x^{-1}) x^{-n} \log^{-n} x.$$

However, this is not a shadow expansion since no shadow field with respect to x^{-1} can contain $1 + x^{-1}$. To obtain a shadow expansion, we must write

$$g = 1 + \sum \log^{-n} x (1 + \log^{-1} x) x^{-n-1}.$$

Now each coefficient $\log^{-n} x(1 + \log^{-1} x) \in \mathcal{S}_2$ and has shadow $\mathbf{t}_1$–expansion $\log^{-n} x + \log^{-n-1} x$. Of course the coefficients of these latter expansions are equal to 1, and are therefore in $\mathcal{S}_1$.

A new consideration arises when we consider the function $(\pi + x^{-1})/(1 - x^{-1} \log^{-1} x)$. If we define the shadow fields in the above fashion, which is the most natural thing to do, then matters are much as in the previous example. This time however we *could* perversely define $\mathcal{S}_2$ to be $\mathbb{Q}(\pi + x^{-1}, \log x)$. The point is that x^{-1} belongs to $\mathbb{Q}(1 + x^{-1})$ but not to $\mathbb{Q}(\pi + x^{-1})$.

Recall that our strategy for computing asymptotics is to exhibit a tower such as (5.1) as a tower of asymptotic fields by giving the scale and the shadow fields attached to each field in the tower. Naturally we shall often obtain the shadow fields for $\mathcal{F}_i = \mathcal{F}_{i-1}(f_i)$ from those for $\mathcal{F}_{i-1}$. When considering $\mathcal{F}_i$ the first question is whether a new scale element needs to be added. Then we need to give a method for expanding f_i in the new scale, keeping closed forms for the coefficients. We need to define new shadow fields, extending those of $\mathcal{F}_{i-1}$, in such a way that the expansion of f_i is a shadow expansion. Then we can obtain shadow expansions of arbitrary members of $\mathcal{F}_i$ by arithmetically combining the expansions of members of $\mathcal{F}_{i-1}$ with the expansion of f_i. This in essence is how we compute expansions for Liouvillian functions. We shall make use of Lemma 17.

5.2.1 Adding Exponentials

It might seem that we have already covered this in Section 5.1, and to a large extent this is true. However we want to define shadow expansions and the rest of the asymptotic–field apparatus, since we may wish to add the exponential of an expression involving integrals for example.

So let $\mathcal{F}$ be an asymptotic field with $f \in \mathcal{F}$ and let $\theta = \exp f$. We define an asymptotic field containing θ. We consider first the case when f tends to a finite limit, l. As in Section 5.1, we can reduce to the case when $f \to 0$. For each i, $1 \leq i \leq n$, we define $\eta_i(\exp f) = \exp(\eta_i(f))$ and similarly $\eta_{i,j}(\exp f) = \exp(\eta_{i,j}(f))$. We extend η_i to $\mathcal{R}_i(\mathcal{F}(\theta))$ in the obvious way, namely for $a_0, \ldots, a_M, b_0, \ldots, b_N \in \mathcal{R}_i(\mathcal{F})$

$$\eta_i\left(\frac{a_0 + a_1\theta + \ldots + a_M\theta^M}{b_0 + b_1\theta + \ldots + b_N\theta^N}\right) = \frac{\eta_i(a_0) + \eta_i(a_1)\eta_i(\theta) + \ldots + \eta_i(a_M)\eta_i(\theta)^M}{\eta_i(b_0) + \eta_i(b_1)\eta_i(\theta) + \ldots + \eta_i(b_N)\eta_i(\theta)^N}$$

and similarly for $\eta_{i,j}$. Note that

$$\xi_i(\exp f) = \exp(\eta_i(f) + \xi_i(f)) - \exp\eta_i(f) = \exp\eta_i(f)\{\exp\xi_i(f) - 1\}.$$

So both shadows and ghosts are given by exponentials. This means that we can add the $\xi_i(\exp f)$, $i = 1, \ldots, n$, in the same way as we added $\exp f$ in order to calculate the next terms in the expansions.

Now we examine the case when $|f| \to \infty$. As in Section 5.1.2 we determine whether θ has the same comparability class as any element of the existing scale $\{\mathbf{t}_1, \ldots, \mathbf{t}_n\}$. If so we divide f by an appropriate power of that scale element and repeat the operation. On the other hand if $\gamma_1(\theta)$ turns out to be a new comparability class, then we remove any terms which have finite limits from the multiseries for f, to give g say and take $\mathbf{t} = \exp(-|g|)$ as a new scale element. As before we can decide where $\mathbf{t}$ should be inserted in the scale, but now we have to define the shadow fields of $\mathcal{F}(\mathbf{t})$ and the maps η_i and $\eta_{i,j}$.

We start with values of i, $1 \leq i \leq n$, for which $\gamma_1(\mathbf{t}_i) < \gamma_1(\mathbf{t})$. There we take $\mathcal{S}_i(\mathcal{F}(\mathbf{t})) = \mathcal{S}_i(\mathcal{F})$. An element, h, of $\mathcal{R}_i(\mathcal{F}(\mathbf{t}))$ is of the form

$$h = \mathbf{t}^r \frac{a_0 + a_1\mathbf{t} + \ldots + a_M\mathbf{t}^M}{b_0 + b_1\mathbf{t} + \ldots + b_N\mathbf{t}^N} \tag{5.11}$$

where $a_0, \ldots, a_M, b_0, \ldots, b_N \in \mathcal{R}_i(\mathcal{F})$, $a_0 b_0 \neq 0$ and for this case $r \geq 0$. If $r > 0$ then $\eta_i(h) = 0$. Otherwise $r = 0$ and we define $\eta_i(h) = \eta_i(a_0)/\eta_i(b_0)$.

When $\gamma_1(\mathbf{t}_i) > \gamma_1(\mathbf{t})$, we take $\mathcal{S}_i(\mathcal{F}(\mathbf{t})) = \mathcal{S}_i(\mathcal{F})(\mathbf{t})$ and then change the indexing to take account of the insertion of $\mathbf{t}$. For h given by (5.11) we define

$$\eta_i(h) = \mathbf{t}^r \frac{\eta_i(a_0) + \eta_i(a_1)\mathbf{t} + \ldots + \eta_i(a_M)\mathbf{t}^M}{\eta_i(b_0) + \eta_i(b_1)\mathbf{t} + \ldots + \eta_i(b_N)\mathbf{t}^N},$$

and again adjust the indexing.

Suppose that $\mathbf{t}_i = \mathbf{t}$ with $i < n+1$ in the new scale, and let h be given by (5.11). Then again $\eta_i(h) = 0$ if $r > 0$ and for $r = 0$ we take $\eta_i(h)$ equal to $\tilde{\eta}_i(a_0)/\tilde{\eta}_i(b_0)$ where $\tilde{\eta}_i$ is the projection of $\mathcal{R}_i(\mathcal{F})$ onto $\mathcal{S}_i(\mathcal{F})$ with i referring to the old scale.

For $\mathbf{t} = \mathbf{t}_{n+1}$ (in the new scale), we put $\mathcal{S}_{n+1}(\mathcal{F}(\mathbf{t})) = \mathcal{F}$ and when $r > 0$ in (5.11), we take $\eta_{n+1}(h) = 0$; otherwise $r = 0$ and we take $\eta_{n+1}(h) = a_0/b_0$.

In all of the above cases, it is easy to see that $h - \eta_i(h) \in \mathcal{I}_i$ and that $\mathcal{S}_i(\mathcal{F}(\mathbf{t}))$ possesses the shadow property with respect to $\mathbf{t}_i$. In particular closure under relative differentiation follows from Lemma 14. It is also clear that η_i is a projection of $\mathcal{R}_i(\mathcal{F}(\mathbf{t}))$ onto $\mathcal{S}_i(\mathcal{F}(\mathbf{t}))$. The definition of the maps $\eta_{i,j}$ is entirely similar but involves more cases. It is left as an exercise for the reader.

Now shadow expansions can be computed using Lemma 17. Alternatively, and more practically, we can observe that what we have done is to show that the expansions of Section 5.1 are shadow expansions. So for example we can use equation (5.6). Theorem 10 ensures that we can always add exponentials to a Hardy field, and hence condition 4 of Definition 21 holds. We have therefore proved the following.

THEOREM **13** *If $\mathcal{F}$ is an asymptotic field and $f \in \mathcal{F}$ then we can define an asymptotic field containing $\mathcal{F}(\exp f)$.*

5.2.2 Adding Integrals

Suppose that in the tower (5.1), $\mathcal{F}_j = \mathcal{F}_{j-1}(g) = \mathcal{F}(g)$ say, where $g = \int f$. Of course g is not uniquely specified by this, and there is a problem in adding an initial condition of the form $g(x_0) = a_0$. Our whole approach has been built around the behaviour of functions for sufficiently large values of the argument, and there is no ready mechanism for encoding initial conditions into this. Indeed since Hardy-field elements are germs at $+\infty$, it makes no sense to ask about values at finite points. Of course elementary functions and other simple functions do have well defined values, but integral extensions may occur at more than one place in the tower (5.1), and moreover integrals may contain singularities. In addition, we are going to want to consider algebraic extensions, and so complex questions about branching may arise.

The line we take is to present the possible forms of asymptotic behaviour. Where integrals are concerned, this means including arbitrary constants in solutions. In other words, our integrals are indefinite. We regard the matter of evaluating the arbitrary constant from an initial condition as a separate question. It is probably quite a difficult one in general. Naturally there are many simple cases when a definite integral can be used without difficulty.

Suppose then that we have a scale $\{\mathbf{t}_1, \ldots, \mathbf{t}_n\}$ in which we can expand elements of $\mathcal{F}_{i-1}$. We need to determine whether or not we need to add to this scale, and to expand g in the new scale keeping closed forms for the coefficients. The following result was given for exp–log functions by Hardy in [40]. The present version is from [103].

THEOREM **14 (Hardy)** *Let f be an element of a Hardy field, and let a be a constant which is sufficiently large for f to be $\mathcal{C}^\infty$ on an open interval containing $[a, \infty)$. Write $F = \int_a^x f$ if this tends to infinity and $F = \int_x^\infty f$ otherwise.*

1. *If $\gamma_1(f) > \gamma_1(x)$ then $F \sim f^2/f'$.*
2. *If $f(x) = x^r h(x)$ with $r \neq -1$ and $\gamma_1(h) < \gamma_1(x)$, then*

$$F(x) \sim \frac{x^{r+1}h}{r+1}.$$

3. *If $f(x) = x^{-1} \log^{-1} x \cdots \log_p^{-1} x h(x)$ with $p \geq 0$, $\gamma_1(h) < \gamma_1(\log_p x)$ and $h \not\sim \log_{p+1}^{-1} x$, then*

$$\begin{aligned} F &\sim \frac{h^2}{h' x \log x \cdots \log_p x}, && \text{if } \gamma_1(h) > \gamma(\log_{p+1} x); \\ &\sim \frac{\log_{p+1}^{s+1} x \bar{h}}{s+1}, && \text{if } h \sim \log_{p+1}^{s} x \bar{h} \text{ with } \gamma_1(\bar{h}) < \gamma_1(\log_{p+1} x) \\ & && \text{and } s \neq -1. \end{aligned}$$

Note that in the second part of case 3, F may be of a new comparability class when $s = 0$.

Since we can compute a multiseries expansion for f, we can in particular obtain a monomial in the scale $\{\mathbf{t}_1, \ldots, \mathbf{t}_n\}$ asymptotic to f. Theorem 14 then allows us to calculate a monomial asymptotic to F after having introduced a new scale element if necessary. We can then see whether or not $|F| \to \infty$ and hence whether F is $\int_a^x f$ or $\int_x^\infty$. If the former, then $g = \int f = F + K$ while if the latter then $g = K - F$, where K is an arbitrary constant.

The next consideration is how to compute the shadows associated with g. For g of the form $K - F$, we would in practice apply the computations below to F. Note that it is not always correct to define $\eta_i(\int f) = \int \eta_i(f)$. For example if $\mathbf{t}_i = x^{-1}$ and $f = x^{-1} \log^{-1} x$ then $\int \eta_i(f) = \int 0 = K$, with K constant, while $\eta_i(\int f) = \eta_i(K - \log^{-2} x) = K - \log^{-2} x$.

Let T be the $\{\mathbf{t}_1, \ldots, \mathbf{t}_n\}$–monomial asymptotic to $\int f$. We can assume that T is positive and not a constant. We define $\eta_i(g/T)$ for the different values of i. Of course we do not have to divide g by all of T; just those factors involving $\mathbf{t}_j$ with $j \geq i$ would suffice. However dividing through by T is probably simpler in practice. As usual, when $i = 1$ the shadow is just the limit. i.e. $\eta_1(g/T) = \lim_{x\to\infty}(g/T)$. For $i > 1$ we distinguish three cases.

Case 1: $\gamma_1(\log T) \geq \gamma_1(\mathbf{t}_i)$

Here we define $\eta_i(g/T) = \eta_i(f/T')$. Since $f/T' \in \mathcal{F}$, we just take $\mathcal{S}_i(\mathcal{F}(g)) = \mathcal{S}_i(\mathcal{F})$ and the shadow property is already guaranteed. We need to know that the difference between g/T and its i–th shadow tends to zero at least as fast as a positive power of $\mathbf{t}_i$.

LEMMA **19** *For each $i = 1, \ldots, n$, there exists a $\delta \in \mathbb{R}^+$ such that*

$$\left|\frac{g}{T} - \eta_i\left(\frac{g}{T}\right)\right| < \mathbf{t}_i^{\delta}.$$

Proof of Lemma 19
We write $f_i = \eta_i(f/T')$ and $g_i = \xi_i(f/T')$. We show that $\xi_i(g/T) \in \mathcal{I}_i$. We have

$$\begin{aligned}\xi_i(g/T) = g/T - \eta_i(g/T) &= T^{-1}\left\{\int f - Tf_i\right\}\\ &= T^{-1}\left\{\int T'(f_i + g_i) - Tf_i\right\}\\ &= T^{-1}\left\{Tg_i - \int T(f_i' + g_i')\right\},\end{aligned} \tag{5.12}$$

as is seen on integrating by parts. Thus

$$|\xi_i(g/T)| \leq |g_i| + \left|T^{-1}\int T(f_i' + g_i')\right|. \tag{5.13}$$

Since $\gamma_1(\mathbf{t}_i) \leq \gamma_1(\log T)$, there exists a $\delta \in \mathbb{R}^+$ such that $\mathbf{t}_i^{-2\delta} < |\log T|$. Then by Lemma 6, $(\mathbf{t}_i^{-2\delta})' < |T'/T|$. Also $f_i' + g_i' < (\mathbf{t}_i^{-\delta})'$ since $f_i + g_i \in \mathcal{R}_i$. Thus

$$|T(f_i' + g_i')| < \left|\frac{(\mathbf{t}_i^{-\delta})'T'}{(\mathbf{t}_i^{-2\delta})'}\right| = \frac{1}{2}\mathbf{t}_i^{\delta}|T'|. \tag{5.14}$$

However, $\gamma_1(\mathbf{t}_i^{\delta}) \leq \gamma_1(\log T)$, and hence by Lemma 7 we have $\gamma_0((\mathbf{t}_i^{\delta})'/\mathbf{t}_i^{\delta}) < \gamma_0(T'/T)$. Therefore $(\mathbf{t}_i^{\delta})'T/\mathbf{t}_i^{\delta}T' \to 0$ and so $(\mathbf{t}_i^{\delta}T)' \sim \mathbf{t}_i^{\delta}T'$. Then (5.14) implies that $|T(f_i'+g_i')| < |(\mathbf{t}_i^{\delta}T)'|$. By making δ smaller if necessary, we may assume that $|g_i| = |\xi_i(f/T')| < \mathbf{t}_i^{\delta}$, and then from (5.13) we have $|\xi_i(g/T)| < 2\mathbf{t}_i^{\delta}$ as required. This completes the proof of Lemma 19.

We also need to express $\xi_i(g/T)$ suitably in closed form, so that in particular we can compute a monomial asymptotic to it. Since $\int T'g_i = Tg_i - \int Tg_i'$, equation (5.12) gives

$$\xi_i(g/T) = T^{-1} \int \{T' g_i - T f_i'\}$$

as required. Note that the arbitrary constant from $\int f$ forms part of the ghost $\xi_i(g/T)$ in this case, since $\gamma_1(K/T) > \gamma_1(\mathbf{t}_i)$.

Case 2: $\gamma_1(\mathbf{t}_i) > \gamma_1(T)$

Now we define

$$\eta_i(g/T) = T^{-1} \int T' \eta_i(f/T'). \tag{5.15}$$

In fact this will also be the definition in the remaining case, $\gamma_1(\log T) < \gamma_1(\mathbf{t}_i) \leq \gamma_1(T)$, but the arbitrary constant of integration will be handled differently there. However we shall combine the cases where possible, as in the next lemma. Note that in Case 2, the integral on the right of (5.15) includes an arbitrary constant.

LEMMA **20** *If* $\gamma_1(\log T) < \gamma_1(\mathbf{t}_i)$ *and* $\phi_i = T^{-1} \int T' \eta_i(f/T')$, *then* $\mathcal{S}_i(\mathcal{F})(\phi_i)$ *has the shadow property SF(i) with respect to* $\mathbf{t}_i$.

Proof of Lemma 20
After Lemma 14, it is sufficient to show that ϕ_i satisfies a differential equation of the form (5.10). It is immediate that

$$\phi_i' = \frac{T'}{T} \left(\eta_i \left(\frac{f'}{T} \right) - \phi_i \right). \tag{5.16}$$

This will be of the required form provided that $\log T \in \mathcal{S}_i(\mathcal{F})$; for then, in the notation of equation (5.10) we can take $g_1 = g_2 = \log T$. Now $T = A\mathbf{t}_1^{c_1}\mathbf{t}_2^{c_2}\cdots\mathbf{t}_j^{c_j}$ for some values of the constants $j, A, c_1, \ldots, c_j$, where $\gamma_1(\log \mathbf{t}_j) < \gamma_1(\mathbf{t}_i)$. So

$$\log T = \log A + c_1 \log \mathbf{t}_1 + \cdots + c_j \log \mathbf{t}_j.$$

According to Definition 10, each $\mathbf{t}_m$, $1 \leq m \leq j$ will either be of the form $\log_k^{-1} x$ or of the form $\exp(-h)$ with every term in the multiseries of h tending to $\pm\infty$. Assuming that $c_1 \neq 0$ we will need to add an extra $\log_{k+1}^{-1} x$ at the beginning of the scale to cater for $\log \mathbf{t}_1$. However Lemma 18 shows that there is no problem in doing so, and clearly this enlargement of the field will not invalidate the conclusion. For $m \geq 2$ with $\mathbf{t}_m$ an inverse of an iterated logarithm, $\log \mathbf{t}_k$ will be the inverse of a scale element already in $\mathcal{S}_i(\mathcal{F})$.

The remaining case is when $\mathbf{t}_m$ is an exponential, say $\mathbf{t}_m = \exp(-h)$. Let the scale element of largest comparability class appearing in the shadow expansion of h be $\mathbf{t}_k$. No positive powers of $\mathbf{t}_k$ can appear in this expansion because the terms must tend to $\pm\infty$. Hence the $\mathbf{t}_k$–expansion of h must be finite. Also $\gamma_1(\mathbf{t}_k) \leq \gamma_1(\log \mathbf{t}_m) < \gamma_1(\mathbf{t}_i)$. The coefficients in the $\mathbf{t}_k$–expansion of h belong to $\mathcal{S}_k(\mathcal{F})$ which is contained in $\mathcal{S}_i(\mathcal{F})$. Since $\mathbf{t}_k \in \mathcal{S}_i(\mathcal{F})$ by condition (2) of Definition 21, we have the conclusion that $h \in \mathcal{S}_i(\mathcal{F})$, as required. Thus SF(i) holds and the proof of Lemma 20 is complete.

The proof that SF(ii) also holds requires slightly different handling in cases two and three, but there is much common ground. In both we use Lemma 16. The following lemma takes us to the point where the paths diverge.

LEMMA **21** *Suppose that* $\gamma_1(\log T) < \gamma_1(\mathbf{t}_i)$ *and let*

$$\phi_i = \eta_i(g/T) = T^{-1}\int T'\eta_i(f/T').$$

Let $p \in \mathcal{S}_i(\mathcal{F})_{alg}$ *and suppose that for some* $\delta \in \mathbb{R}^+$, $|\phi_i - p| < \mathbf{t}_i^\delta$. *Then* ϕ_i *and* p *both satisfy the differential equation*

$$y' = \frac{T'}{T}\left(\eta_i\left(\frac{f'}{T}\right) - y\right). \tag{5.17}$$

Proof of Lemma 21
Equation (5.16) shows that ϕ_i satisfies equation (5.17). On the other hand, by Lemma 6 parts (5) and (6), $|\phi_i - p| < \mathbf{t}_i^\delta$ implies that $|\phi_i' - p'| < |(\mathbf{t}_i^\delta)'|$. So

$$\begin{aligned}\left|\eta_i\left(\frac{f}{T'}\right) - p - \frac{Tp'}{T'}\right| &\le \left|\eta_i\left(\frac{f}{T'}\right) - \phi_i - \frac{T\phi_i'}{T'}\right| + |\phi_i - p| + \left|\frac{T}{T'}\right| |\phi_i' - p'| \\ &< \left|\frac{T(\mathbf{t}_i^\delta)'}{T'}\right| + |\mathbf{t}_i^\delta|.\end{aligned} \tag{5.18}$$

Now since $\gamma_1(\log T) < \gamma_1(\mathbf{t}_i)$, we have $|\log T| > \mathbf{t}_i^{\delta/2}$. Hence $|T'/T| > |(\mathbf{t}_i^{\delta/2})'|$ and so

$$\left|\frac{T(\mathbf{t}_i^\delta)'}{T'}\right| < \left|\frac{(\mathbf{t}_i^\delta)'}{(\mathbf{t}_i^{\delta/2})'}\right| = 2|\mathbf{t}_i^{\delta/2}|.$$

It then follows from (5.18) that

$$\left|\eta_i\left(\frac{f}{T'}\right) - p - \frac{Tp'}{T'}\right| < 3|\mathbf{t}_i^{\delta/2}|. \tag{5.19}$$

However, we already know from the proof of Lemma 20 that $\log T \in \mathcal{S}_i(\mathcal{F})$ and since $\mathcal{S}_i(\mathcal{F})_{alg}$ is closed under relative differentiation, the left–hand side of (5.19) belongs to $\mathcal{S}_i(\mathcal{F})_{alg}$. Hence it must be zero by Lemma 15, and so p also satisfies (5.17). This completes the proof of Lemma 21.

It is now easy to see that in Case 2, $\mathcal{S}_i(\phi)$ has the shadow property SF(ii) with respect to $\mathbf{t}_i$, for two solutions of (5.17) must differ by KT^{-1} for some constant K. But then $|\phi_i - p| < \mathbf{t}_i^\delta$ gives $|KT^{-1}| < \mathbf{t}_i^\delta$ and since $\gamma_1(\mathbf{t}_i) > \gamma_1(T)$ this implies that $K = 0$. Thus $|\phi_i - p| < \mathbf{t}_i^\delta$ implies that $\phi_i = p$, and Lemma 16 gives the desired conclusion.

We now show that $gT^{-1} - \phi_i \in \mathcal{I}_i$ in Case 2. We have

$$\frac{g}{T} - \phi_i = T^{-1}\int f - T^{-1}\int T'\eta_i\left(\frac{f}{T'}\right) = T^{-1}\int\left\{T'\xi_i\left(\frac{f}{T'}\right)\right\} \tag{5.20}$$

As in Lemma 19, we write $g_i = \xi_i(f/T')$. Then $|g_i| < \mathbf{t}_i^\delta$ for some $\delta \in \mathbb{R}^+$. Since $\gamma_1(T) < \gamma_1(\mathbf{t}_i)$, this gives $2|\log T| < |\log g_i|$, and so

$$2\left|\frac{T'}{T}\right| < \left|\frac{g_i'}{g_i}\right|.$$

Hence $|T'g_i| < |Tg_i'|/2$, which in turn implies that

$$|(Tg_i)'| \geq |Tg_i'| - |T'g_i| \geq |T'g_i|/2.$$

On substituting into (5.20), we obtain

$$\left|\frac{g}{T} - \phi_i\right| < 2T^{-1}\left|\int (Tg_i)'\right| < 2\mathbf{t}_i^\delta, \tag{5.21}$$

as required.

Now (5.20) tells us that $\xi_i(g/T) = T^{-1}\int T'g_i$ for some determination of the integral. In fact we must have $\xi_i(g/T) = -T^{-1}\int_x^\infty T'g_i$ since (5.21) shows that the integral tends to zero at infinity. From the expression for $\xi_i(g/T)$ we can then compute later terms shadow expansions of elements of $\mathcal{F}(g)$ as indicated earlier.

Case 3: $\gamma_1(\log T) < \gamma_1(\mathbf{t}_i) \leq \gamma_1(T)$

Again we define

$$\eta_i(g/T) = T^{-1}\int T'\eta_i(f/T'),$$

but now the arbitrary constant must belong to the ghost. For otherwise two different determinations of the constant would give values of $\eta_i(g/T)$ differing by KT^{-1} with K constant, and this would violate the shadow property since $\gamma_1(\mathbf{t}_i) \leq \gamma_1(T)$. It follows that the integral defining $\eta_i(g/T)$ must be a definite one, but which?

Clearly if there is such an integral which is already in $\mathcal{S}_i(\mathcal{F})$, we must choose that one, and similarly if there is one in $\mathcal{S}_i(\mathcal{F})_{alg}$. Note once more that by SF(ii), there cannot be two different integrals in this field. This apparently leaves the rather difficult looking problems of how to decide whether there is an integral in the algebraic closure of the existing shadow field, and if not, how to define the limits of integration in $\int T'\eta_i(f/T')$. In fact these problems can very largely be circumvented.

Recall that the reason why it is so important to have closed forms for the coefficients is that we have to be able to detect whether a coefficient is functionally equivalent to zero. The zero–equivalence algorithms of Sections 2.2 and 2.3 are recursive in nature. So when we need to know about the zero–equivalence of an expression involving $h = \int T'\eta_i(f/T')$, we will be considering a polynomial, $P(h)$, over the field $\mathcal{S}_i(\mathcal{F})$. If h is not algebraic over $\mathcal{S}_i(\mathcal{F})$ for any determination of the integral, then $P(h)$ can only be equivalent to zero if all its coefficients are. On the other hand if there is a value of the constant of integration for which h is algebraic over $\mathcal{S}_i(\mathcal{F})$, it is unique and is the value we choose. Thus the question we have to answer is whether for any value of the constant of integration, our polynomial is equivalent to zero. Happily the method of Section 2.2 can easily be adapted to do this. For in Algorithm 2 of that section, we reduce P to a square–free form and compute $\gcd_{\mathcal{F}}(P, P^*)$. If this turns out to have degree zero then by Lemma 1, $P(h)$ will not be equivalent to zero for any determination of the constant of integration. On the other hand if the degree of the gcd is positive, it contains a root satisfying the same differential equation as h. In other words there is an integral which makes P zero. In that case our integral is that one and $P(h) \equiv 0$. If we never meet a P for which $\gcd_{\mathcal{F}}(P, P^*)$ has positive degree, we never need to know precisely which integral is the shadow of g/T since we can still expand the integral in the usual way.

Of course once we have an arbitrary constant in our expansion, all the integrals that follow are definite ones of the form $\int_x^\infty \cdots$.

It remains to prove that in Case 3, $\gamma_1(T^{-1}g - \phi_i) \le \gamma_1(\mathbf{t}_i)$. If $\gamma_1(g_i) > \gamma_1(T)$, we obtain (5.21) as in Case 2. Similarly, if $\gamma_1(g_i) < \gamma_1(T)$, then by Lemma 7, $|T'g_i|/2 > |Tg_i'|$ and again we deduce that $|T'g_i| < 2|(Tg_i)'|$, which gives (5.21) as before. If $\log g_i \sim a \log T$ with a a constant not equal to -1, then $g_i'/g_i \sim aT'/T$. Hence $T'g_i \sim (Tg_i)'/(a+1)$, and as above $|T^{-1}g - \phi_i| < 2|g_i|/(a+1)$. Finally if $\log g_i \sim -\log T$, then $|g_i| < 1/\sqrt{T}$. Hence

$$|\psi_i| < T^{-1} \int \frac{T'}{\sqrt{T}} = \frac{2}{\sqrt{T}}.$$

Thus in all cases, $\gamma_1(T^{-1}g - \phi_i) \le \gamma_1(\mathbf{t}_i)$ as required.

Finally, note that we can write $\xi_i(g/T)$ in terms of an integral exactly as in Case 2.

Example
We illustrate the use of asymptotic fields by calculating an asymptotic expansion for the function

$$H = -x \exp\left(\frac{1}{x}\right) + \int \frac{x-1}{x} \exp\left(\frac{1}{x} + e^{-x}\right).$$

We take $\mathbb{K}$ as the field of exp–log constants and build a tower of fields whose top field contains H. We have $\mathcal{F}_0 = \mathbb{K}$, $\mathcal{F}_1 = \mathbb{K}(x)$, $\mathcal{F}_2 = \mathbb{K}(x, e^{-x})$,

$\mathcal{F}_3 = \mathcal{F}_2(\exp(x^{-1}), \mathcal{F}_4 = \mathcal{F}_3(\exp(x^{-1} + e^{-x}))$ and $\mathcal{F}_5 = \mathcal{F}_4(g)$, where $g = \int \frac{x-1}{x} \exp(x^{-1} + e^{-x}) = \int f$ say.

Our next objective is to build the corresponding tower of asymptotic fields. For $\mathbb{K}(x)$, we have scale $\{x^{-1}\}$. Also $\mathcal{R}_1(\mathbb{K}(x))$ consists of the quotients of polynomials in $\mathbb{K}[x^{-1}]$ whose denominator has non–zero constant term. Here the projection η_1 is obtained by mapping x^{-1} to zero. At the next stage we add the exponential e^{-x}. It is immediately clear that $-x$ is not asymptotic to a multiple of the logarithm of the only existing scale function, and so e^{-x} needs to be added to the scale. Thus the scale for $\mathcal{F}_2$ is $\{\mathbf{t}_1, \mathbf{t}_2\}$, where $\mathbf{t}_1 = x^{-1}$ and $\mathbf{t}_2 = e^{-x}$. Then $\mathcal{R}_2$ and $\mathcal{R}_1$ are defined in the obvious way, and the map η_2 is then obtained by projecting e^{-x} to zero. Similarly η_1 is obtained by projecting both e^{-x} and x^{-1} to zero.

To get to $\mathcal{F}_3$ we again add an exponential, $\exp(x^{-1})$, but this time its argument tends to zero and so there are no new scale elements to add. According to Section 5.2.1, we take

$$\eta_2(\exp(x^{-1})) = \exp(\eta_2(x^{-1})) = \exp(x^{-1})$$

and

$$\eta_1(\exp(x^{-1})) = \exp(\eta_1(x^{-1})) = \exp 0 = 1.$$

Things are very similar when it comes to $\mathcal{F}_4$. There

$$\eta_2(\exp(x^{-1} + e^{-x})) = \exp(\eta_2(x^{-1} + e^{-x})) = \exp(x^{-1}),$$

and

$$\eta_1(\exp(x^{-1} + e^{-x})) = \exp(\eta_1(x^{-1} + e^{-x})) = 1.$$

Now to $\mathcal{F}_5$ and the nub of the matter. We calculate that

$$f = \frac{x-1}{x} \exp\left(\frac{1}{x} + e^{-x}\right) \sim 1,$$

and so Theorem 14 tells us that $g = \int f \sim x$. Thus we take $T = x$ and compute shadows of g/T. For η_2, we are in Case 2 since $\gamma_1(T) < \gamma_1(\mathbf{t}_2)$. We define

$$\eta_2\left(\frac{g}{T}\right) = \frac{1}{T}\int T'\eta_2\left(\frac{f}{T'}\right) = \frac{1}{x}\int \eta_2\left(\frac{x-1}{x}\exp\left(\frac{1}{x} + e^{-x}\right)\right)$$
$$= \frac{1}{x}\int\left(\frac{x-1}{x}\exp x^{-1}\right).$$

Note that the last integral still includes an arbitrary constant. Actually $\eta_2(g/T) = \exp x^{-1} + Kx^{-1}$, but we do not need to know this fact.

The calculation of $\eta_1(g/T)$ brings us into Case 3. We have

$$\eta_1\left(\frac{g}{T}\right) = \frac{1}{x}\int \eta_1\left(\frac{x-1}{x}\exp\left(\frac{1}{x} + e^{-x}\right)\right) = \frac{1}{x}\int 1.$$

This time the integral is a definite one, and clearly we want $\eta_1(g/T)$ to be equal to 1. However with other examples we might not have a closed form available for the integral, and so to illustrate the method we proceed as if in ignorance of the integral.

Our next step then, is to obtain suitable expressions for the ghosts of g/T. As in (5.20) we have

$$\xi_2\left(\frac{g}{T}\right) = \frac{1}{T}\int T'\xi_2\left(\frac{f}{T'}\right) = \frac{1}{x}\int \xi_2\left(\frac{x-1}{x}\exp\left(\frac{1}{x}+e^{-x}\right)\right)$$
$$= \frac{1}{x}\int \frac{x-1}{x}\exp x^{-1}(\exp(e^{-x})-1),$$

and similarly

$$\xi_1\left(\frac{g}{T}\right) = \frac{1}{x}\int\left\{\frac{x-1}{x}\exp(x^{-1}+e^{-x})-1\right\}.$$

Finally we are in a position to deal with $H = g - x\exp x^{-1}$. This belongs to $\mathcal{R}_2(\mathcal{F}_5)$ because both g and $x\exp x^{-1}$ do. Now

$$\eta_2\left(\frac{H}{x}\right) = \eta_2\left(\frac{g}{T}\right) - \eta_2(\exp x^{-1}) = \frac{1}{x}\int\left(\frac{x-1}{x}\exp x^{-1}\right) - \exp x^{-1}.$$

On testing the right–hand side for zero equivalence we discover that it is zero for some determination of the integral. Hence $\eta_2(H) = K$ where K is the constant of integration. Thus

$$H = K + x\xi_2\left(\frac{g}{T}\right) = K + \int\frac{x-1}{x}\exp x^{-1}(\exp(e^{-x})-1). \tag{5.22}$$

By applying the same methods to the integral on the right of (5.22), one can obtain

$$\begin{aligned} H \sim K &- \int_x^\infty\left(1-\frac{1}{t}\right)e^{1/t}e^{-t}dt - \int_x^\infty\left(1-\frac{1}{t}\right)e^{1/t}\frac{e^{-2t}}{2}dt \\ &- \int_x^\infty\left(1-\frac{1}{t}\right)e^{1/t}\frac{e^{-3t}}{3!}dt - \cdots \\ \sim K &- e^{-x}\left(1-\frac{1}{2x^2}+\frac{2}{3x^3}+\cdots\right) + \frac{e^{-2x}}{4}\left(1-\frac{1}{2x^2}-\frac{5}{6x^3}+\cdots\right) \\ &+ O\left(e^{-3x}\right). \end{aligned} \tag{5.23}$$

Note that the terms of the second series have a meaning by virtue of the closed forms in (5.23).

Exercise
Use the methods of the present section to show that if

$$f = \sqrt{\log x - 1} + \frac{1}{2\sqrt{\log x - 1}} + \frac{1}{x}$$

then $(\int f)^2 - x^2(\log x - 1) \sim 2x \log^{3/2} x$, and check the result using a closed form for the integral; c.f. [103].

5.2.3 Algebraic Equations

We now consider the problem of adding a function defined by an algebraic equation to an asymptotic field. As mentioned, there is a problem in distinguishing different solutions here too. Before grappling with that, we show that the most natural way of obtaining shadows actually works with algebraic extensions.

THEOREM **15** *Let $\mathcal{F}$ be an asymptotic field with scale $\{\mathbf{t}_1, \ldots, \mathbf{t}_n\}$ and suppose that $1 \leq i \leq n$. Let P be a polynomial over $\mathcal{R}_i(\mathcal{F})$ and let f be an element of $\mathcal{R}_i(\mathcal{F}_{alg})$ satisfying $P(f) = 0$. Then there exists an element ϕ_i of $\mathcal{S}_i(\mathcal{F})_{alg}$ such that the following hold.*

1. *$\tilde{\eta}_i(P)(\phi_i) = 0$, where $\tilde{\eta}_i(P)$ is obtained from P by applying η_i to the coefficients.*

2. *$f - \phi_i \in \mathcal{J}_i = \mathcal{J}_i(\mathcal{F}_{alg})$.*

3. *$\mathcal{S}_i(\mathcal{F})(\phi_i)$ has the shadow property with respect to t_i.*

Proof of Theorem 15
Write $\tilde{P} = \tilde{\eta}_i(P)$. The main task in the proof is to show that $\tilde{P}$ has a root sufficiently close to f. To this end, we use induction on the degree to show that if Q is a polynomial over $\mathcal{S}_i(\mathcal{F})$ and $Q(f) \in \mathcal{I}_i$ then there is a $g \in \mathcal{S}_i(\mathcal{F})_{alg}$ such that $f - g \in \mathcal{I}_i$ and $Q(g) = 0$.

If Q is of degree one, this is a triviality. Suppose then that the statement holds for polynomials of degree less than n, and let Q be of degree n with $Q(f) \in \mathcal{I}_i$. We begin by observing that if $\alpha \in \mathcal{I}_i$ and $\partial Q/\partial y$ has a zero at a point y_0 in the interval $(f - \alpha, f + \alpha)$, then $Q(y_0)$ is the zero function. For here y_0 must be an element of $\mathcal{S}_i(\mathcal{F})_{alg}$ and hence so must $Q(y_0)$. However $Q(y_0) = (Q(y_0) - Q(f)) + Q(f)$, and $Q(y_0) - Q(f)$ belongs to $\mathcal{I}_i$ since it is divisible by $y_0 - f$. By hypothesis $Q(f)$ belongs to this ideal and hence so does $Q(y_0)$. However by Lemma 15, zero is the only element of $\mathcal{S}_i(\mathcal{F})_{alg} \cap \mathcal{I}_i$.

That done, we return to the problem of finding a root of Q sufficiently close to f. If $(\partial Q/\partial y)(f) \in \mathcal{I}_i$ then by induction there is a $g \in \mathcal{S}_i(\mathcal{F})_{alg}$ such that $f - g \in \mathcal{I}_i$ and $(\partial Q/\partial y)(g) = 0$. But then $Q(g) = 0$ by the above argument and we have found a suitable root of Q. Suppose then that $(\partial Q/\partial y)(f) \notin \mathcal{I}_i$. Then for $\alpha \in \mathcal{I}_i$,

$$Q(f + \alpha) = Q(f) + \alpha \left\{ \frac{\partial Q}{\partial y}(f) + \beta(\alpha) \right\},$$

where $\beta(\alpha) \in \mathcal{I}_i$. Since $Q(f)$ belongs to $\mathcal{I}_i$ and $(\partial Q/\partial y)(f)$ does not, there must exist values of $\alpha \in \mathcal{I}_i$ such that $Q(f+\alpha) > 0$ and $Q(f-\alpha) < 0$. By the intermediate-value property of real-closed fields, [120], Q has a root in $\mathcal{S}_i(\mathcal{F})_{alg}$ between $f+\alpha$ and $f-\alpha$, as required.

Now $\tilde{P}(f) = \tilde{\eta}_i(P(f)) \in \mathcal{I}_i$ and so $\tilde{P}$ has a root in an interval of the form $(f-\alpha, f+\alpha)$ with $\alpha \in \mathcal{I}_i$. Clearly $\tilde{P}$ cannot have two distinct roots in such an interval or their difference would be a non-zero element of $\mathcal{S}_i(\mathcal{F})_{alg} \cap \mathcal{I}_i$. So we may define $\phi_i = \eta_i(f)$ to be the unique root of $\tilde{P} = 0$ whose difference with f lies in $\mathcal{I}_i$. Now properties 1 and 2 of the theorem hold by construction and 3 follows from Lemma 15. Thus we have proved Theorem 15.

Our aim now is to show that under the hypotheses of Theorem 15, $\mathcal{F}(f)$ is an asymptotic field. Naturally we intend to define $\eta_i(f)$ as the root of $\tilde{\eta}_i(P)$ close to f, but we have a problem in distinguishing f from the other roots of P. The arguments we put concerning arbitrary constants attached to integrals apply equally well here. If we want to distinguish roots by their values at some finite point, x_0, then we will need to know about branch points in the interval $[x_0, \infty)$. Some progress has been made with such problems over the past few years; for example see [69, 57]. However we take a different line of approach here. Our algorithm lists the different asymptotic representations of the roots, with roots being distinguished by their asymptotic behaviour. In a sense this amounts to taking $+\infty$ as the evaluation point. It is not always possible to make such a distinction by naive expansion in series, and in the general case considerable care is needed with the handling of shadows and ghosts.

Let P be a square–free polynomial over an asymptotic field, $\mathcal{F}$. We seek expansions of the roots of P, and moreover we would like to create an asymptotic field containing them and $\mathcal{F}$. Clearly we do not need to add any new scale elements to the scale of $\mathcal{F}$ because the roots of P are algebraic over $\mathcal{F}$. Let $\{\mathbf{t}_1, \ldots, \mathbf{t}_n\}$ be the scale.

We calculate a $\{\mathbf{t}_1, \ldots, \mathbf{t}_n\}$–monomial asymptotic to each root by trial substitution. Thus we put $f = \mathbf{t}_n^{r_n} \cdots \mathbf{t}_1^{r_1} \lambda$, with the understanding that λ is asymptotic to a non–zero constant, and consider the leading terms in $P(f)$. We first look at the powers of $\mathbf{t}_n$ and use the fact that the leading monomials of $P(f)$ must cancel. The exponents will be of the form $\alpha_i + r_n\beta_i$. Working by hand, we can then draw a Newton polygon by plotting the points (α_i, β_i), see for example [8]. In computer implementation, the same result is achieved by considering pairs (α_i, β_i) and (α_j, β_j) and solving the equation $\alpha_i + r_n\beta_i = \alpha_j + r_n\beta_j$. Then we check that with this value of r_n, these particular pairs were indeed dominant. The result is a list of possible values of r_n.

We consider each element of the list in turn. The value of r_n is substituted and the possible values of r_{n-1} obtained in similar fashion. These are then substituted in each case, and values of r_{n-2} sought, and so on, finishing with the calculation of possible asymptotic values for λ. In practice it might be more economical to roll the calculations of the r_is together. At the end of this stage we have a list of monomials representing the possible asymptotic behaviour of the roots of P. Possibly some monomials in the list might not correspond to any actual root.

We can determine how many roots there are asymptotic to a given monomial by using a version of Sturm's Theorem for polynomials over Hardy fields. Note first that, by Theorem 9, there is a Hardy field containing $\mathcal{F}$ and all the roots of $P = 0$. Suppose that P is of degree n. Define $A_0 = P$, $A_1 = \partial P/\partial y$ and for $i = 1, \dots, n-1$ define $-A_{i+1}$ to be the remainder on dividing A_{i-1} by A_i. Note that $A_n(y) \in \mathcal{F}$; i.e. A_n is of degree zero. For any element c of $\mathcal{F}$, we define $V(c)$, the *variation* of P at c, to be the number of changes in sign (with zeros ignored) in the sequence $A_0(c)$, $A_1(c)$,..., $A_n(c)$. The usual proof (see, for example, [48]) now yields the following.

THEOREM **16 (Sturm)** *Let a and b belong to $\mathcal{F}$ with $a < b$. Suppose that P is square free and that $P(a)P(b) \neq 0$. Then the number of zeros of P between a and b is equal to $V(a) - V(b)$.*

To find out how many roots there are asymptotic to each possible monomial we choose other monomials, $\mathbf{s}_i$, separating them. We then calculate the variation at each $\mathbf{s}_i$ and also at the points $\pm\infty$. Note that the sign of a polynomial at $+\infty$ is the sign of the leading coefficient, and likewise the sign at $-\infty$ is $(-1)^{degree}$ times the sign of the leading coefficient.

To compute expansions, we proceed as follows. For each monomial, m, we replace f by $\bar{f} = f/m$ in the equation $P = 0$. Then we divide the equation through by the largest monomial occurring in a coefficient. This ensures in particular that the coefficients belong to $\mathcal{R}_i(\mathcal{F})$ for all i. Now for each i, we can compute the $\mathbf{t}_i$–shadow of $\bar{f}$ as a root of the equation $\tilde{\eta}_i(P) = 0$.

We may summarize the above as follows.

ALGORITHM **7**

1. *Compute the gcd of P and $\partial P/\partial y$ over $\mathcal{F}$ and divide P by the result. P is then square free with the same roots as before.*
2. *Substitute $y = \mathbf{t}_1^{r_1}\mathbf{t}_2^{r_2}\cdots\mathbf{t}_n^{r_n}\lambda$ in P, where $r_1, \dots, r_n$ are real numbers and λ tends to a non–zero constant. Determine possible values for $r_n, r_{n-1}, \dots, r_1$ and the limit of λ by trial substitution.*
3. *Choose evaluation points separating the possible monomials found during the previous stage. Compute the Sturm sequence for P and the variation at each separation point and at $\pm\infty$. Sturm's Theorem then gives the number of roots asymptotic to each of the $\{\mathbf{t}_1, \dots, \mathbf{t}_n\}$–monomials found.*
4. *To obtain the shadow of f/m with f a root asymptotic to m, let $\bar{f} = f/m$ and substitute $y \longrightarrow \bar{f}$ in $P = 0$. Divide the resulting equation through by the largest coefficient. The i–th shadow of $\bar{f}$ is then obtained as a root of $\tilde{\eta}_i(P) = 0$.*

A number of matters remain. We show below how to continue the expansions of the roots. However the shadows and ghosts will be defined as roots of algebraic equations over suitable function fields, and since the algebraic equations will in general have several roots, we will not know a priori which ghost goes with which shadow. We give a method for deciding that question.

First observe that $\xi_i(\bar{f})$ satisfies the equation

$$\operatorname{res}_z\{P(z), \tilde{\eta}_i(P)(z-y)\} = 0, \tag{5.24}$$

where res_z is the resultant with z taken as the variable. For if $y = \xi_i(\bar{f})$ then $P(z)$ and $\tilde{\eta}_i(P)(z-y)$ have a root $z = \bar{f}$ in common. Conversely if a particular value of y gives a common root, then y must be of the form

$$\text{root\_of}\,\{P = 0\} - \text{root\_of}\,\{\tilde{\eta}_i(P) = 0\}. \tag{5.25}$$

We can then obtain monomials asymptotic to the solutions of (5.24). Some of the roots of (5.24) will not be i–th ghosts of roots of P since the roots in (5.25) need not correspond. However we can detect these phantom solutions since they will not belong to $\mathcal{I}_i$. At this stage we have algebraic equations for shadows and ghosts and monomials for the roots of these equations. However we still have to link each ghost monomial with its corresponding shadow monomial.

We have separation points between the shadows. Let s be the one between the greatest shadow, α, and the rest, and let the roots of $P = 0$ having i–th shadow α be $\alpha + \delta_1, \ldots, \alpha + \delta_J$. We may assume that s has been chosen so that $\alpha + \delta_j > s$ for all $j = 1, \ldots, J$. Consider the equation

$$\operatorname{res}_z\{P(z^2 + s), \tilde{\eta}_i(P)((z-y)^2 + s)\} = 0. \tag{5.26}$$

If y is a root of this equation, then there is a z such that $z^2 + s$ is a zero, $\bar{f}$, of P and $(z-y)^2 + s$ is a zero of $\tilde{\eta}_i(P)$. However only roots with the largest possible shadow will give real values of z. If we eliminate z between the simultaneous equations

$$z^2 + s = \alpha + \delta_j, \qquad (z-y)^2 + s = \alpha$$

and solve for y, we obtain

$$y = \pm\sqrt{\alpha + \delta_j - s} \pm \sqrt{\alpha - s}. \tag{5.27}$$

Each y of this form will satisfy (5.26). However to recover values of the δ_j we use only values of y in $\mathcal{I}_i$, these being the ones for which the signs on the right of (5.27) are opposite. The others would not give any new values of δ_j and would in fact lead to equations for δ_j exhibiting cancellation. We would be unable to obtain asymptotic information about δ_j from the latter since we know only monomials asymptotic to $\alpha - s$ and y. In addition we might as well restrict attention to positive values of y.

In the case when δ_j is positive, it will satisfy

$$\delta_j = y^2 + 2y\sqrt{\alpha - s}, \tag{5.28}$$

while if it is negative it will satisfy

$$\delta_j = y^2 - 2y\sqrt{\alpha - s}. \tag{5.29}$$

Of course we do not know the sign's of the δ_j in advance, but we do have, from equation (5.24), a list of the ghosts associated with some shadow. This may rule out one (or both!) of the cases (5.28) and (5.29). If we are unlucky, and both possibilities remain, we can use the fact that the roots of

$$\mathrm{res}_z\{P(z^2 + s), \tilde{\eta}_i(P)((z - y^2)^2 + s)\} = 0 \tag{5.30}$$

will be given by solutions of

$$y^2 = \sqrt{\alpha - s + \delta_j} - \sqrt{\alpha - s},$$

$j = 1, \ldots, J$. So only positive δ_j will give solutions of (5.30).

Now if we want further terms in the expansions, we can apply the same processes to (5.24) as we applied to $P = 0$. The method for linking shadows and ghosts may be summarized as follows.

ALGORITHM **8**

1. *Choose a separation point, s, lying between the largest shadow, α, and the rest. Then apply the previous method to the equation*

   $$\mathrm{res}_z\{P(z^2 + s), \tilde{\eta}_i(P)((z - y)^2 + s)\} = 0.$$

   A real root, p, of this equation which belongs to $\mathcal{I}_i$ corresponds to a ghost, $\pm 2p\sqrt{\alpha - s} + p^2$, with shadow equal to α. The asymptotic forms of these expressions can then be checked against those of the ghosts obtained from (5.24). The ambiguity of sign can if necessary be resolved by considering the equation

   $$\mathrm{res}_z\{P(z^2 + s), \tilde{\eta}_i(P)((z - y^2)^2 + s)\} = 0. \tag{5.31}$$

   If the expression (5.25) *contains a power of y as a factor, then some roots of $P = 0$ are equal to their shadows. We can discover which roots by computing* $\gcd\{P, \tilde{\eta}_i(P)\}$. *This factor can then be divided out from P for the rest of the calculation.*
2. *Now place s between the second largest i–th shadow and those below it and repeat the computation. We obtain the ghosts corresponding to the top two i–th shadows, and since we already know those corresponding to the top shadow, this tells us which to link to the second.*
3. *Now place s between the third largest i–th shadow and those below it, and so on.*
4. *There may be roots of $P = 0$ with the same shadow and having ghosts asymptotic to the same $\{\mathbf{t}_1, \ldots, \mathbf{t}_n\}$–monomial. In that case compute shadows of the ghosts and obtain the multiplicities of roots from the Sturm theorem. If* these *roots have different shadows, separate them by asymptotic expansion as above. If not, consider* their *ghosts, i.e. replace P by* $\mathrm{res}_z\{P(z), \tilde{\eta}_i(P)(z - y)\}$ *and apply the above methods again.*

5. *If necessary go on to consider further shadows and so on.*

Example
Consider the polynomial P over the field $\mathbb{K}(x, e^x)$, where P is given by

$$P(y) = e^{3x}y^2 - (2xe^{2x} - 1)y + (x^2 - 1)e^x, \tag{5.32}$$

and $\mathbb{K}$ is a field of constants containing all the end and evaluation points we shall need. Our scale is $\{x^{-1}, e^{-x}\}$, so we put $y = e^{cx}x^dK$. We first compare powers of e^x in the terms of P. The largest powers occurring are $e^{(3+2c)x}$, $e^{(2+c)x}$ and e^x, and for two at least of these to cancel, we must have $c = -1$. Next we substitute for c and compare powers of x to obtain $d = 1$. Finally we see that $K \sim 1$. Thus the asymptotic expansions of all solutions begin $y \sim e^{-x}x + \cdots$. On writing $y = xe^{-x} + y_1$ we get

$$y_1^2 + e^{-3x}y_1 - e^{-2x} + xe^{-4x} = 0.$$

Now if we treat y_1 in the way we treated y, we see that $y_1 \sim \pm e^{-x}$. To see how many roots there are of each type, we use the Sturm Theorem. We have

$$\begin{aligned} A_0 &= P = e^{3x}y^2 - (2xe^{2x} - 1)y + (x^2 - 1)e^x, \\ A_1 &= \partial P/\partial y = 2e^{3x}y - 2xe^{2x} + 1. \end{aligned}$$

A_2 is then minus the remainder obtained from dividing A_0 by A_1. A short calculation gives

$$A_2 = e^x - xe^{-x} + \frac{e^{-3x}}{4}.$$

We choose xe^{-x} as our intermediate evaluation point, and arrive at the following table.

	A_0	A_1	A_2	Variation
At ∞	+	+	+	0
At xe^{-x}	−	+	+	1
At $-\infty$	+	−	+	2

The Sturm Theorem then tells us that there is one root asymptotic to $xe^{-x} + e^{-x}$ and another asymptotic to $xe^{-x} - e^{-x}$.

Now we substitute $y_1 = e^{-x} + y_2$ to get another term in the first expansion. This gives

$$y_2^2e^{3x} + y_2(2e^{2x} + 1) + (x + 1)e^{-x} = 0.$$

For the first and second term on the left to be dominant and cancel, we must have $y_2 \sim -2e^{-x}$. This must be rejected because it violates the criterion $y_2 = o(e^{-x})$. (Note that this value of y_2 in fact corresponds to the *other* root). For the first and third term to cancel y_2^2 would have to be negative. The remaining possibility is that the second and third terms dominate and cancel, and this yields $y_2 \sim -xe^{-3x}/2$. After a similar calculation with the second root, we see that the two roots have expansions $xe^{-x} + e^{-x} - xe^{-3x}/2 + \cdots$ and $xe^{-x} - e^{-x} + xe^{-3x}/2 + \cdots$.

Of course this example is relatively simple and the conclusions could be obtained in other ways, but it illustrates the general method.

Exercise

Show similarly that the equation

$$(x+1)^2y^3 - \{e^x(x+1)^2 + 2x(x+1)\}y^2 + \{2x(x+1)e^x + x^2\}y - x^2e^x + e^{-x} = 0$$

has roots asymptotic to e^x and $(x \pm e^{-x})(x+1)^{-1}$. Note that in this case the last two roots will not be separated by a power–series expansion. Details are given in [103].

5.3 Compositions with Meromorphic Functions

If f is an element of an asymptotic field and G is meromorphic at the limit of f, then at least in principle we can use the Laurent series of G to obtain a multiseries for $G \circ f$. However G has to be finitely presented in some way and we have to able to handle zero equivalence and cancellation problems. It turns out that for all of these it is sufficient for G to be given by an algebraic differential equation. Without loss of generality, we can take the limit of f to be zero. Similarly we can reduce to the case when G is analytic at the limit of f by replacing G by $x^M G$ for a suitable value of M. We then have the following result.

THEOREM **17** *Let $\mathcal{F}$ be an asymptotic field and let f be an element of $\mathcal{F}$ which tends to zero. Let G be a C^∞ function defined in a neighbourhood of the origin with $G(0), \ldots, G^{(N-1)}(0)$ equal to given constants. Suppose that G satisfies the differential equation*

$$\frac{d^N y}{dx^N} = R\left(y, \frac{dy}{dx}, \ldots, \frac{d^{N-1}y}{dx^{N-1}}\right) \tag{5.33}$$

in some neighbourhood of the origin, where $R(y_0, \ldots, y_{N-1})$ is a rational function over $\mathbb{R}(x)$, analytic at the point $(0, G(0), \ldots, G^{(N-1)}(0))$.

Then $\mathcal{E} =_{def} \mathcal{F}(G \circ f, G' \circ f, \ldots, G^{(N-1)} \circ f)$ is an asymptotic field.

Note that under the hypotheses of the theorem, G will be analytic at the origin since it satisfies an analytic differential equation there; see [8] for example.

The main part of the proof of Theorem 17 is concerned with establishing the shadow property. The definitions are as might be expected. If ϕ_i is the i–th shadow of f, then for $k = 1, \ldots, N-1$ we define

$$\eta_i(G^{(k)} \circ f) = G^{(k)} \circ \phi_i \tag{5.34}$$

and for $0 \leq i < j \leq N-1$,

$$\eta_{i,j}(G^{(k)} \circ \phi_j) = G^{(k)} \circ \phi_i. \tag{5.35}$$

It is then straightforward to establish SF(i).

LEMMA **22** *Write* $\Sigma_i = \mathcal{S}_i(\mathcal{F})(G \circ \phi_i, \ldots, G^{(N-1)} \circ \phi_i)$. *Then* Σ_i *is closed under relative differentiation.*

Proof of Lemma 22
The proof follows similar lines to that of Lemma 14. Let

$$P \in \mathcal{S}_i(\mathcal{F})[G \circ \phi_i, \ldots, G^{(N-1)} \circ \phi_i].$$

Then P may be written as a finite sum.

$$P = \sum_{\mathbf{J}} P_{\mathbf{J}} (G \circ \phi_i)^{j_0} \ldots (G^{(N-1)} \circ \phi_i)^{j_{N-1}}.$$

where $\mathbf{J}$ is the multi-index $(j_0, \ldots, j_{N-1})$. For $0 \leq l < N-1$, we have

$$(G^{(l)} \circ \phi_i)' = (G^{(l+1)} \circ \phi_i)\phi_i',$$

while

$$(G^{(N-1)} \circ \phi_i)' = (R(G, \ldots, G^{(N-1)}) \circ \phi_i)\phi_i'.$$

Moreover, for any $\mathbf{J}$, $P_{\mathbf{J}}'/\phi_i' \in \mathcal{S}_i(\mathcal{F})$ since both $P_{\mathbf{J}}$ and ϕ_i belong to $\mathcal{S}_i(\mathcal{F})$. Hence $P'/\phi_i' \in \Sigma_i$. Now let P_1/Q_1 and P_2/Q_2 be two elements of $\mathcal{S}_i(\mathcal{F})(G \circ \phi_i, \ldots, G^{(N-1)} \circ \phi_i)$. Then

$$\frac{(P_1/Q_1)'}{(P_2/Q_2)'} = \frac{\{(P_1'/\phi_1')Q_1 - (Q_1'/\phi_1')P_1\}Q_2^2}{\{(P_2'/\phi_1')Q_2 - (Q_2'/\phi_1')P_2\}Q_1^2} \in \Sigma_i,$$

and so Σ_i is closed under relative differentiation. Lemma 22 is therefore established.

It is more difficult to establish SF(ii). Our basic strategy is to add elements $G \circ \phi_i, \ldots, G^{(N-1)} \circ \phi_i$ to $\mathcal{S}_i(\mathcal{F})$ one at a time, using induction and Lemma 16. The first case is the simplest and introduces some of the main techniques.

LEMMA **23** *Let* $\mathcal{F}$ *and* G *be as above and let* $\mathbf{t}_i$ *be a scale element. Suppose that* $\mathcal{S} = \mathcal{S}_i(\mathcal{F})$ *contains* $G^{(j)}(0)$ *for* $j = 0, 1, \ldots, N-1$. *Let* $q \in \mathcal{S}_{alg}$ *and let* $\phi \in \mathcal{S}$ *with* $\phi \to 0$. *If there is a* $\delta \in \mathbb{R}^+$ *such that* $|G \circ \phi - q| < \mathbf{t}_i^\delta$ *then* $G \circ \phi = q$.

Proof of Lemma 23
We prove that $G \circ \phi$ and q are identical by showing that $q \circ \phi^{-1}$ and G both satisfy the differential equation (5.33) with the same initial conditions.

If ϕ is zero, then $G \circ \phi = G(0) \in \mathcal{S}$, and the shadow property of $\mathcal{S}$ implies that $G \circ \phi = q$. Otherwise ϕ is non-constant (since it tends to zero), and on differentiating the inequality $|G \circ \phi - q| < \mathbf{t}_i^\delta$ we obtain

$$\left|G' \circ \phi - \frac{q'}{\phi'}\right| < \frac{\delta \mathbf{t}_i^\delta \mathbf{t}_i'}{\mathbf{t}_i \phi'}. \tag{5.36}$$

Now $\gamma_0(\mathbf{t}_i^{\delta/3}) < \gamma_0(\log \phi)$, since $\mathbf{t}_i^{\delta/3} \to 0$ and $|\log \phi| \to \infty$. Hence by Lemma 6(5),

$$\gamma_0 \left(\frac{\delta}{3} \mathbf{t}_i^{\delta/3} \frac{\mathbf{t}_i'}{\mathbf{t}_i}\right) < \gamma_0 \left(\frac{\phi'}{\phi}\right).$$

Since ϕ belongs to a shadow field with respect to $\mathbf{t}_i$, $\mathbf{t}_i^{\delta/3}/\phi \to 0$, and so

$$\frac{\mathbf{t}_i^\delta \mathbf{t}_i'}{\mathbf{t}_i \phi'} = \mathbf{t}_i^{\delta/3} \frac{\mathbf{t}_i^{\delta/3}}{\phi} \frac{\mathbf{t}_i^{\delta/3} \mathbf{t}_i' \phi}{\mathbf{t}_i \phi'} = o(\mathbf{t}_i^{\delta/3}).$$

Hence $|G' \circ \phi - q'/\phi'| < \mathbf{t}_i^{\delta/3}$.

Rather than accumulating ever more complicated fractions of δ in the exponents, we make the convention that for the rest of the proof of Theorem 17, δ will denote a positive constant but not necessarily the same one every time. Such conventions have a long and honourable history in Analysis, and no confusion should be caused here. We also write Δq for the relative derivative q'/ϕ'.

Our conclusions from above can then be stated as

$$|G \circ \phi - q| < \mathbf{t}_i^\delta \Rightarrow |G' \circ \phi - \Delta q| < \mathbf{t}_i^\delta.$$

On iterating we obtain that

$$|G^{(j)} \circ \phi - \Delta^j q| < \mathbf{t}_i^\delta \tag{5.37}$$

for $j = 0, \ldots, N$, where $\Delta^0 q = q$. We define R_ϕ to be the differential polynomial obtained from R by composing the coefficients with ϕ. Then the triangle inequality gives

$$\begin{aligned} |\Delta^N q - R_\phi(q, \Delta q, \ldots, \Delta^{N-1} q)| &\le |\Delta^N q - G^{(N)} \circ \phi| \\ &+ |R_\phi(G \circ \phi, \ldots, G^{(N-1)} \circ \phi) - R_\phi(q, , \ldots, \Delta^{N-1} q)|, \end{aligned} \tag{5.38}$$

since G satisfies (5.33). However

$$\begin{aligned} R_\phi(G \circ \phi, \ldots, G^{(N-1)} \circ \phi) - R_\phi(q, , \ldots, \Delta^{N-1} q) &= (G \circ \phi - q) R_{\phi,0} \\ + (G' \circ \phi - \Delta q) R_{\phi,1} + \cdots + (G^{(N-1)} \circ \phi &- \Delta^{N-1} q) R_{\phi,N-1}, \end{aligned} \tag{5.39}$$

where $R_{\phi,0}, \ldots, R_{\phi,N-1} \in \mathcal{S}_i(G \circ \phi, \ldots, G^{(N-1)} \circ \phi)$. Now since $\phi \in \mathcal{S}_i$ and G and its derivatives are analytic at 0, $|G^{(j)} \circ \phi| < \mathbf{t}_i^{-\epsilon}$ for every positive real constant ϵ and every $j = 0, \ldots, N-1$. Likewise $|R_{\phi,j}| < \mathbf{t}_i^{-\epsilon}$ and then from (5.37) and (5.39)

$$|R_\phi(G \circ \phi, \ldots, G^{(N-1)} \circ \phi) - R_\phi(q, , \ldots, \Delta^{N-1} q)| < \mathbf{t}_i^\delta.$$

Hence from (5.38) we have that

$$|\Delta^N q - R_\phi(q, \Delta q, \ldots, \Delta^{N-1} q)| < \mathbf{t}_i^\delta.$$

But $\Delta^N q - R_\phi(q, \Delta q, \ldots, \Delta^{N-1} q) \in \mathcal{S}_i$ and so the shadow property there gives

$$\Delta^N q = R_\phi(q, \Delta q, \ldots, \Delta^{N-1} q). \tag{5.40}$$

Now ϕ will be monotonic for large values of x and so will have a compositional inverse ψ. Let us write $Q = q \circ \psi$. A simple induction then shows that $Q^{(j)} = \Delta^j q$ for $j = 1, \ldots, N$. We can then compose both sides of equation (5.40) with ψ to give

$$Q^{(N)} = R(Q, Q', \ldots, Q^{(N-1)}).$$

So Q satisfies the same differential equation as G.

To complete the proof of Lemma 23 we must show that Q and G also satisfy the same initial conditions at 0. Essentially this follows by taking limits in (5.37) for $j = 0, \ldots, N-1$, but we need some care since it is not a priori clear that the derivatives of Q exist at the origin. Fortunately it is sufficient for the right–hand derivatives to exist, and this we can prove as we go along. Thus (5.37) with $j = 0$ gives $|(G - Q) \circ \phi| < \mathbf{t}_i^\delta$, and on letting $x \to \infty$ we get that the right–hand limit of Q at 0 exists and is equal to $G(0)$. Suppose inductively that the right–hand derivative, $Q_+^{(j)}(0)$, of $Q^{(j-1)}$ at 0 exists and is equal to $G^{(j)}(0)$, where $0 \leq j \leq N-2$. Then

$$\left| \frac{G^{(j)}(\phi) - G^{(j)}(0)}{\phi} - \frac{Q^{(j)}(\phi) - Q_+^{(j)}(0)}{\phi} \right| = \left| \frac{(G - Q)(\phi)}{\phi} \right| < \frac{\mathbf{t}_i^\delta}{\phi},$$

by (5.37). Since the right–hand side tends to zero as $x \to \infty$, we see that the right–hand derivative of $Q^{(j)}$ exists at 0 and is equal to $G^{(j+1)}(0)$. By induction Q and G satisfy the same initial conditions at 0 and so by the uniqueness theorem from [8], for example, $Q = G$. Hence $G \circ \phi = q$ and the proof of Lemma 23 is complete.

Corollary *$\mathcal{S}_i(G \circ \phi)$ has the shadow property with respect to $\mathbf{t}_i$.*

This results from Lemmas 22, 23 and 16.

Lemma **24** *For $j = 0, \ldots, N-1$, $\mathcal{S}_i(G \circ \phi, \ldots, G^{(j)} \circ \phi)$ has the property SF(ii) with respect to $\mathbf{t}_i$.*

Proof of Lemma 24

We already have this for $j = 0$, and it is natural to use induction. If we could just replace $\mathcal{S}_i$ by $\mathcal{S}_i(G \circ \phi, \ldots, G^{(j-1)} \circ \phi)$ in Lemma 23 this would now be simple enough. However in order to deduce (5.40) we needed to use the fact that $\mathcal{S}_i$ is closed under relative differentiation, and this is not true of $\mathcal{S}_i(G \circ \phi, \ldots, G^{(j-1)} \circ \phi)$. More specifically we used the fact that the relative derivatives $\Delta^i q$ belong to a shadow field. Now it turns out that we can adapt the approach of Lemma 23 but we have to do some more work. The key idea is to approximate the $\Delta^i q$ by elements of $\mathcal{S}_i(G \circ \phi, \ldots, G^{(j-1)} \circ \phi)$.

Suppose then that $\mathcal{S}_i(G \circ \phi, \ldots, G^{(j-1)} \circ \phi)$ has the property SF(ii) with respect to $\mathbf{t}_i$. We aim to show the same for $\mathcal{S}_i(G \circ \phi, \ldots, G^{(j)} \circ \phi)$. It follows from Lemma 15 that

$$\bar{\mathcal{S}} =_{def} \mathcal{S}_i(G \circ \phi, \ldots, G^{(j-1)} \circ \phi)_{alg}$$

also has property SF(ii) with respect to $\mathbf{t}_i$. Let $q \in \bar{\mathcal{S}}$ and suppose that $|G^{(j)} \circ \phi - q| < \mathbf{t}_i^\delta$. Exactly as in the proof of the previous lemma, we can show that for $k = 0, \ldots, N - j$

$$|G^{(j+k)} \circ \phi - \Delta^k q| < \mathbf{t}_i^\delta, \tag{5.41}$$

and hence that

$$|\Delta^{N-j} q - R_\phi(G \circ \phi, \ldots, G^{(j-1)} \circ \phi, q, \Delta q, \ldots, \Delta^{N-j-1} q)| < \mathbf{t}_i^\delta. \tag{5.42}$$

Let $h \in \bar{\mathcal{S}}$. Then since h is given as a root of a polynomial over $\mathcal{S}_i(G \circ \phi, \ldots, G^{(j-1)} \circ \phi)$, we can regard h as a function of the $j+1$ variables x, $G \circ \phi, \ldots, G^{(j-1)} \circ \phi$; thus $h = h(x, G \circ \phi, \ldots, G^{(j-1)} \circ \phi)$. We write $\partial_k h$ for the partial derivative of h with respect to its k–th variable, $k = 1, \ldots, j+1$. Then

$$\Delta h = \frac{\partial_1 h}{\phi'} + \partial_2 h \cdot G'(\phi) + \cdots + \partial_j h \cdot G^{(j-1)}(\phi) + \partial_{j+1} h \cdot G^{(j)}(\phi).$$

The problem is that $G^{(j)}(\phi)$ does not belong to $\bar{\mathcal{S}}$, but (5.41) gives us an approximation that does, namely q. We embody this idea in the following definition.

$$h^\star =_{def} \frac{\partial_1 h}{\phi'} + \partial_2 h \cdot G'(\phi) + \cdots + \partial_j h \cdot G^{(j-1)}(\phi) + \partial_{j+1} h \cdot q.$$

We can apply this with $h = q$, $h = q^\star$, and so on.

LEMMA **25** *Let $q^{\star 1} = q^\star$ and for $k = 2, \ldots,$ let $q^{\star k} = (q^{\star(k-1)})^\star$. Then for $k = 1, \ldots, N - j$,*

$$|q^{\star k} - \Delta^k q| < \mathbf{t}_i^\delta. \tag{5.43}$$

Proof of Lemma 25
We prove by induction that

$$\Delta^k q - q^{\star k} = \sum_{r=0}^{k-1} a_{k,r}(G^{(j+r)} \circ \phi - \Delta^r q), \tag{5.44}$$

where each $a_{k,r} \in \bar{\mathcal{S}}[G^{(j)} \circ \phi, \ldots, G^{(N-1)} \circ \phi]$. The lemma will then follow from (5.41).

Suppose first that $k = 1$. We have

$$\Delta q = q'/\phi' = (\partial_1 q)/\phi' + (\partial_2 q)G' \circ \phi + \ldots + (\partial_{j+1} q)G^{(j)} \circ \phi$$

and so $\Delta q - q^\star = (\partial_{j+1} q)(G^{(j)} \circ \phi - q)$, as required. Now suppose that the conclusion holds for a certain value of k. Then from the definitions

$$\Delta(q^{\star k}) = q^{\star(k+1)} + \partial_{j+1}(q^{\star k})(G^{(j)} \circ \phi - q),$$

and hence

$$\Delta^{k+1} q - q^{\star(k+1)} = \Delta\left(\Delta^k q - q^{\star k}\right) + \partial_{j+1}\left(q^{\star k}\right)\left(G^{(j)} \circ \phi - q\right).$$

The induction hypothesis then gives

$$\begin{aligned}
&\Delta^{k+1} q - q^{\star(k+1)} \\
&= \Delta\left(\sum_{r=0}^{k-1} a_{k,r}(G^{(j+r)} \circ \phi - \Delta^r q)\right) + \partial_{j+1}\left(q^{\star k}\right)\left(G^{(j)} \circ \phi - q\right) \\
&= \sum_{r=0}^{k-1} \Delta a_{k,r}(G^{(j+r)} \circ \phi - \Delta^r q) + \sum_{r=0}^{k-1} a_{k,r}(G^{(j+r+1)} \circ \phi - \Delta^{r+1} q) \\
&\qquad + \quad \partial_{j+1}\left(q^{*^i}\right)\left(G^{(j)} \circ \phi - q\right),
\end{aligned}$$

since $\Delta(G^{(j+r)} \circ \phi) = (G^{(j+r)} \circ \phi)'/\phi' = G^{(j+r+1)} \circ \phi$. This expression for $\Delta^{k+1} q - q^{\star(k+1)}$ is of the required form because $a_{k,r}$ and $\partial_{j+1}(q^{\star k})$ belong to $\bar{\mathcal{S}}[G^{(j)} \circ \phi, \ldots, G^{(N-1)} \circ \phi]$ and moreover

$$\Delta a_{k,r} = a^\star_{k,r} + \partial_{j+1} a_{k,r}(G^{(j)} \circ \phi - q) \in \bar{\mathcal{S}}[G^{(j)} \circ \phi, \ldots, G^{(N-1)} \circ \phi].$$

Thus (5.44) follows by induction, and Lemma 25 is proved.

Continuation of the Proof of Lemma 24
From (5.42) and (5.43), we get that

$$|q^{\star N-j} - R_\phi(G(\phi), \ldots, G^{(j-1)}(\phi), q, q^\star, \ldots, q^{\star N-j-1})| < \mathbf{t}_i^\delta.$$

However $q, q^\star, \ldots, q^{\star N-j} \in \mathcal{S}_i \subset \bar{\mathcal{S}}$ and by the induction hypothesis SF(ii) is valid there. Hence

$$q^{\star N-j} = R_\phi(G(\phi), \ldots, G^{(j-1)}(\phi), q, q^\star, \ldots, q^{\star N-j-1}). \tag{5.45}$$

Now if we use (5.44) to substitute for $q, q^\star, \ldots, q^{\star N-j}$ in (5.45), we get

$$\begin{aligned}
&\Delta^{N-j} q + \sum_{r=0}^{N-j-1} a_{N-j,r}(\Delta^r q - G^{(j+r)} \circ \phi) \\
&\quad = R_\phi(G \circ \phi, \ldots, G^{(j)} \circ \phi, q, \Delta q + a_{1,1}(q - G^{(j+1)}), \ldots, \\
&\qquad\quad \Delta^{N-j-1} q + \sum_{r=0}^{N-j-1} a_{N-j-1,r}(\Delta^r q - G^{(j+r)} \circ \phi)).
\end{aligned}$$

On composing with ψ (where ψ is the inverse of ϕ), we see that both $G^{(j)}$ and $Q = q \circ \psi$ satisfy the differential equation

$$\begin{aligned} y^{(N-j)} &+ \sum_{r=0}^{N-j-1} (a_{N-j,r} \circ \psi)(y^{(r)} - G^{(j+k)}) \\ &= R(G, \ldots, G^{(j)}, y, y' + (a_{1,1} \circ \psi)(y - G^{(j)}), \ldots, \\ &\quad y^{(N-j-2)} + \sum_{r=0}^{N-j-1} (a_{N-j-1,k} \circ \psi)(y^{(r)} - G^{(j+r)})). \end{aligned} \tag{5.46}$$

The method that was used at the end of Lemma 23 applies again here, and gives that $Q_+^{(r)}(0)$ exists and is equal to $G^{(j+r)}(0)$ for $r = 0, \ldots, N-j-1$. Therefore by the uniqueness theorem, $G^{(j)} = Q$ and hence $G^{(j)} \circ \phi = q$. Now Lemma 16 gives the conclusion of Lemma 24.

Next we need the following result in order to be able to obtain expansions of ghosts.

LEMMA **26** *Let ϕ_1 and ϕ_2 be elements of a Hardy field, both tending to zero, such that $\gamma_1(\phi_2) > \gamma_1(\phi_1)$. Let G be analytic in a neighbourhood of the origin. Then for $r = 1, 2, \ldots,$ we have*

$$\left| G \circ (\phi_1 + \phi_2) - \sum_{s=0}^{r} \frac{\phi_2^s G^{(s)} \circ \phi_1}{s!} \right| < |\phi_2|^{r+1-\epsilon} \tag{5.47}$$

for every $\epsilon \in \mathbb{R}^+$.

Proof of Lemma 26
For $n \in \mathbb{N}$ we have

$$\begin{aligned} |{}_nC_{r+1}\phi_2^{r+1}\phi_1^{n-r-1} + \cdots + \phi_2^n| &< |\phi_2|^{r+1}|\phi_1|^{n-r-1}({}_nC_{r+1} + \cdots + 1) \\ &< |\phi_2|^{r+1}|\phi_1|^{n-r-1}2^n. \end{aligned}$$

Let the power series for G at the origin be $G(y) = \sum c_n y^n$. Then $G^{(s)}(y) = \sum_{n=s}^{\infty} c_n s! {}_nC_s y^{n-s}$. Now

$$\begin{aligned} \left| G \circ (\phi_1 + \phi_2) - \sum_{s=0}^{r} \frac{\phi_2^s G^{(s)} \circ \phi_1}{s!} \right| &= \left| \sum_{s=r+1}^{\infty} \frac{\phi_2^s G^{(s)} \circ \phi_1}{s!} \right| \\ &= \left| \sum_{s=r+1}^{\infty} \sum_{n=s}^{\infty} c_n \phi_2^s \, {}_nC_s \phi_1^{n-s} \right| \\ &= \left| \sum_{n=r+1} c_n \sum_{s=r+1}^{n} {}_nC_s \phi_2^s \phi_1^{n-s} \right| \\ &< \sum_{n=r+1}^{\infty} |c_n||\phi_2|^{r+1}|\phi_1|^{n-r-1}2^n \end{aligned}$$

$$= \left|\frac{\phi_2}{\phi_1}\right|^{r+1} \sum |c_n||2\phi_1|^n$$
$$< |\phi_2|^{r+1-\epsilon},$$

for every $\epsilon \in \mathbb{R}^+$, since the series converges uniformly to a bounded analytic function on an interval of the form (a, ∞). This completes the proof of Lemma 26.

COROLLARY *For each $j = 0, \ldots, N-1$ there exists a $\delta \in \mathbb{R}^+$ such that*

$$\left|G^{(j)} \circ f - G^{(j)} \circ \phi\right| < \mathbf{t}_i^\delta \tag{5.48}$$

This is obtained by applying the Lemma to $G^{(j)}$ with $\phi_1 = \phi$ and $\phi_2 = f - \phi$.

Completion of the proof of Theorem 17

We first note that zero equivalence calculations may be handled by the methods of Chapter 2. Also we have conditions (1) and (2) of Definition 21.

Next we observe that for any $j = 0, \ldots, N-1$, we can compute a $\{\mathbf{t}_1, \ldots, \mathbf{t}_n\}$–monomial asymptotic to $G^{(j)} \circ f$. For let T be a $\{\mathbf{t}_1, \ldots, \mathbf{t}_n\}$–monomial asymptotic to f and let the first non–zero term of the power series of $G^{(j)}(X)$ be $b_s X^s$. Then $G^{(j)}(f) \sim b_s T^s$.

In order to be able to compute a $\{\mathbf{t}_1, \ldots, \mathbf{t}_n\}$–monomial asymptotic to an arbitrary element of $\mathcal{E} = \mathcal{F}(G \circ f, \ldots, G^{(N-1)} \circ f)$ it is sufficient to to this for elements of $E =_{def} \mathcal{F}[G \circ f, \ldots, G^{(N-1)} \circ f]$. If g is such an element then we may write g as a finite sum

$$g = \sum_{\mathbf{m}} a_{\mathbf{m}} G(f)^{m_0} G'(f)^{m_1} \cdots G^{(N-1)}(f)^{m_{N-1}}. \tag{5.49}$$

We can find $\{\mathbf{t}_1, \ldots, \mathbf{t}_n\}$–monomials asymptotic to each $a_{\mathbf{m}}$ and each $G^{(j)}(f)$. By dividing out by the appropriate least common multiple we can reduce to the case when $g \in \mathcal{R}_1(E)$. Then if $\eta_1(g) \not\equiv 0$, g is asymptotic to this constant. Otherwise we may find a $j < n$ such that $\eta_i(g) \equiv 0$ for $i = 1, \ldots, j$ but $\eta_{j+1}(g) \not\equiv 0$. Alternatively, if $\eta_i(g) \equiv 0$ for all i we take $j = n$. We compute $\mathbf{t}_j$–expansions of all the terms in (5.49). This can be done for the $a_{\mathbf{m}}$ since they belong to $\mathcal{F}$ and Lemma 26 guarantees the expansions of the $G^{(i)}(f)$. A non–zero term must be found after a finite number of steps since $\eta_{j+1}(g) \not\equiv 0$ (or in the case $j = n$ since $g \not\equiv 0$). The coefficient of this term will be an algebraic combination of j–th shadows, for which $\mathbf{t}_{j-1}$–expansions may be calculated, and so on. So we can obtain a $\{\mathbf{t}_1, \ldots, \mathbf{t}_n\}$–monomial asymptotic to g.

The above techniques can obviously be extended to calculate shadow expansions of elements of E and hence of elements of $\mathcal{E}$. Similarly expansions of the various shadows and ghosts may be obtained. It follows that no non–zero element of the field generated by $\mathcal{E}$ and the various Σ_i can have arbitrarily large zeros or poles. Since this field is clearly closed under differentiation, it is then a Hardy field, and this is enough to establish Theorem 17.

Example
Consider the function

$$f(x) = \mathrm{sn}(x^{-1} + e^{-x}) - \mathrm{sn}(x^{-1}), \tag{5.50}$$

where $\mathrm{sn}(y) = \mathrm{sn}(y, 1/2)$ is of course the Jacobi elliptic function; see [123] for example. It may be defined as the solution of the differential equation

$$2y'y'' + 2yy'(1 - y^2/4) + y(1 - y^2)y'/2 = 0, \tag{5.51}$$

subject to the initial conditions $y(0) = 0, y'(0) = 1$. Its series development begins

$$\mathrm{sn}(t) = t - \frac{5}{24}t^3 + \frac{73}{1920}t^5 - \frac{79}{9216}t^7 + \frac{24487}{13271040}t^9 - \frac{461287}{1167851520}t^{11} + O\left(t^{12}\right).$$

We build the tower of asymptotic fields

$$\mathbb{R} \subset \mathbb{R}(x) \subset \mathbb{R}(x, e^x) \subset \mathbb{R}(x, e^x, \mathrm{sn}(x^{-1})) \subset \mathbb{R}(x, e^x, \mathrm{sn}(x^{-1}), \mathrm{sn}(x^{-1} + e^{-x})).$$

Our scale is $\{\mathbf{t}_1 = x^{-1}, \mathbf{t}_2 = e^{-x}\}$. We define the maps η_1 and η_2 as in the proof of Theorem 17, and discover that $\eta_1(f) = \eta_2(f) = 0$. We then obtain the $\mathbf{t}_2$-expansion of $\mathrm{sn}(x^{-1} + e^{-x})$ from the Taylor series for sn. For this we need to use the relations

$$\frac{d\mathrm{sn}(u)}{du} = \mathrm{cn}(u)\mathrm{dn}(u), \quad \frac{d\mathrm{cn}(u)}{du} = -\mathrm{sn}(u)\mathrm{dn}(u), \quad \frac{d\mathrm{dn}(u)}{du} = \frac{-1}{4}\mathrm{sn}(u)\mathrm{cn}(u),$$

where cn and dn are also Jacobi elliptic functions, having respective expansions

$$\mathrm{cn}(\mathbf{t}_1) = \left(1 - \frac{1}{2}{\mathbf{t}_1}^2 + \frac{1}{12}{\mathbf{t}_1}^4 + O\left({\mathbf{t}_1}^6\right)\right)$$

and

$$\mathrm{dn}(\mathbf{t}_1) = \left(1 - \frac{1}{8}{\mathbf{t}_1}^2 + \frac{17}{384}{\mathbf{t}_1}^4 + O\left(\mathbf{t}_1^6\right)\right).$$

Thus we obtain that $\mathrm{sn}(\mathbf{t}_1 + \mathbf{t}_2)$ is equal to

$$\begin{aligned}&\mathrm{sn}(\mathbf{t}_1) + \mathrm{cn}(\mathbf{t}_1)\mathrm{dn}(\mathbf{t}_1)\mathbf{t}_2 + \left(-1/2\,\mathrm{dn}^2(\mathbf{t}_1)\mathrm{sn}(\mathbf{t}_1) - 1/8\,\mathrm{cn}^2(\mathbf{t}_1)\mathrm{sn}(\mathbf{t}_1)\right){\mathbf{t}_2}^2\\ &+ \left(1/6\,\mathrm{dn}(\mathbf{t}_1)\mathrm{sn}^2(\mathbf{t}_1)\mathrm{cn}(\mathbf{t}_1) - 1/6\,\mathrm{dn}^3(\mathbf{t}_1)\mathrm{cn}(\mathbf{t}_1) - 1/24\,\mathrm{cn}^3(\mathbf{t}_1)\mathrm{dn}(\mathbf{t}_1)\right){\mathbf{t}_2}^3\\ &+O(\mathbf{t}_2^4).\end{aligned}$$

On substituting these and rearranging, we get the multiseries for f

$$\begin{aligned}&\left(1 - \frac{5}{8x^2} + \frac{73}{384x^4} + O(x^{-6})\right)e^{-x} + \left(\frac{-5}{8x} + \frac{73}{192x^3} + O(x^{-5})\right)e^{-2x}\\ &+\left(\frac{-5}{24} + \frac{73}{192x^2} + \frac{2765}{9216x^4} + O(x^{-6})\right)e^{-3x} + O(e^{-4x}).\end{aligned}$$

Exercise. Obtain similar multiseries expansions for the functions

$$\exp(x^{-1}) - 1 - \sin(\exp(x^{-1} + e^{-x}) - 1)$$

and

$$\sin(\exp((x^{-1} + e^{-x}) - 1) - \sin(\exp(x^{-1}) - 1).$$

Details are in [101].

5.4 Cartesian Representations

The algorithms we have described earlier in this chapter for calculating multiseries have been based on finding an asymptotic scale, $\{\mathbf{t}_1 \ldots, \mathbf{t}_k\}$, and expanding in terms of its elements. In [116], Joris van der Hoeven points out that this method may be highly inefficient when very large powers are present in the input expression. For example, consider the expression

$$\frac{1}{1-x^{-1}} - \frac{1}{1-x^{-1}-x^{-N}},$$

with N large. If the algorithm of Section 5.1 is applied, $N+1$ terms will have to be examined before a non–zero term is encountered. While the present author does not share the view that this means that the complexity of that algorithm is worse than any iterated exponential, it does indicate a possible drawback.

Van der Hoeven's idea is to choose a finite set of monomials in the scale elements $\{\mathbf{t}_1 \ldots, \mathbf{t}_k\}$ and to expand in terms of these. So in the example above, we would use monomials $z_1 = x^{-1}$ and $z_2 = x^{-N}$. The selected monomials are treated as different 'variables', just as different scale elements are with the classical algorithm, except that now we have to take into account the possibility of cancellation between products of monomials.

Having obtained a scale for our expression as in Section 5.1, we obtain a set $\mathcal{M}$ of monomials recursively. Thus for c constant $\mathcal{M}(c) = \emptyset$ and $\mathcal{M}(\{x^c\}) = \{x^c\}$. Then

$$\mathcal{M}(E_1 \pm E_2) = \mathcal{M}(E_1) \cup \mathcal{M}(E_2),$$

and

$$\mathcal{M}(E_1 \times E_2) = \{m_1 m_2;\ m_1 \in \mathcal{M}(E_1)\ \&\ m_2 \in \mathcal{M}(E_2)\}.$$

Given an expression E_1/E_2, we write $E_2 = L + R$, where L is a product of monomials and $R/L \to 0$. Then, bearing in mind that $E_2^{-1} = L^{-1}\sum_{n=0}^{\infty}(-R/L)^n$, we have

$$\mathcal{M}(E_1/E_2) = \{m_1 L^{-1};\ m_1 \in \mathcal{M}(E_1)\} \cup \{m_2 L^{-1};\ m_2 \in \mathcal{M}(E_2)\}.$$

To handle logarithms, we first write the logarithm of a product as a sum of logarithms. Then if E consists of a single monomial (so that $\mathcal{M}(E) = \{E\}$)

then $\mathcal{M}(\log E) = \{\log E\}$. Otherwise E may be written in the form $E = L+R$ as above, and we have $\log E = \log L+\log(1+R/L)$. Since the second logarithm can be given by a series, we put

$$\mathcal{M}(\log E) = \mathcal{M}(\log L) \cup \mathcal{M}(R/L).$$

Similarly if $c \in \mathbb{R} \setminus \{0\}$,

$$\mathcal{M}((L+R)^c) = \{L^c\} \cup \{mL^{-1}; m \in \mathcal{M}(R)\}.$$

For the exponential of E, we write E in the form $G + l + H$ where l is a constant, all the multiseries terms of G tend to infinity and all of those of H tend to zero. Here $\exp G$ will be a monomial in the scale elements $\{\mathbf{t}_1 \ldots, \mathbf{t}_k\}$, and we write

$$\mathcal{M}(\exp E) = \{\exp G\} \cup \mathcal{M}(H).$$

H may be given as an infinite series, but if this causes difficulty, we may replace the above formula by

$$\mathcal{M}(\exp E) = \{\exp G\} \cup \mathcal{M}(E),$$

the only disadvantage being that we may have introduced unnecessary extra monomials into $\mathcal{M}$.

Having thus obtained the set of monomials $\mathcal{M} = \mathcal{M}(f)$, we replace any element of $\mathcal{M}$ which tends to infinity by its inverse. We arrive at a set of monomials $\mathcal{M} = \{z_1, z_2, \ldots, z_r\}$ such that each z_i tends to zero and f can be written as a polynomial, P_f, in $\{z_1, z_1^{-1}, \ldots, z_r^{-1}\}$ and the *z-functions* of such polynomials. The z–functions are $\mathrm{zexp}(t) = \exp t - 1$, $\mathrm{zlog}(t) = \log(1+t)$ and $\mathrm{zpow}(t, c) = (1+t)^c - 1, c \in \mathbb{R} \setminus \mathbb{N}$. They are applied only to t which tend to zero, [97].

Let $S_{\mathcal{M}}$ denote the set of power products of elements in $\mathcal{M}$. Thus

$$S_{\mathcal{M}} = \{z_1^{p_1} \cdots z_r^{p_r};\ p_1, \ldots, p_r \in \mathbb{Z}\}.$$

We partially order the set $S_{\mathcal{M}}$ by setting

$$z_1^{p_1} \cdots z_r^{p_r} \le z_1^{q_1} \cdots z_r^{q_r} \iff (p_1 \le q_1) \wedge (p_2 \le q_2) \wedge \cdots \wedge (p_r \le q_r).$$

We refer to this ordering as the *monomial ordering*. Note that the monomial ordering is quite different from the Hardy–field ordering on the corresponding functions. For example if $z_1 = x^{-1}$ and $z_2 = e^{-x}$ then $z_1 z_2^2$ and $z_1^2 z_2$ are incomparable in the monomial ordering where as, of course, in the Hardy–field ordering $x^{-1}e^{-2x} < x^{-2}e^{-x}$. However, since each z_i tends to zero, if $m_1 < m_2$ in the monomial ordering then $m_2 m_1^{-1} \to 0$. Another point to note is that there cannot be a monomial which is larger than an increasing infinite sequence of other monomials.

Since the z–functions have power series expansions, we see that f may be given by a multiple Laurent series, u, in the elements of M. Van der Hoeven refers to such a Laurent series as a *Cartesian representation*.

To obtain a multiseries from the Cartesian representation, we begin by computing a set, $G = idm(P_f)$ of *intermediate dominant monomials.* These are a set of monomials in elements of $\mathcal{M}$ which may be the dominant ones of u, or alternatively may cancel out in u when replaced by the functions they represent. If they do cancel out they are removed and a new set of intermediate dominant monomials is calculated.

More formally, we say that G is a set of intermediate dominant monomials if each element of G is minimal in the monomial ordering and f is given by the series obtained from u by removing monomials less than all those in G.

The following example is given in [116]. Let

$$f(x) = \frac{1}{z_1(1-z_1-z_2)} - \frac{1}{z_1} - 1 - z_1 - 3z_2 - \frac{z_2}{z_1}.$$

On expanding $(1-z_1-z_2)^{-1}$ as far as the term in $(z_1+z_2)^3$, we obtain

$$f(x) = \frac{z_2^2}{z_1} - z_2 + z_1^2 + 3z_1z_2 + 3z_2^2 + \frac{z_2^3}{z_1} + \cdots .$$

The minimal monomials are $z_1^{-1}z_2^2$, z_2 and z_1^2, and these make up G, provided that we do not get cancellation; i.e. if the functions represented by z_1 and z_2 are different and likewise those represented by z_1^2 and z_2. Note that there cannot be such cancellation between $z_1^{-1}z_2^2$ and z_1^2 since their coefficients are both positive. If z_1 and z_2 represent the same function then we have a cancellation $z_1^{-1}z_2^2 - z_2 \equiv 0$, and we recompute G as $G = \{z_1^2, z_1z_2, z_1^{-1}z_2^3\}$. Similarly if z_1^2 and z_2 represent the same function, we obtain $G = \{z_1^{-1}z_2^2, z_1z_2, z_1^3\}$. Assuming that f is not equivalent to zero, the process of recomputing G cannot continue indefinitely since any monomial would become minimal in the monomial ordering after a finite number of steps.

For any monomial, m, we write $\bar{m}$ for the function which m represents. Let f be a non–zero exp–log function with Cartesian representation u. A set of intermediate dominant monomials for u may then be computed as follows:

1. For $i = 1, \ldots, r$, let $z_i^{p_i}$ be the lowest power of z_i occurring in u.
2. Set $G = \{z_1^{p_1}, z_2^{p_2}, \ldots, z_r^{p_r}\}$.
3. Let

$$M = \{m \in G;\ \bar{m} \text{ is maximal in the Hardy–field ordering}\}.$$

4. If the sum of the coefficients in u of the elements of M is not functionally equivalent to zero, return G.
5. Otherwise let $G \setminus M = \{\phi_1, \ldots, \phi_q\}$ with $\bar{\phi}_1 \geq \bar{\phi}_2 \geq \cdots \geq \bar{\phi}_q$ in the Hardy–field ordering, and for $1 \leq i \leq q$ denote by $u_{\phi_i,\ldots,\phi_q}$ the sum of the terms in u whose monomials are greater than or equal to one of $\phi_i, \ldots, \phi_q$ in the monomial ordering.

6. If for some i, $\bar{u}-\bar{u}_{\phi_i,\ldots,\phi_q} \equiv 0$ choose i minimal and put $G=\{\phi_i,\ldots,\phi_q\}$. If there is no such i, put $G := (G\setminus M)\cup M\{z_1, z_2, \ldots, z_k\}$. In either case, remove non–minimal elements from G and return to stage 3.

Here $M\{z_1, z_2, \ldots, z_k\} = \{mz_i;\ m\in M,\ i=1,\ldots,k\}$.

To see that the algorithm is correct, we note firstly that G starts with minimal monomials in u and this is maintained at stage 6. Secondly we only remove monomials from G whose contributions to u have been found to cancel out. Note also that since $u - u_{\phi_i,\ldots,\phi_q}$ contains only finitely many terms we can apply a zero–equivalence procedure to it.

To prove termination, suppose the contrary. Let G_1, G_2, ... be the successive values of G at the beginning of the main loop, and let M_1, M_2, ... be the corresponding values for M. We write $((G_j))$ for set of monomials which are greater than or equal to some element of G_j in the monomial ordering. So if Φ is a subset of the set, S_Z of monomials in $Z=\{z_1, z_2, \ldots, z_k\}$ then

$$((\Phi)) = \{\psi\in S_Z\ \exists\phi\in\Phi \text{ such that } \phi\leq_Z\psi\}.$$

We observe that $((G_1))\supseteq((G_2))\supseteq((G_3))\supseteq\cdots$ and set $F=\bigcap_{j\geq 1}((G_j))$.

Let u_F be the series obtained from u by using only the monomials in F. Then since we only remove monomials corresponding to cancelling terms, we have

$$f(u)=f(u_{((G_1))})=f(u_{((G_2))})=f(u_{((G_3))})=\cdots=f(u_F),$$

where $f(u)$ is the function defined by the series u. By Dickson's Lemma, [1], F is finitely generated, say by Γ. Now we can never have $G_{j+1}=G_j$ because G_{j+1} is either $G_j\setminus M_j$ or $(G_j\setminus M_j)\cup M_j\{z_1,\ldots,z_k\}$ and $M_j\neq\emptyset$. So after a finite number of steps G_j cannot contain any monomial which is less than an element of Γ. Fix this j.

Clearly $\Gamma\subsetneq G_{j+1}$ and so nothing in M_j can be in Γ. Hence $\Gamma\subsetneq\{\phi_i,\ldots,\phi_q\}$ when we pass through the loop for the j-th time. On the other hand $f(u)=f(u_{((\Gamma))})=f(u_{((\{\phi_i,\ldots,\phi_q\}))})$, and so $G_{j+1}=\{\phi_i,\ldots,\phi_q\}$. Hence $\Gamma=\{\phi_i,\ldots,\phi_q\}=G_{j+1}$, which forces G_{j+2} to equal G_{j+1}.

This contradiction proves the termination.

6

Algebraic Differential Equations

Asymptotics have been much used in the study of differential equations. The method of *undetermined coefficients* is one common technique. At its most basic, this consists of substituting a general power series into the equation and then comparing terms in order to find the coefficients. The same idea can be made to work with series in other base functions, and also series where several different base functions may appear, for example x and $\log x$. A drawback is that one has to know in advance which base functions are going to be needed. There is a great fund of expertise on such questions, especially for linear ordinary differential equations; see [119] for example. In this chapter, we look at what can be done for non–linear equations using nested expansions.

It should be said immediately that one cannot expect too much. As innocuous–looking an equation as

$$\frac{dy}{dx} = 1 + y^2, \tag{6.1}$$

which by most reckoning is not very far from being linear, already shows very pathological asymptotics. The solution $\tan(x + k)$, $k \in \mathbb{R}$, has poles at every argument of the form $(n + \frac{1}{2})\pi - k$. So we can hardly expect an asymptotic representation at infinity by power series[1]. Yet from a naive viewpoint, there is little in the equation to suggest this, nor to explain the appearance of the mysterious number π.

We have also seen in previous chapters that oscillation can cause problems in asymptotics, so we commence by looking for solutions to our differential equations which lie in a Hardy field. This allows us to use the machinery of the previous chapters. However it does not require us to guess in advance what sort of growth orders solutions might have.

[1] See Chapter 10 however

6.1 Nested Forms of Hardy–Field Solutions

Suppose that we have an ordinary algebraic differential equation,

$$P\langle y\rangle = P(y, y', \ldots, y^{(n)}) = 0, \tag{6.2}$$

and seek the growth of solutions $f(x)$ which belong to some Hardy field. Of course there will often be many such solutions, and as explained in Chapter 5, distinguishing them by their values at a finite point does not fit in with our approach. Therefore our question becomes 'how can we describe the set of all possible asymptotic behaviours of Hardy–field solutions to (6.2)?'.

We consider first the cases when $f \to \infty$. We recall from Theorem 11 that either $\mathbb{K}\langle\langle f\rangle\rangle$ contains infinitely–many different comparability classes, or else f has a pnf $\exp_s(\log_m^d x\phi_1(x))$. However since f satisfies (6.2), $f^{(n)}$ is algebraic over $\mathbb{K}(f, f', \ldots, f^{(n-1)})$. The same applies to $f^{(r)}$ with $r > n$, and so the rank of $\mathbb{K}(f, f', \ldots, f^{(n-1)})$ cannot exceed $n+1$. This simple but important fact was pointed out by M. Rosenlicht in [79]. It follows immediately that cases [2], [3] and [4] of Theorem 11 are excluded for functions satisfying algebraic differential equations.

Moreover if the pnf of f is $\exp_s(\log_m^d x\phi(x))$, Theorem 11 asserts that the Rosenlicht field $\mathbb{K}\langle\langle f\rangle\rangle$ (see Chapter 3.2) contains an element asymptotic to $\phi(x)$. But then the same argument may be applied to ϕ as was to f. The following result is from [99].

THEOREM **18** *Let f be an element of a Hardy field with $f \to \infty$, which satisfies an algebraic differential equation of order n over a Hardy field, $\mathcal{F}$, of finite rank, r. Then f has a nested form $\{(\epsilon_i, s_i, m_i, d_i, \phi_i),\ i = 1, 2, \ldots, k\}$ such that*

$$\sum_{j=1}^{k} s_j + \delta_k + m_k \le n + r, \tag{6.3}$$

where $\delta_k = 0$ if $d_k = 1$ and $m_k = 0$ and otherwise $\delta_k = 1$. Moreover there exist elements $\Gamma_0, \ldots, \Gamma_\Sigma$ of the Rosenlicht field $\mathcal{F}\langle\langle f\rangle\rangle$ such that the following properties hold:

1. *$\Sigma = \sum_{j=1}^{k} s_j + \delta_k + m_k$.*
2. *Each Γ_i may be expressed as a rational function of a finite set of real powers of $f,\ f', \ldots, f^{(i)}$ with real coefficients. Conversely, $f^{(i)}$ may be similarly expressed as a rational function of a finite set of real powers of $\Gamma_0,\ \Gamma_1,\ \ldots,\ \Gamma_i$. Both of these rational functions are computable.*
3. *For $i = 0, \ldots, \Sigma - 1$, we have $\gamma_1(\Gamma_i) > \gamma_1(\Gamma_{i+1})$. Also $\gamma_1(\Gamma_\Sigma) = \gamma_1(1)$.*
4. *The set of comparability classes of the Γ_is is equal to*

$$\{\gamma_1(1)\} \cup \bigcup_{i=0}^{m_k+\delta_k-1} \{\gamma_1(\log_i)\} \cup \bigcup_{j=1}^{k}\bigcup_{t=1}^{s_j}\{\gamma_1(\exp_t(\log_{m_j}^{d_j}\phi_j))\}.$$

Of course if f tends to $-\infty$ or to a finite limit A, we can apply Theorem 18 to $-f$ or to $\pm(A-f)^{-1}$ respectively. It is also worth pointing out that a similar theorem can be obtained even when f does not satisfy a differential equation provided that the rank of $\mathcal{F}\langle\langle f\rangle\rangle$ is finite, [99]. The proof is identical to the one we now give.

Proof of Theorem 18
The first part of the statement is already more or less clear from Chapter 3, and will become even more so once we define the Γ_is. We proceed with this, concentrating initially on the first pnf $f = \exp_s(\log_m^d x\phi(x))$; c.f. [99, 79]. As explained in Section 4.4, if $s > 0$ and $\log_m^d x\phi(x)$ is asymptotic to $A\log_m x$ with A constant, we can rewrite the pnf in the form $f = \exp_{s-1}(\log_{m+1}^A x\psi(x))$ and repeat this procedure if applicable. So we may assume that our pnf is the first part of a nested form.

For $f, g \in \mathcal{F}^\infty$, we write $f \approx g$ if $\gamma_1(f/g) < \gamma_1(f)$; i.e. $\log f \sim \log g$. Lemma 6 then implies that $f^\triangle \sim g^\triangle$, where we recall from Section 3.4 that $f^\Delta = f'/f$. As in Proposition 4, we make considerable use of this logarithmic derivative. However our aim there was to deduce the existence of a pnf, whereas here we want to find suitable Γ_is.

The generic method of defining the Γ_is is as follows. Suppose we already have Γ_j with $\Gamma_j \approx \exp_t(\log_m^d x\phi_1)$, $t \geq 2$. Then $\log \Gamma_j \sim \exp_{t-1}(\log_m^d x\phi_1)$ and so

$$\begin{aligned}\Gamma_j^\triangle &\sim (\exp_{t-1}(\log_m^d x\phi_1))' \\ &= \exp_{t-1}(\log_m^d x\phi_1)\ldots\exp_1(\log_m^d x\phi_1)\{(\log_m^d x)'\phi_1 + \log_m^d x\phi_1'\}.\end{aligned}$$

Since $\gamma_1(\phi_1) < \gamma_1(\log_m x)$, Lemma 7 implies that $\gamma_0(\phi_1^\Delta) < \gamma_0((\log_m^d x)^\Delta)$. So $\log_m^d x\phi_1' = o((\log_m^d x)'\phi_1)$, and

$$\Gamma_j^\triangle \sim \exp_{t-1}(\log_m^d x\phi_1)\ldots\exp(\log_m^d x\phi_1)L_m^{-1}\log_m^{d-1} x\phi_1 d, \qquad (6.4)$$

where as in Section 3.4, $L_m = x\log x\log_2 x\cdots\log_{m-1} x$. If $\gamma_1(\Gamma_j^\triangle) > \gamma_1(x)$, we take $\Gamma_{j+1} = \Gamma_j^\triangle$. Otherwise $r = m - t + 1 \geq 0$, and we suppose that we have already defined an element, η_r, of $\mathcal{F}$ which is asymptotic to L_r; we take $\eta_0 = 1$. We set $\Gamma_{j+1} = \Gamma_j^\triangle\eta_r$, and then by Lemma 11, either $\Gamma_{j+1} \approx \log_r^{-1} x$ if $\gamma_1(\log_r x) > \gamma_1(\exp_{t-1}(\log_m^d x\phi_1))$ or $\Gamma_{j+1} \approx \exp_{t-1}(\log_m^d x\phi_1)$ if $\gamma_1(\log_r x) < \gamma_1(\exp_{t-1}(\log_m^d \phi_1))$. Note that except in the relatively trivial case $t = 1$, the two comparability classes cannot be equal because of our assumptions concerning the pnf.

If $\Gamma_{j+1} \approx \log_r^{-1} x$, then $\Gamma_{j+1}^\triangle \sim -L_{r+1}^{-1}$ and so we take $\eta_{r+1} = -\left(\Gamma_{j+1}^\triangle\right)^{-1}$, giving $\eta_{r+1} \sim L_{r+1}$. We then define $\Gamma_{j+2} = \Gamma_j^\triangle\eta_{r+1}$, so that

$$\begin{aligned}\Gamma_{j+2} \sim \exp_{t-1}(\log_m^d x\phi_1)\ldots\exp(\log_m^d x\phi_1)\log_{r+1}^{-1} x\ldots \\ \ldots\log_{m-1}^{-1} x\log_m^{d-1} x\phi_1 d.\end{aligned}$$

Then $r = m - t + 1 < r + 1$ implies that $\Gamma_{j+2} \approx \exp_{t-1}(\log_m^d x\phi_1)$.

If on the other hand $\Gamma_{j+1} \approx \exp_{t-1}(\log_m^d x\phi_1)$ and $t > 2$, we take $\Gamma_{j+2} = -\Gamma_{j+1}^{\triangle}\eta_r$. Since $r < m-t+2$ it follows that $\gamma_1(\log_r x) > \gamma_1(\exp_{t-2}(\log_m^d x\phi_1))$, and hence that $\Gamma_{j+2} \approx \log_r^{-1} x$. Then η_{r+1} is defined as before.

In practice we shall need to worry about various special situations. We also employ some slight variations to make more efficient use of the Γ_is already defined. We now give the details for the various cases.

Case 1: $m = s = 0$

Here $f = x^d\phi_1$. We put $\Gamma_0 = f$. Then

$$f^{\triangle} = \frac{d}{x} + \phi_1^{\triangle}.$$

By Lemma 7, $\gamma_0(\phi_1^{\triangle}) < \gamma_0(d/x)$ and so $f^{\triangle} \sim d/x$. We therefore take $\eta_1 = f^{\triangle}/d$ and $\Gamma_1 = \Gamma_0\eta_1^d \sim \phi_1$.

Case 2: $m = 0, s > 0$

We set $\Gamma_0 = f = \exp_s(x^d\phi_1)$ and for $i = 1, \ldots, s$, we let $\Gamma_i = \Gamma_{i-1}^{\triangle}$. Suppose, inductively, that $\Gamma_{i-1} \approx \exp_{s-i+1}(x^d\phi_1)$. Then if $i < s$, (6.4) gives

$$\Gamma_i \sim \exp_{s-i}(x^d\phi_1)\ldots\exp(x^d\phi_1)x^{d-1}\phi_1 d \approx \exp_{s-i}(x^d\phi_1).$$

For the case $i = s$, we have

$$\Gamma_s = (\log \Gamma_{s-1})' \sim (x^d\phi_1)' = dx^{d-1}\phi_1 + x^d\phi_1' \sim dx^{d-1}\phi_1,$$

by Lemma 7 (as in (6.4)). If $d = 1$ this means that $\Gamma_s \sim \phi_1$. Otherwise $\Gamma_s \approx x^{d-1}$ and hence $\Gamma_s^{\triangle} \sim (d-1)x^{-1}$. We then put

$$\Gamma_{s+1} = \frac{\Gamma_s}{d}\left(\frac{\Gamma_s^{\triangle}}{(d-1)}\right)^{d-1} \sim \phi_1.$$

Case 3: $m > 0, s = 0$

In this case $f = \log_m^d x\phi_1$ with $\gamma_1(\phi_1) < \gamma_1(\log_m x)$. So

$$f^{\triangle} \sim dx^{-1}\log^{-1} x \ldots \log_m^{-1} x = dL_{m+1} \approx x^{-1}.$$

We therefore take $\Gamma_0 = f^{\triangle}$. Since $\Gamma_0 \approx x^{-1}$, we have $\Gamma_0^{\triangle} \sim -x^{-1}$. We define $\Gamma_1 = -\Gamma_0/\Gamma_0^{\triangle}$, so that $\Gamma_1 \sim d\log^{-1} x \ldots \log_m^{-1} x \approx \log^{-1} x$.

Suppose that $1 < i < m$ and that Γ_{i-1} has been defined with $\Gamma_{i-1} \approx \log_{i-1}^{-1} x$. Then $\Gamma_{i-1}^{\triangle} \sim -L_i^{-1}$ and so we take $\Gamma_i = -\Gamma_0/\Gamma_{i-1}^{\triangle}$. This gives $\Gamma_i \sim d\log_i^{-1} x \ldots \log_m^{-1} x \approx \log_i^{-1} x$. Hence by induction, $\Gamma_i \sim d\log_i^{-1} x \cdots \log_m^{-1} x \approx \log_i^{-1} x$ for $i = 0, \ldots, m$. For the case $i = m$, we obtain $\Gamma_m \sim d\log_m^{-1} x$, and we take

$$\Gamma_{m+1} = f(d^{-1}\Gamma_m)^d \sim \phi_1.$$

Case 4: $s > m > 0$

By Lemma 11, $\gamma_1(f) = \gamma_1(\exp_s(\log_m^d x\phi_1)) > \gamma_1(x)$. We therefore take $\Gamma_0 = f$. For $i = 1, \ldots, s-m$, we take $\Gamma_i = \Gamma_{i-1}^{\triangle}$. Suppose inductively that $1 \le i \le s-m-1$ and that $\Gamma_{i-1} \approx \exp_{s-i+1}(\log_m^d x\phi_1)$; of course for $i=1$, we know this. By (6.4),

$$\Gamma_i \sim \exp_{s-i}(\log_m^d x\phi_1)\ldots\exp(\log_m^d x\phi_1)x^{-1}\ldots\log_{m-1}^{-1} x \log_m^{d-1} x\phi_1 d,$$

and from Lemma 11, $\gamma_1(\exp_{s-i}(\log_m^d x\phi_1)) > \gamma_1(x)$, since $s-i > m$. Hence $\Gamma_i \approx \exp_{s-i}(\log_m^d x\phi_1)$ and so by induction, this holds for $i = 0, \ldots, s-m-1$. From (6.4) again we have

$$\Gamma_{s-m} \sim \exp_m(\log_m^d x\phi_1)\ldots\exp(\log_m^d x\phi_1)L_m^{-1}\log_m^{d-1} x\phi_1 d. \tag{6.5}$$

We now consider two subcases.

<u>Case 4a: $d > 1$ or $d = 1$ and $\phi_1 \to \infty$</u>

Here, Lemma 11 implies that $\Gamma_{s-m} \approx \exp_m(\log_m^d x\phi_1)$. We take $\Gamma_{s-m+1} = \Gamma_{s-m}^{\triangle}$, so that

$$\Gamma_{s-m+1} \sim \exp_{m-1}(\log_m^d x\phi_1)\ldots\exp(\log_m^d x\phi_1)L_m^{-1}\log_m^{d-1} x\phi_1 d. \tag{6.6}$$

From Lemma 11 again, $\Gamma_{s-m+1} \approx x^{-1}$, and it follows that $\Gamma_{s-m+1}^{\triangle} \sim -x^{-1}$. Hence we take $\Gamma_{s-m+2} = -\Gamma_{s-m+1}/\Gamma_{s-m+1}^{\triangle}$. Then

$$\begin{aligned}\Gamma_{s-m+2} &\sim \exp_{m-1}(\log_m^d x\phi_1)\ldots\exp(\log_m^d x\phi_1)\log^{-1} x\ldots\\ &\quad\ldots\log_{m-1}^{-1} x\log_m^{d-1} x\phi_1 d\\ &\approx \exp_{m-1}(\log_m^d x\phi_1).\end{aligned}$$

Suppose $0 < i < m-1$ and that $\Gamma_{s-m+2i-1}$ and Γ_{s-m+2i} have been defined with $\Gamma_{s-m+2i-1} \approx \log_{i-1}^{-1} x$ and $\Gamma_{s-m+2i} \approx \exp_{m-i}(\log_m^d x\phi_1)$. Then

$$\Gamma_{s-m+2i-1}^{\triangle} \sim -L_i^{-1} \tag{6.7}$$

and

$$\Gamma_{s-m+2i}^{\triangle} \sim \exp_{m-i-1}(\log_m^d x\phi_1)\ldots\exp(\log_m^d x\phi_1)L_m^{-1}\log_m^{d-1} x\phi_1 d. \tag{6.8}$$

We let

$$\Gamma_{s-m+2i+1} = -\frac{\Gamma_{s-m+2i}^{\triangle}}{\Gamma_{s-m+2i-1}^{\triangle}}. \tag{6.9}$$

It then follows from (6.7) and (6.8) that

$$\begin{aligned}\Gamma_{s-m+2i+1} &\sim \exp_{m-i-1}(\log_m^d x\phi_1)\ldots\exp(\log_m^d x\phi_1)\log_i^{-1} x\ldots\\ &\quad\ldots\log_{m-1}^{-1} x\log_m^{d-1} x\phi_1 d\\ &\approx \log_i^{-1} x.\end{aligned}$$

Then $\Gamma^{\triangle}_{s-m+2i+1} \sim -L^{-1}_{i+1}$ and so we put

$$\Gamma_{s-m+2i+2} = -\frac{\Gamma^{\triangle}_{s-m+2i}}{\Gamma^{\triangle}_{s-m+2i+1}}.$$

This gives

$$\begin{aligned}\Gamma_{s-m+2i+2} &\sim \exp_{m-i-1}(\log^d_m x\phi_1)\ldots\exp(\log^d_m x\phi_1)\log^{-1}_{i+1} x\ldots\\ &\quad\ldots\log^{-1}_{m-1} x\log^{d-1}_m x\phi_1 d\\ &\approx \exp_{m-i-1}(\log^d_m x\phi_1).\end{aligned}$$

Thus by induction we can define $\Gamma_{s-m+1},\ldots\Gamma_{s+m-2}$ with the desired properties. If we take $i = m-1$ in (6.9) we get $\Gamma_{s+m-1} \approx \log^{-1}_{m-1} x$. We then set

$$\Gamma_{s+m} = -\frac{\Gamma^{\triangle}_{s+m-2}}{d\Gamma^{\triangle}_{s+m-1}} \sim \log^{d-1}_m x\phi_1,$$

by (6.8). If $d = 1$, then $\Gamma_{s+m} \sim \phi_1$. Otherwise $\Gamma_{s+m} \approx \log^{d-1}_m$ and hence $\Gamma^{\triangle}_{s+m} \sim (d-1)L^{-1}_{m+1}x$. Then

$$\frac{\Gamma^{\triangle}_{s+m}}{(d-1)\Gamma^{\triangle}_{s+m-1}} \sim -\log^{-1}_m x.$$

We therefore take

$$\Gamma_{s+m-1} = \Gamma_{s+m}\left(\frac{\Gamma^{\triangle}_{s+m}}{(1-d)\Gamma^{\triangle}_{s+m-1}}\right)^{d-1} \sim \phi_1.$$

Case 4b: $d < 1$ or $d = 1$ and $\phi_1 \to 0$

In this case, (6.5) and Lemma 11 give $\Gamma_{s-m} \approx x^{-1}$. Hence $\Gamma^{\triangle}_{s-m} \sim -x^{-1}$ and so we take $\Gamma_{s-m+1} = -\Gamma_{s-m}/\Gamma^{\triangle}_{s-m}$. From (6.5) and from the Lemma again, we have that $\Gamma_{s-m+1} \approx \exp_m(\log^d_m x\phi_1)$.

Suppose that $0 \le i < m-1$ and that Γ_{s-m+2i} and $\Gamma_{s-m+2i+1}$ have been defined with $\Gamma_{s-m+2i} \approx \log^{-1}_i x$ and $\Gamma_{s-m+2i+1} \approx \exp_{m-i}(\log^d_m x\phi_1)$.

We take

$$\Gamma_{s-m+2i+2} = -\frac{\Gamma^{\triangle}_{s-m+2i+1}}{\Gamma^{\triangle}_{s-m+2i}}.$$

As before $\Gamma_{s-m+2i+2} \approx \log^{-1}_{i+1} x$, and we let

$$\Gamma_{s-m+2i+3} = -\frac{\Gamma^{\triangle}_{s-m+2i+1}}{\Gamma^{\triangle}_{s-m+2i+2}} \approx \exp_{m-i-1}(\log^d_m x\phi_1).$$

We have thus defined $\Gamma_{s-m+1,\ldots,}\Gamma_{s+m-1}$. Next we put

$$\Gamma_{s+m} = -\frac{\Gamma^{\triangle}_{s+m-1}}{d\Gamma^{\triangle}_{s+m-2}} \sim \log_m^{d-1} x\phi_1 .$$

If $d = 1$, then of course $\Gamma_{s+m} \sim \phi_1$. Otherwise $\Gamma_{s+m} \approx \log_m^{d-1} x$ and we take

$$\Gamma_{s+m-1} = \Gamma_{s+m}\left(\frac{\Gamma^{\triangle}_{s+m}}{(1-d)\Gamma^{\triangle}_{s+m-1}}\right)^{d-1} \sim \phi_1 .$$

Case 5: $m > s > 0$

We let $\Gamma_0 = f^{\triangle} \approx x^{-1}$ and for $i = 1, \ldots, m-s-1$ we take $\Gamma_i = -\Gamma_0/\Gamma^{\triangle}_{i-1}$. Then as in Case 3,

$$\begin{aligned}\Gamma_i &\sim \exp_{s-1}(\log_m^d x\phi_1)\ldots\exp x(\log_m^d x\phi_1)\log_i^{-1} x\ldots \\ &\quad\ldots\log_{m-1}^{-1} x \log_m^{d-1} x\phi_1 d \\ &\approx \log_i^{-1} x.\end{aligned}$$

Case 5a: $d > 1$ or $d = 1$ and $\phi_1 \to \infty$

In this case, Lemma 11 implies that $\gamma_1(f) > \gamma_1(\log_{m-s} x)$ and hence we take $\Gamma_{m-s} = f$ and $\Gamma_{m-s+1} = -\Gamma_0/\Gamma^{\triangle}_{m-s-1} \approx \log_{m-s}^{-1} x$. Similarly we let $\Gamma_{m-s+2} = -\Gamma_0/\Gamma^{\triangle}_{m-s+1} \approx \exp_{s-1}(\log_m^d x\phi_1)$. For $i = 1, \ldots, s-2$, we define

$$\Gamma_{m-s+2i+1} = -\frac{\Gamma^{\triangle}_{m-s+2i}}{\Gamma^{\triangle}_{m-s+2i-1}} \quad\text{and}\quad \Gamma_{m-s+2i+2} = -\frac{\Gamma^{\triangle}_{m-s+2i}}{\Gamma^{\triangle}_{m-s+2i+1}}. \tag{6.10}$$

Then $\Gamma_{m-s+2i+1} \approx \log_{m-s+i+1}^{-1} x$ and $\Gamma_{m-s+2i+2} \approx \exp_{s-i-1}(\log_m^d x\phi_1)$. We define Γ_{m+s-1} by taking $i = s-1$ in (6.10), so that $\Gamma_{m+s-1} \approx \log_{m-1}^{-1}$. Then we set

$$\Gamma_{m+s} = -\frac{\Gamma^{\triangle}_{m+s-2}}{d\Gamma^{\triangle}_{m+s-1}} \sim \log_m^{d-1} x\phi_1 .$$

If $d \neq 1$, we take

$$\Gamma_{m+s+1} = \Gamma_{m+s}\left(\frac{\Gamma^{\triangle}_{m+s}}{(1-d)\Gamma^{\triangle}_{m+s-1}}\right)^{d-1} \sim \phi_1 .$$

Case 5b: $d < 1$ or $d = 1$ and $\phi_1 \to 0$

Now $\gamma_1(f) < \gamma_1(\log_{m-s} x)$ and so we take $\Gamma_{m-s} = -\Gamma_0/\Gamma^{\triangle}_{m-s-1} \approx \log_{m-s}^{-1} x$ and $\Gamma_{m-s+1} = f$. Then for $i = 1, \ldots, s-1$, we let

$$\Gamma_{m-s+2i} = -\frac{\Gamma^{\triangle}_{m-s+2i-1}}{\Gamma^{\triangle}_{m-s+2i-2}}$$

and

$$\Gamma_{m-s+2i+1} = -\frac{\Gamma^{\triangle}_{m-s+2i-1}}{\Gamma^{\triangle}_{m-s+2i}}.$$

Then we set

$$\Gamma_{m+s} = -\frac{\Gamma^{\triangle}_{m+s-1}}{d\Gamma^{\triangle}_{m+s-2}} \sim \log_m^{d-1} x\phi$$

and if $d \neq 1$, we let

$$\Gamma_{m+s+1} = \Gamma_{m+s}\left(\frac{\Gamma^{\triangle}_{m+s}}{(1-d)\Gamma^{\triangle}_{m+s-1}}\right)^{d-1} \sim \phi_1.$$

Case 6: $s = m > 0$

Here again, we consider two sub-cases.

Case 6a: $s = m > 0$ and either $d > 1$ or $d = 1$ and $\phi_1 \to \infty$

We take $\Gamma_0 = f$, $\Gamma_1 = f^{\triangle}$ and for $i = 1, \ldots, s-1$,

$$\Gamma_{2i} = -\frac{\Gamma^{\triangle}_{2i-2}}{\Gamma^{\triangle}_{2i-1}} \approx \exp_{s-i}(\log_s^d x\phi_1)$$

and

$$\Gamma_{2i+1} = -\frac{\Gamma^{\triangle}_{2i}}{\Gamma^{\triangle}_{2i-1}} \approx \log_i^{-1} x.$$

As before

$$\Gamma_{2s} = \frac{\Gamma^{\triangle}_{2s-1}}{d\Gamma^{\triangle}_{2s-2}} \sim \log_s^{d-1} x\phi_1 \tag{6.11}$$

and if $d \neq 1$,

$$\Gamma_{2s+1} = \Gamma_{2s}\left(\frac{\Gamma^{\triangle}_{2s}}{(1-d)\Gamma^{\triangle}_{2s-1}}\right)^{d-1}. \tag{6.12}$$

Case 6b: $s = m > 0$ and either $d < 1$ or $d = 1$ and $\phi_1 \to 0$

Now $\gamma_1(f) < \gamma_1(x)$ and so we take $\Gamma_0 = f^{\triangle} \approx x^{-1}$. We let $\Gamma_1 = f$ and for $i = 1, \ldots, s-1$, set

$$\Gamma_{2i} = -\frac{\Gamma^{\triangle}_{2i-1}}{\Gamma^{\triangle}_{2i-2}} \approx \log_i^{-1} x$$

and

$$\Gamma_{2i+1} = \frac{-\Gamma^{\triangle}_{2i-1}}{\Gamma^{\triangle}_{2i}} \approx \exp_{s-i}(\log_s^d x\phi_1).$$

Γ_{2s} is defined by (6.11) and if $d \neq 1$, Γ_{2s+1} is defined by (6.12). Thus we have dealt with the pnfs.

For $j = 1, \ldots, k$ we define δ_j to be 1 if $d_j \neq 1$ or if $s_j = 0$, and to be 0 otherwise. We write $t_j = s_j + \delta_j$. We have s_1, m_1, d_1 and ϕ_1 such that $f = \exp_{s_1}(\log_{m_1}^{d_1} \phi_1)$. If $\gamma_0(\phi_1) = 0$ then $\{(+1, s_1, m_1, d_1, \phi_1)\}$ is a nested form for f. Theorem 11 then implies the existence of $\Gamma_0, \ldots, \Gamma_{t_1+m_1}$ with suitable properties.

If $\gamma_0(\phi_1) \neq 0$, we may apply Theorem 11 to either $\Gamma_{t_1+m_1}$ or $\Gamma_{t_1+m_1}^{-1}$ according to whether $\gamma_0(\Gamma_{t_1+m_1}) > 0$ or not. Thus $\Gamma_{t_1+m_1} = \exp_{s_2}^{\pm 1}(\log_{m_2}^{d_2} \psi_2)$ for suitable choices of s_2, m_2, d_2 and ψ_2. The method given above then guarantees the existence of elements $\Gamma_0^1, \ldots, \Gamma_{t_2+m_2}^1$ of $\mathcal{F}$ as required. The comparability classes of these elements will include $\gamma_1(x), \ldots, \gamma_1(\log_{m_1+\delta_1-1} x)$ but will otherwise be distinct from the comparability classes of the Γ_is, since $\gamma_0(\phi_1) > \gamma_0(\log_{m_1} x)$. Hence we may relabel $\Gamma_{m_1+\delta_1}^1, \ldots, \Gamma_{t_2+m_2}^1$ as $\Gamma_{t_1+m_1+1}, \ldots, \Gamma_{s_1+t_2+m_2}$ and then $\gamma_1(\Gamma_0) > \gamma_1(\Gamma_1) > \ldots > \gamma_1(\Gamma_{s_1+t_2+m_2})$. Since $\Gamma_{t_1+m_1} \sim \phi_1$, there is a $\phi_2 \sim \psi_2$ such that $\phi_1 = \exp_{s_2}^{\pm 1}(\log_m^{d_2} x\phi_2)$. If $\gamma_0(\phi_2) = 0$, then $\{(+1, s_1, m_1, d_1, \phi_1), (\pm 1, s_2, m_2, d_2, \phi_2)\}$ is a nested form for f. Otherwise we apply the above method to $\Gamma_{s_1+t_2+m_2}$ and continue as before. In this way we obtain a sequence $\{(\epsilon_j, s_j, m_j, d_j, \phi_j), \, j = 1, \ldots, k\}$ and the process will continue until some $\gamma_0(\phi_k)$ is zero. It is clear that this must happen after a finite number of steps since (6.3) must continue to hold and $\{m_j\}$ is strictly increasing; moreover $t_j \geq 1$. So f has a nested form. The inequality (6.3), the existence of the Γ_j's and properties required of a nested form are then clear from Theorem 11 and the above method of construction. Finally, we see that δ_k can be zero only when m_k is zero. This completes the proof of Theorem 18.

What we have done above applies to any algebraic differential equation, and makes no distinction between different equations of the same order. For a particular equation, many of the nested forms whose parameters satisfy (6.3) will turn out not to be possible. We illustrate this by looking in more detail at the case of second order equations over $\mathbb{R}$; c.f. [99].

Suppose then that f satisfies such an equation. Now $\delta_k = 1$ unless $m_k = 0$ and from the definition of a nested form $m_i \geq i - 1$. Hence the condition $\sum_i s_i + m_k + \delta_k \leq n$ implies that $k \leq 2$. Possibilities are therefore

$$\{k = 2,\ s_1 = s_2 = 0,\ m_1 = 0,\ m_2 = 1\} \quad \text{or} \quad \{k = 1,\ s_1 = 0,\ m_1 \leq 1\}$$
$$\text{or} \quad \{k = 1,\ s_1 \leq 2,\ m_1 = 0\}.$$

Note in particular that we cannot have $k = s_1 = m_1 = 1$ since then $\delta_1 = 1$ and the condition $\sum_i s_i + m_k + \delta_k \leq n$ will be violated. Moreover the case $k = 1$, $s_1 = 2$, $m_1 = 0$ is only possible if $d_1 = 1$. Let A stand for a non-zero constant and C for any constant. The possible asymptotic forms for f are then:

(a) $C + (A + \varepsilon)x^{d_1} \log^{d_2} x$ with $C = 0$ if $d_1 > 0$ or if $d_1 = 0$ and $d_2 > 0$;
(b) $C \pm \exp(x^{d_1}(A + \varepsilon))$ with $C = 0$ if $A > 0$;
(c) $\pm \exp_2(x(A + \varepsilon))$ with $A > 0$;

(d) $C \pm \exp_2^{-1}(x(A+\varepsilon))$ with $A > 0$.

We calculate the Γ_is in each case.

Case (a)(i): $f = C + (A+\varepsilon)x^{d_1}\log^{d_2} x$, $d_1 \neq 0$, $d_2 \neq 0$

We take $\Gamma_0 = f - C$. Then as in the proof of Theorem 18, Case 1, we take

$$\Gamma_1 = \Gamma_0 \left(\frac{\Gamma_0^\triangle}{d_1}\right)^{d_1} \sim A \log^{d_2} x. \tag{6.13}$$

Then as in Case 3,

$$\Gamma_2 = \Gamma_1 \left(\frac{\Gamma_1^\triangle/d_2}{\Gamma_0^\triangle/d_1}\right)^{d_2} \sim A. \tag{6.14}$$

Next, we calculate f, f' and f'' in terms of Γ_0, Γ_1, and Γ_2. Of course

$$f = \Gamma_0 + C \tag{6.15}$$

and therefore

$$f' = \Gamma_0' = d_1 \Gamma_0^{1-1/d_1} \Gamma_1^{1/d_1}, \tag{6.16}$$

from (6.13). Differentiating (6.16) gives

$$f'' = (d_1 - 1)(\Gamma_1/\Gamma_0)^{1/d_1}\Gamma_0' + (\Gamma_1/\Gamma_0)^{1/d_1 - 1}\Gamma_1'$$

and then substituting for Γ_0' and Γ_1' from (6.13) and (6.14) yields

$$f'' = d_1(d_1-1)\Gamma_0^{1-2/d_1}\Gamma_1^{2/d_1} + d_2 \Gamma_0^{1-2/d_1}\Gamma_1^{2/d_1 - 1/d_2}\Gamma_2^{1/d_2}. \tag{6.17}$$

Case (a)(ii): $f = C + x^{d_1}(A+\varepsilon)$, $d_1 \neq 0$

Here the rank of $\mathbf{R}\langle f\rangle$ (i.e. $\mathbb{R}(f, f', f'' \ldots)$) is equal to 1. Accordingly, we define $\tilde{\Gamma}_0 = f - C$, $\tilde{\Gamma}_1 = \tilde{\Gamma}_0'$ and $\tilde{\Gamma}_2 = \tilde{\Gamma}_1'$.

Case (a)(iii): $f = C + (A+\varepsilon)\log^{d_2} x$, $d_2 \neq 0$

We follow the proof of Theorem 18, Case 3 and take

$$\Gamma_0 = \frac{f'}{d_2(f-C)} \sim x^{-1}\log^{-1} x, \tag{6.18}$$

$$\Gamma_1 = -\Gamma_0^2/\Gamma_0' \sim \log^{-1} x \tag{6.19}$$

and

$$\Gamma_2 = (f-C)\Gamma_1^{d_2} \sim A.$$

Then

$$f = C + \Gamma_2 \Gamma_1^{-d_2}, \tag{6.20}$$

and from (6.18)

$$f' = d_2\Gamma_0(f - C) = d_2\Gamma_0\Gamma_1^{-d_2}\Gamma_2, \tag{6.21}$$

by (6.20). From (6.19) we see that , $\Gamma_0' = -\Gamma_0^2\Gamma_1^{-1}$. On the other hand, Γ_0' is equal to $(f'/d_2(f-C))'$ by (6.18). Hence

$$\frac{f''}{d_2(f-C)} - \frac{1}{d_2}\left(\frac{f'}{(f-C)}\right)^2 = -\Gamma_0^2\Gamma_1^{-1}$$

and so

$$f'' = \frac{f'^2}{f-C} - d_2(f-C)\Gamma_0^2\Gamma_1^{-1}.$$

On substituting, using (6.21) and (6.20), we get

$$f'' = d_2^2\Gamma_0^2\Gamma_1^{-d_2}\Gamma_2 - d_2\Gamma_0^2\Gamma_1^{-1-d_2}\Gamma_2. \tag{6.22}$$

Case (b): $f = C \pm \exp(x^{d_1}(A+\varepsilon))$, $d_1 > 0$

Here, as in Case 2 of Theorem 18, we take $\Gamma_0 = \pm(f - C)$. If $d_1 = 1$, we define $\tilde{\Gamma}_1 = \Gamma_0'$ and $\tilde{\Gamma}_2 = \tilde{\Gamma}_1'$. Otherwise, we take

$$\Gamma_1 = \Gamma_0'\Gamma_0^{-1}d_1^{-1} \sim Ax^{d_1-1}. \tag{6.23}$$

and

$$\Gamma_2 = \Gamma_1\left(\frac{\Gamma_1'}{(d_1-1)\Gamma_1}\right)^{d_1-1} \sim A. \tag{6.24}$$

To obtain the expressions for f and its derivatives (in the case $d_1 \neq 1$), we start from

$$f = \pm\Gamma_0 + C. \tag{6.25}$$

Then (6.23) gives

$$f' = \pm d_1\Gamma_0\Gamma_1. \tag{6.26}$$

Hence $f'' = \pm(d_1\Gamma_0'\Gamma_1 + d_1\Gamma_0\Gamma_1')$, and on substituting for Γ_0' and Γ_1' using (6.23) and (6.24) we obtain

$$f'' = \pm\Gamma_0\Gamma_1\left(d_1^2\Gamma_1 + d_1(d_1-1)\left(\frac{\Gamma_2}{\Gamma_1}\right)^{\frac{1}{d_1-1}}\right). \tag{6.27}$$

It is understood that one must have a consistent choice of + or – sign in (6.25), (6.26) and (6.27).

Case(c): $f = \pm \exp_2(x(A+\varepsilon))$

Here also we follow Case 2 of the proof of Theorem 18. We take $\Gamma_0 = \pm f$, $\Gamma_1 = \Gamma_0^{\triangle}$, and $\Gamma_2 = \Gamma_1^{\triangle}$.

Then

$$f' = \pm \Gamma_0' = \pm \Gamma_0 \Gamma_1$$

and

$$\begin{aligned} f'' &= \pm(\Gamma_0' \Gamma_1 + \Gamma_0 \Gamma_1') \\ &= \pm \Gamma_0 \Gamma_1 (\Gamma_1 + \Gamma_2). \end{aligned}$$

Case (d): $f = C \pm \exp_2^{-1}(x(A+\varepsilon))$

Take $\Gamma_0 = \pm(f-C)^{-1} = \exp_2(x(A+\varepsilon))$. Then, as in Case (c), $\Gamma_1 = \Gamma_0^{\triangle}$ and $\Gamma_2 = \Gamma_1^{\triangle}$. Then

$$f = C \pm \Gamma_0^{-1} \tag{6.28}$$

and so

$$f' = \pm \Gamma_0' \Gamma_0^{-2} = \pm \Gamma_0^{-1} \Gamma_1. \tag{6.29}$$

Differentiation of (6.29) gives $f'' = \pm \Gamma_0^{-2}(\Gamma_0 \Gamma_1' - \Gamma_0' \Gamma_1)$, and hence

$$f'' = \pm \Gamma_0^{-1} \Gamma_1 (\Gamma_2 - \Gamma_1). \tag{6.30}$$

6.1.1 Examples

We first apply the results of the above analysis to the equation $f'' + \lambda f = 0$, $\lambda \in \mathbb{R}$; c.f. [99]. Of course, the solutions of this equation are exceedingly well known, but it nonetheless illustrates the method quite well.

We start with Case (a)(i) and substitute (6.15) and (6.17) into the differential equation. This gives us the Γ–equation

$$d_1(d_1-1)\Gamma_0^{1-2/d_1}\Gamma_1^{2/d_1} + d_2 \Gamma_0^{1-2/d_1}\Gamma_1^{2/d_1-1/d_2}\Gamma_2^{1/d_2} + \lambda(\Gamma_0 + C) = 0. \tag{6.31}$$

If $d_1 > 0$, then $C = 0$ and the leading term is $\lambda \Gamma_0$. Hence $\lambda = 0$. If $d_1 \neq 1$, the leading term becomes $d_1(d_1-1)\Gamma_0^{1-2/d_1}\Gamma_1^{2/d_1}$, (since $\Gamma_1^{-1/d_2} \sim \log^{-1} x$). If $d_1 = 1$, the only remaining term is $d_2 \Gamma_0^{-1}\Gamma_1^{2-1/d_2}\Gamma_2^{1/d_2}$. In neither case can (6.31) be satisfied.

Similarly if $d_1 < 0$ the leading term is λC which must therefore be zero. The next largest term is $\lambda \Gamma_0$ and hence $\lambda = 0$. The leading term then becomes again $d_1(d_1-1)\Gamma_0^{1-2/d_1}\Gamma_1^{2/d_1}$ and as before the Γ–equation cannot be satisfied. Thus Case (a)(i) cannot occur.

In Case (a)(ii), the rank, r, of $\mathbf{R}\langle f \rangle$ is one less than the order of the differential equation. Hence we work with $\tilde{\Gamma}_0, \ldots, \tilde{\Gamma}_r$. Here $r = 1$ and $\tilde{\Gamma}_0 =$

$f-C=x^{d_1}(A+\varepsilon)$. Thus $\tilde{\Gamma}_1=\Gamma_0'=d_1x^{d_1-1}(A+\varepsilon)\sim d_1x^{-1}\Gamma_0$ and $\tilde{\Gamma}_2=\tilde{\Gamma}_1'\sim d_1(d_1-1)x^{-2}\Gamma_0$. On substituting into the differential equation, we obtain

$$\tilde{\Gamma}_2+\lambda(\tilde{\Gamma}_0+C)=0.$$

On using the fact that $\tilde{\Gamma}_2\sim d_1(d_1-1)x^{-2}\tilde{\Gamma}_0$, we see that $\lambda=0$ and hence $d_1(d_1-1)x^{-2}\tilde{\Gamma}_0=0$. This implies $d_1=1$. So we obtain a possible asymptotic form for f when $\lambda=0$, namely $f=x(A+\varepsilon)$.

Turning our attention now to (a)(iii), we obtain from (6.20) and (6.22), the Γ-equation

$$d_2^2\Gamma_0^2\Gamma_1^{-d_2}\Gamma_2-d_2\Gamma_0^2\Gamma_1^{-1-d_2}\Gamma_2+\lambda(\Gamma_1^{-d_2}\Gamma_2+C)=0. \tag{6.32}$$

As before, whether C is zero or not, we must have $\lambda=0$. The leading term is then $-d_2\Gamma_0^2\Gamma_1^{-1-d_2}\Gamma_2$ whose coefficient, $-d_2$, cannot be asymptotically zero. So (6.32) is unsatisfiable and Case (I)(iii) cannot occur.

Now for Case (b). Substitution into $f''+\lambda f=0$ from (6.25) and (6.27), in the case when $d_1\neq 1$, gives the Γ-equation

$$d_1^2\Gamma_0\Gamma_1^2+d_1(d_1-1)\Gamma_0\Gamma_1^{1-\frac{1}{d_1-1}}\Gamma_2^{\frac{1}{d_1-1}}+\lambda(\Gamma_0+C)=0, \tag{6.33}$$

where we have absorbed the $\pm$ sign into C. Note first that Γ_1 tends to ∞ if $d_1>1$, and to 0 if $d_1<1$. Hence, for positive A, the leading term is $d_1^2\Gamma_0\Gamma_1^2$ if $d_1>1$, and either $\lambda\Gamma_0$ or (if $\lambda=0$) $d_1^2\Gamma_0\Gamma_1^2$, otherwise. If A is negative, we see at once that C must be zero and the leading terms are then as for positive A. In no case can the leading coefficient be zero and so (6.33) cannot be satisfied.

If $d_1=1$, we work with Γ_0 and its derivatives. We have $\tilde{\Gamma}_1=\tilde{\Gamma}_0'\sim A\Gamma_0$ and $\tilde{\Gamma}_2=f_0''\sim A^2\Gamma_0$. The Γ-equation is then

$$A^2\Gamma_0+\lambda(\Gamma_0+C)\sim 0.$$

If A is negative, we see that C must be zero. Hence, whatever the sign of C, we require $A^2+\lambda=0$. This cannot hold if $\lambda\geq 0$, but if $\lambda<0$ we obtain two possible asymptotic forms for f, namely $\exp(\pm x(\sqrt{-\lambda}+\varepsilon))$.

Turning next to Case (c) we obtain the Γ-equation

$$\Gamma_0\Gamma_1^2+\Gamma_0\Gamma_1\Gamma_2+\lambda\Gamma_0=0,$$

which is easily seen to be unsatisfiable.

Similarly for Case (d), (6.28) and (6.30) give

$$\Gamma_0^{-1}\Gamma_1\Gamma_2-\Gamma_0^{-1}\Gamma_1^2+\lambda(C+\Gamma_0^{-1}).$$

On examining the leading terms, we discover successively that $\lambda C=0$ and $\Gamma_0^{-1}\Gamma_1^2=0$. The latter is clearly impossible and thus neither Case (c) nor Case (d) can occur.

We have thus rediscovered the following facts:

(i) If $\lambda > 0$, $f'' + \lambda f$ has no non–zero solutions lying in a Hardy field.
(ii) If $\lambda = 0$, the possible asymptotic form for f is $x(A + \varepsilon)$.
(iii) If $\lambda < 0$, two possible asymptotic forms are $\exp(x(\sqrt{-\lambda}+\varepsilon))$ and $\exp(-x(\sqrt{-\lambda} + \varepsilon))$.

Next we take a brief look at how the techniques apply to the equation

$$y''y - y' + 1 = 0. \tag{6.34}$$

Starting with case (a)(i), we substitute (6.15), (6.16) and (6.17) into (6.34) to obtain

$$\{d_1(d_1 - 1)\Gamma_0^{1-2/d_1}\Gamma_1^{2/d_1} + d_2\Gamma_0^{1-2/d_1}\Gamma_1^{2/d_1-1/d_2}\Gamma_2^{1/d_2}\}\{\Gamma_0 + C\} \\ -d_1\Gamma_0^{1-1/d_1}\Gamma_1^{1/d_1} + 1 = 0.$$

If d_1 is positive, then of course we take $C = 0$. On inspecting the exponents of Γ_0, we obtain the equation $1 - 1/d_1 = 0$, and so $d_1 = 1$. Then the exponents of Γ_1 give $2 - 1/d_2 = 1$, and hence $d_2 = 1$. Finally we see that $\Gamma_2 \sim 1$, which implies that $A = 1$. Thus we have a possible solution asymptotic to $x \log x$. In fact this is an exact solution, as may easily be checked, but the algorithm in no way needs this information. Taking d_1 to be negative rapidly leads to a contradiction.

Turning next to case (a)(ii), we could use the substitutions given above, but it is just as easy to work directly with the expression $f(x) = C+x^{d_1}(A+\epsilon)$. We get

$$d_1(d_1-1)Cx^{d_1-2}(A+\epsilon)+d_1(d_1-1)x^{2d_1-2}(A+\epsilon)^2 - d_1x^{d_1-1}(A+\epsilon)+1 = 0.$$

Comparing powers of x leads first to $d_1 = 1$, and then to $A = 1$. Thus another possible asymptotic solution is $x(1 + o(1))$. Again $y = x$ is in fact an exact solution.

The substitutions for case (a)(iii) give

$$\{d_2^2\Gamma_0^2\Gamma_1^{-d_2}\Gamma_2 - d_2\Gamma_0^2\Gamma_1^{-1-d_2}\Gamma_2\}\{C + \Gamma_2\Gamma_1^{-d_2}\} - d_2\Gamma_0\Gamma_1^{-d_2}\Gamma_2 + 1 = 0.$$

Comparing powers of Γ_0 gives a contradiction.

Similarly cases (b), (c) and (d) are easily eliminated, and so the possible asymptotic solutions are $x \log x(1 + o(1))$ and $x(1 + o(1))$. A slightly different example is similarly treated in [99].

We finish this section by looking at a second–order equation over $\mathbb{R}(x)$, namely

$$x^2g'' - x^3(g')^2 - 2x^2gg' + (3x + 1)g' - xg^2 + g = 0. \tag{6.35}$$

Rather than considering all the possibilities, we look for solutions of the form $g(x) = Ax^d \log_2^{1/c} x$. Even this one case is sufficiently complicated to make the use of an algebra system highly desirable. MAPLE was therefore used for the more intricate parts of the calculation.

We start by substituting $h = A^{-1}x^{-d}g$, so that $h \sim \log_2^{1/c} x$. We obtain $g' = Adx^{d-1}h + Ax^d h'$ and

$$g'' = Ad(d-1)x^{d-2}h + 2Adx^{d-1}h' + Ax^d h''.$$

Substitution in (6.35) yields

$$\begin{aligned} &x^d Ad^2 h + 2\,x^d Adh + 2\,x^{1+d} Adh' + x^{2+d} Ah'' - x^{1+2\,d} A^2 d^2 h^2 \\ &-2\,x^{2+2\,d} A^2 dhh' - x^{3+2\,d} A^2 h'^2 - 2\,x^{1+2\,d} A^2 h^2 d - 2\,x^{2+2\,d} A^2 hh' \\ &+3\,Ax^{1+d} h' + Ax^{d-1} dh + Ax^d h' - x^{1+2\,d} A^2 h^2 + Ax^d h = 0. \end{aligned} \tag{6.36}$$

We take $\Gamma_2 = h$. Then $h^{\triangle} \sim c^{-1}x^{-1}\log^{-1} x \log_2^{-1} x$, and so we set

$$\Gamma_1 = xh^{\triangle}h^c, \tag{6.37}$$

yielding $\Gamma_1 \sim c^{-1}\log^{-1} x$. Finally we put

$$\Gamma_0 = -c^{-1}\Gamma_1'\Gamma_1^{-2}, \tag{6.38}$$

which gives $\Gamma_0 \sim x^{-1}$. So at this stage we obtain a comparability class that is already in the coefficient field, allowing the possibility of a solution.

Next we express the derivatives of h in terms of the Γ. Of course we allow x to appear in coefficients. A rearrangement of (6.37) yields $h' = x^{-1}\Gamma_1\Gamma_2^{1-c}$, and similarly from (6.38) we get

$$h'' = -x^{-2}\Gamma_1\Gamma_2^{1-c} - cx^{-1}\Gamma_0\Gamma_1^2\Gamma_2^{1-c} + (1-c)x^{-2}\Gamma_1^2\Gamma_2^{1-2c}.$$

Substitution of these and $h = \Gamma_2$ into (6.36), followed by simplification, gives us

$$\begin{aligned} x^d &\left\{\Gamma_2 A\left(d^2 + 2d + 1 + \frac{d}{x}\right) + \Gamma_1\Gamma_2^{1-c}A\left(2d + 2 + \frac{1}{x}\right)\right. \\ &\left.+\Gamma_1^2 A(\Gamma_2^{1-2c}(1-c) - \Gamma_2^{1-c}\Gamma_0 xc)\right\} - x^{2d+1}\{\Gamma_2^2(d^2 + 2d + 1) \\ &+\Gamma_1\Gamma_2^{2-c}A^2(2d+2) + \Gamma_1^2\Gamma_2^{2-2c}A^2\} = 0. \end{aligned} \tag{6.39}$$

If $d > -1$, then because x and Γ_0 are of a larger comparability class than Γ_1 and Γ_2, the terms in x^{2d+1} must dominate. Hence the coefficient of x^{2d+1}, namely $\Gamma_2^2(d^2 + 2d + 1) + \Gamma_1\Gamma_2^{2-c}A^2(2d+2) + \Gamma_2^{2-2c}A^2$ must cancel asymptotically. But Γ_1 is the largest comparability class of this coefficient, and so $\Gamma_2^2(d^2 + 2d + 1) + \Gamma_2^{2-2c}A^2$ must similarly cancel. Since $c \neq 0$ and $A \neq 0$, $\Gamma_2^2(d^2 + 2d + 1)$ must be the dominant term here. This cannot exhibit cancellation, and so this case is eliminated.

Similarly if $d < -1$, the coefficient of x^d must cancel, and hence the coefficient of $x^d\Gamma_1^0$ also. Again we get a contradiction, and so $d = -1$.

Substituting $d = -1$ into (6.39) yields the equation

$$x^{-1}\Gamma_1^2 A\{\Gamma_2^{1-2c}(1-c) - \Gamma_0 x \Gamma_2^{1-c} c - \Gamma_2^{2-2c} A\} + x^{-2} A\{\Gamma_1 \Gamma_2^{1-c} - \Gamma_2\} = 0. \tag{6.40}$$

$x^{-1}\Gamma_1^2$ dominates the left–hand side of (6.40) unless its coefficient shows cancellation. The cases $c < 1$ and $c > 1$ are eliminated easily. Hence $c = 1$ and (6.40) becomes

$$-x^{-1}\Gamma_1^2 A\{\Gamma_0 x + A\} + x^{-2} A\{\Gamma_1 - \Gamma_2\} = 0. \tag{6.41}$$

This forces $A = -1$ and so our asymptotic form is $g \sim -x^{-1}\log_2 x$.

6.2 The Number of Cases

The techniques described in the previous section give a method for computing, term by term, all the possible nested forms of Hardy–field solutions of a given algebraic differential equation. If we want to obtain the subsequent forms in the nested expansions one possible line of approach might be to regard the unknown part of the expansion as a new function and substitute into the differential equation. So suppose we have the nested expansion

$$f = N(x, \phi(x))$$

for a possible solution of (6.2). We can substitute into (6.2), but ϕ may occur inside nested exponentials in N. In that case it will be necessary to differentiate further and eliminate in order to obtain a differential equation for ϕ. Moreover the various $\log_i x$ in N will still appear in the coefficients and further differentiation and elimination will be required if we want an equation over $\mathbb{R}$. To illustrate the second point consider again the equation (6.34). As we have seen, a possible nested form is $y = x \log x + \phi$. Differentiation yields $y' = \log x + 1 + \phi'$ and $y'' = x^{-1} + \phi''$. On substituting into (6.34) and simplifying, we get

$$\phi''\phi + x \log x \phi'' + x^{-1}\phi - \phi' = 0. \tag{6.42}$$

If we wanted an equation over $\mathbb{R}$, we would have to differentiate twice more and eliminate x and $\log x$ between the three equations. The resulting equation is fourth–order and quite large, as calculation with an algebra system will show. For this reason, it is better to work with (6.42) itself.

It is not hard to see that in general we shall have to differentiate once to remove each exponential from the nested form. So if the parameters of the nested form are $\{(\epsilon_i, s_i, m_i, d_i), i = 1, \ldots, k\}$ (there is no need to make the ϕ_is explicit here), a total of $\sum_1^k s_i$ differentiations will be needed. The resulting differential equation will have coefficients in the field $\mathbb{R}[x, \log x, \ldots, \log_{m_k}]$, so to get an equation over $\mathbb{R}$, a further $m_k + 1$ differentiations will be needed. Since our bound on the parameters of the nested form is $m_k + \sum_1^k s_i + \delta \leq n$, we can expect that the order of the new equation may be as large as $2n$. It is therefore rather important to get some measure of the growth of the method of the previous section as a function of the order, n, of the differential equation. One way of doing this is to calculate the number of nested forms which need to be considered, and that is what we do in this section.

The first thing we have to decide is when two cases are to be treated as different. Clearly the exponents, d_i, have to remain as parameters, but, for example, should $\log^{d_1} x(A+o(1))$ be counted a case in its own right, or as a special case of $x^{d_0}\log^{d_1} x$? There is no unique right answer to this, but the examples above suggest treating it as separate. That is what was done in [105], on which this section is based, and we make the same choice here. In any event the order of magnitude of the growth is not greatly affected one way or the other.

Let $N(r)$ denote the number of distinct nested forms possible for a Rosenlicht field of rank r. The first three values are $N(1) = 3$, $N(2) = 7$ and $N(3) = 19$. The corresponding forms are:

$$\begin{aligned}
r=1\ & x^{d_1}(A+\mathrm{o}(1)), \quad e^{\pm x(A+\mathrm{o}(1))},\\
r=2\ & x^{d_1}(A+\mathrm{o}(1)), \quad \log^{d_1} x(A+\mathrm{o}(1)), \quad x^{d_1}\log^{d_2} x(A+\mathrm{o}(1)),\\
& \exp\{\pm x^{d_1}(A+\mathrm{o}(1))\}, \quad \exp\{\pm\exp\{x(A+\mathrm{o}(1))\}\},\\
r=3\ & x^{d_1}(A+\mathrm{o}(1)), \quad \log^{d_1} x(A+\mathrm{o}(1)), \quad \log_2^{d_1} x(A+\mathrm{o}(1)),\\
& x^{d_1}\log^{d_2} x(A+\mathrm{o}(1)), \quad x^{d_1}\log_2^{d_2} x(A+\mathrm{o}(1)),\\
& \log^{d_1} x\log_2^{d_2} x(A+\mathrm{o}(1)), \quad x^{d_1}\log^{d_2} x\log_2^{d_3} x(A+\mathrm{o}(1)),\\
& e^{\pm x^{d_1}(A+\mathrm{o}(1))}, \quad e^{\pm x^{d_1}\log^{d_2} x(A+\mathrm{o}(1))},\\
& e^{\pm\log^{d_1} x(A+\mathrm{o}(1))}, \quad e^{\pm e^{x^{d_1}(A+\mathrm{o}(1))}},\\
& e_3^{\pm 1}(x(A+\mathrm{o}(1))), \quad x^{d_1}e^{\pm\log^{d_2} x(A+\mathrm{o}(1))}.
\end{aligned}$$

To obtain a general expression for $N(r)$, we use a classical technique from combinatorics. We encode the constraints on the parameters s_i and m_i in the language of regular expressions . We then obtain a regular expression which completely describes the possible choices for the s_i and m_i. The next step is to compute the generating function of this regular expression. $N(r)$ can then be obtained from the coefficients of the power–series expansion of the generating function. Here we only give the details required for our specific problem. For more about generating functions see [33] or [93].

We begin by defining $q_i = m_i - m_{i-1} - s_i$ for $i \geq 2$. It is easier to work with q_i than m_i for technical reasons. Note that once s_i is fixed (and of course m_{i-1} also), there are as many choices for q_i as for m_i. We can now give the conditions for a set of parameters to represent a valid nested form as

$$i \geq 2 \ \&\ s_i = 0 \quad \Rightarrow \quad q_i \geq 1, \tag{6.43}$$

$$i \geq 2 \ \& \ s_i \neq 0 \quad \Rightarrow \quad q_i \geq 0, \tag{6.44}$$

$$m_1 + s_1 + \sum_{i=2}^{k}(q_i + 2s_i) + \delta \ \leq \ r. \tag{6.45}$$

Note that q_i is non–negative because of (4.14).

We require an alphabet to describe a nested form $\{(\epsilon_i, s_i, m_i, d_i), i = 1, \ldots, k\}$. The letter '$z$' will be used to denote the start of a new quadruple. Then if $s_i \neq 0$, the letters P for 'plus' and M for 'minus' will be used to encode the value of $\epsilon_i = \pm 1$. If $s_i = k$, with $i > 1$, then we concatenate the letter 's' a total of k times. Similarly, $q_i = k$ will be encoded by the k-fold concatenation of 'q'. The case $i = 1$ requires special consideration, as witnessed by condition (6.45). We therefore use separate letters 's_1' and 'm_1' to denote increments of s_1 and m_1. Finally we use the letter 'δ' every time we have to account for $\delta = 1$ instead of 0 in (6.45). Our alphabet is thus $\{z, P, M, q, s, s_1, m_1, \delta\}$. We can use it to associate a unique word with each nested form (where the d_i and ϕ_k are unspecified parameters). For example, the nested form

$$\exp_2\{x^\alpha \log^\beta x \exp^{-1}(\log_4^\gamma x)\} \tag{6.46}$$

is given by parameter sequence $\{(1, 2, 0, \alpha), (1, 0, 1, \beta), (-1, 1, 4, \gamma)\}$ and represented by the word

$$zPs_1s_1zPqzMsqq\delta. \tag{6.47}$$

Next we want to encode the constraints (6.43)-(6.45) as a regular expression. Only regular expressions will be permitted as words in our language. We introduce a slight variation on the standard notation for regular expressions. We write $\emptyset$ for the empty word. $|$ denotes an alternative and $^+$ denotes repetition an arbitrary, but strictly positive, number of times. The notation * indicates unrestricted repetition; so that $m_1^\star = \emptyset | m_1^+$, for example. Various sorts of brackets {, }, (,), [,], are used as separators in the normal way.

To simplify the expressions, we put $s_1^\ddagger := \emptyset | (P|M)s_1^+$. Then for $k = 1$ and $m_1 = 0$ we have the description

$$z[\emptyset | (P|M)s_1^+](\emptyset | \delta) = zs_1^\ddagger(\emptyset | \delta).$$

This translates as follows: Start a new word; either do nothing, or else put a sign (+ or −) and exponentiate a positive number of times; either stop or put $d_1 = 1$ (the significance of δ here). Note that the case when $s_1 = 0$ is different from the others, because there we can take $\epsilon_1 = 1$ and regard any negative sign as being subsumed into the value of d_1. If $k = 1$ and $m_1 \neq 0$ then $\delta = 1$, and so this case is described by $zs_1^\ddagger m_1^+ \delta$. Thus for $k = 1$ and arbitrary m_1, we have the description

$$[zs_1^\ddagger(\emptyset | \delta)] \mid [zs_1^\ddagger m_1^+ \delta] = zs_1^\ddagger[\emptyset | (m_1^\star \delta)],$$

where the two main alternatives on the left correspond to the cases $m_1 = 0$ and $m_1 \neq 0$ respectively.

We now move on to the case $k > 1$. Then the possible first steps are encoded by $zs_1^\ddagger m_1^\star$, as before. Those that follow are given by $z\{q^+|(P|M)s^+q^\star\}$, which translates conditions (6.43) and (6.44). The possible choices of s_i and m_i for an arbitrary value of r are given by the regular expression

$$zs_1^\ddagger\left(\emptyset|m_1^\star\left(z\{q^+|(P|M)s^+q^\star\}\right)^\star\delta\right). \tag{6.48}$$

Thus we have a one–to–one correspondence between nested forms and the words belonging to the language defined by (6.48).

Now for the generating function of our language. If $c_{k,r}$ denotes the number of nested forms of length k obeying conditions (6.43)-(6.45), the generating function of $c_{k,r}$ is defined to be

$$\sum_{k,r\geq 0} c_{k,r}z^k u^r, \tag{6.49}$$

We now compute a closed form for $\sum c_{k,r}z^k u^r$. This is done by first computing the generating function of $g_{k,r}$ the number of words such that (6.45) holds with an equality. For instance, the word (6.47) will be enumerated by z^3u^8 because the length of the nested form is 3, and by (6.45), the least order an of equation having a solution of asymptotic growth (6.46) is 8.

The generating function we are seeking is the sum of the contributions of all the words of the language (6.48). We obtain this by the general method, as given in Eilenberg (1974) pp. 196–198 for example. When two sublanguages $\mathcal{A}$ and $\mathcal{B}$ with empty intersection have generating functions $A(z,u)$ and $B(z,u)$, then it is a fact that the alternative $\mathcal{A}|\mathcal{B}$ has generating function $A(z,u)+B(z,u)$. Similarly the unambiguous concatenation $\mathcal{AB}$ has generating function $A(z,u)B(z,u)$. Our task is therefore reduced to computing the generating functions of the letters of the language. Because they all increase r by 1 in (6.45), s_1, q, m_1 and δ have generating function u, while the generating function of s is u^2 because s_i occurs with coefficient 2 on the left of (6.45). Obviously the generating function of z is z and that of $\emptyset$, P and M is 1. Combining the rules on alternative and concatenation yields classical rules for repetition:

$$\begin{aligned}
\mathcal{A}^\star &\mapsto \sum_{j=0}^{\infty} A^j(z,u) = \frac{1}{1-A(z,u)}\\
\mathcal{A}^+ &\mapsto \frac{1}{1-A(z,u)} - 1 = \frac{A(z,u)}{1-A(z,u)}\\
\mathcal{A}^\ddagger &\mapsto 1+(1+1)\mathcal{A}^+ = 1+\frac{2A(z,u)}{1-A(z,u)}.
\end{aligned}$$

Note that the $(1+1)$ appearing on the last line is to account for the $P|M$ in the expression for $\mathcal{A}^\ddagger$. On using these rules, we see that firstly the generating function for the language defined by the expression $z\{q^+|(P|M)s^+q^\star\}$ is

$$P(z,u) = z\left(\frac{u}{1-u} + 2\frac{u^2}{(1-u^2)(1-u)}\right).$$

Hence that for (6.48) is

$$z\left(1+2\frac{u}{1-u}\right)\left(1+\frac{u}{(1-u)(1-P(z,u))}\right)$$

$$= z\left(1+2\frac{u}{1-u}\right)\left(1+\frac{u}{(1-u)\left(1-z\left(\frac{u}{1-u}+2\frac{u^2}{(1-u^2)(1-u)}\right)\right)}\right).$$

So far we have only considered the case when we have equality in (6.45). To account for the $\leq$ sign there we have to sum all these contributions, except that the cases when $m_k = 0$ and $d_k = 1$ should only be counted when $s_1 = r$. This is because for $s_1 < r$ such a case will be included without the restriction on d_k. The regular expression for the case $s_1 < r$ is therefore obtained by suppressing the $\emptyset$ in (6.48). The corresponding generating function is therefore reduced to

$$z\left(1+2\frac{u}{1-u}\right)\frac{u}{(1-u)\left(1-z\left(\frac{u}{1-u}+2\frac{u^2}{(1-u^2)(1-u)}\right)\right)}.$$

Then summation of the contributions over r is obtained by dividing the new generating function by $1-u$. To see this note that

$$\frac{\sum c_{k,r}z^k u^r}{1-u} = \sum c_{k,r}z^k u^r \sum u^n = \sum_{k,N}\left(z^k u^N \sum_{n+r=N} c_{k,r}\right).$$

Finally, the extra case is recovered by adding back the generating function of $zs_1^\dagger$, namely $z[1+2u/(1-u)]$. After simplification we get the function

$$C(z,u) = \frac{z(1+u)(1-zu-2zu^2+u^3+zu^3)}{(1-u-u^2+u^3-zu-2zu^2+zu^3)(1-u)}. \tag{6.50}$$

We have therefore proved the following result.

THEOREM **19** *The number of nested forms of length k obeying (6.3) is the coefficient of $z^k u^r$ in the Taylor expansion at the origin of the function $C(z,u)$.*

We could actually have carried out the above analysis without using the variable z, but it allows a simple check to be made by computing some of the coefficients of $C(z,u)$. Thus if we use the classical notation $[z^k]f(z)$ to denote the kth Taylor coefficient of $f(z)$ at the origin, we obtain

$$[z^1]C(z,u) = 1 + 3u + 6u^2 + 11u^3 + 18u^4 + O(u^5),$$
$$[z^2]C(z,u) = u^2 + 7u^3 + 24u^4 + O(u^5),$$
$$[z^3]C(z,u) = u^3 + 10u^4 + O(u^5).$$

From these, we can obtain the first values of $N(r)$ by summing the coefficients of the various powers of u. Thus $N(1)$ is given by the coefficient of u, which is 3. Similarly $N(2) = 6 + 1$, which is the sum of the coefficients of u^2, and $N(3) = 11 + 7 + 1$. These values agree with those obtain above by listing the nested forms. Note that the use of z allows us to get the lengths of the nested forms from the coefficients of the monomials. Thus for $r = 3$, we see that we have eleven forms of length 1, seven of length 2 and one of length 3, which again can be checked against the list.

Now we are finally in a position to estimate $N(r)$. The result from [105] is the following.

THEOREM **20** *The number of possible positive nested forms of solutions of algebraic differential equations of order r over* $\mathbb{R}$ *which tend to zero or infinity is*

$$N(r) = \frac{99 + 109\lambda - 34\lambda^2}{158}\lambda^{-r} + \frac{99 + 109\mu - 34\mu^2}{158}\mu^{-r} + \frac{99 + 109\nu - 34\nu^2}{158}\nu^{-r},$$

where λ, μ *and* ν *are the roots of* $2x^3 - 3x^2 - 2x + 1$, *and* $\lambda \approx 0.355$, $\mu \approx -0.745$, $\nu \approx 1.889$.

A numerical version of this result is

$$N(r) = 0.845\, e^{1.034\, r} + O(e^{0.295\, r}), \quad r \to \infty.$$

Proof of Theorem 20
Setting $z = 1$ in $C(z, u)$ corresponds to summing over all possible values of k, as we can see by inspecting (6.49). Thus we get the generating function of $N(r)$ as

$$C(u) = \sum_{r \geq 0} N(r) u^r = \frac{(1 + u)(1 - 2u^2)}{1 - 2u - 3u^2 + 2u^3}. \tag{6.51}$$

Since the denominator is irreducible, the partial fraction decomposition of C is obtained by reducing $(1+u)(1-2u^2)/(6u^2-6u-2)$ modulo $1-2u-3u^2+2u^3$. This technique is well known, and can be seen to give the correct answer as follows. Suppose that $P/Q \in \mathbb{C}(u)$ and has the partial fraction decomposition

$$\frac{P(u)}{Q(u)} = \frac{p_1}{u - \alpha_1} + \frac{p_2}{u - \alpha_2} + \cdots + \frac{p_k}{u - \alpha_k}.$$

Then for $1 \leq i \leq k$,

$$p_i = \lim_{u \to \alpha_i} \frac{P(u)(u - \alpha_i)}{Q(u)} = \frac{P(\alpha_i)}{Q'(\alpha_i)}.$$

In the present case, this gives

$$C(u) = \sum_{1 - 2\alpha - 3\alpha^2 + 2\alpha^3 = 0} \frac{99 + 109\alpha - 34\alpha^2}{158(1 - \frac{u}{\alpha})}. \tag{6.52}$$

Let λ, μ and ν be the three roots of $2x^3 - 3x^2 - 2x + 1$. Then on extracting the coefficients from (6.52), we get Theorem 20.

6.3 Reducing the Complexity

Theorem 20 has important consequences for the method of section 6.1. We have already seen that if f belongs to a Hardy field and satisfies an algebraic differential equation of order n over $\mathbb{R}$, then the second term of the nested expansion will satisfy a similar equation of order at most $2n$. We can use the method of section 6.1 to obtain the possible nested forms for the second term, and then repeat the process to obtain the third term, and so on. Unfortunately the order of the differential equation might double at each stage. So to compute the rth term we might need to consider a differential equation of order $n2^{r-1}$. Alas the bound of Theorem 20 then allows the possibility of more than $\exp(n2^{r-1})$ cases.

If the situation is this bad then the method can only be used for the first two or three terms at best, but can this combinatorial explosion really happen? It is clear that for first term, all the possibilities given by the theorem can occur. However the bounds on the subsequent terms are another matter. The rth term cannot just be any Hardy–field solution of the differential equation of order $n2^{r-1}$ obtained. The terms have to fit together to give a solution of the original equation! So the upper bound of $\exp(1.035n2^{r-1})$ may be a vast over–estimate for $r > 1$. Support for such a point of view can be drawn from the fact that an explosion of complexity in the terms has not been found classically, although probably nobody looked for it! Another point is that the search for the rth term can be bounded from the parameters of the $r-1$ terms *already obtained*, instead of what the above analysis tells us they might, at worst, have been. So if the $r-1$ terms generate a Rosenlicht field of rank R, the rth term will lie in a Rosenlicht field of rank at most $2R$.

The most important point however is that there are theorems which greatly limit the number of cases that need to be considered for the later terms in the nested expansion. Firstly, it is proved in [102] that later terms cannot be of larger comparability class than would be allowed for the first term. Secondly a theorem of Strodt and Wright, [110] gives a bound on the value of m for which $\log_m x$ can appear in the expansion of a solution of a first–order equation; this bound depends only on the degree. The theorem assumes that there are no exponentials present in the expansion.

There are computational problems with the method of section 6.1 apart from the possible number of cases to consider. The process of substituting a finite nested expansion into the original differential equation and differentiating out the existing terms will undoubtedly produce equations which are much more complicated as well as being of higher order. It is therefore worth looking to see what can be done to alleviate this.

Some saving is made by not differentiating out all the logarithms. Here we rely on the fact that we only need to obtain a differential equation for the new term with coefficients in some Hardy field generated by the previous terms in order for Theorem 18 to apply. So it is only the exponentials enclosing the new term which have to be differentiated out. It is then worth taking terms out of exponentials when this can be done. For example if we are considering $\exp_2(x(1+\phi))$, this might be better written as $\exp\{\exp x \cdot \psi\}$ with $\psi = \exp(x\phi)$ as the new unknown. In other words a multiseries or a star product approach might be better here. Moreover for cases where $x\phi \to 0$, we can do better still by letting $\nu = \exp(\exp x(\exp(x\phi) - 1))$. Then $\exp_2(x(1+\phi)) = \exp_2 x \cdot \nu$.

We next look at another way in which the complexity can sometimes be reduced.

6.3.1 A Theorem for Sparse Differential Equations

Suppose that we are given an algebraic differential equation over $\mathcal{F}$ of order n whose coefficients are $c_1, \dots, c_J$, and let f be a solution which lies in a Hardy field. We have seen with regard to iterated logarithms that we can reduce the number of cases to be considered by taking account of the fact that it is $\mathcal{F}\langle f \rangle$, rather than $\mathbf{R}\langle f \rangle$, that has to be of rank at most $n + \operatorname{rank}(\mathcal{F})$. Thus if $\Gamma_0, \dots, \Gamma_\Sigma$ are as in Theorem 18, then there must be a rational function of the c_is and their derivatives which is of the same comparability class as some Γ_j, $j \le n$. In fact we can do better than this.

THEOREM **21** *Let $c_1, \dots, c_J$, $\Gamma_0, \dots, \Gamma_\Sigma$ and f be as above. Then there exist m, with $0 \le m \le n$, and real numbers $\alpha_1, \dots, \alpha_J, \beta_1, \dots, \beta_J$, such that either $\sum_{j=1}^J \beta_j c_j \ne 0$ and*

$$\gamma_1(\Gamma_m) = \gamma_1\left(\sum_{j=1}^J \alpha_j c_j \Big/ \sum_{j=1}^J \beta_j c_j\right)$$

or else

$$\gamma_1(\Gamma_m) = \gamma_1\left(\sum_{j=1}^J \alpha_j c_j\right).$$

Proof of Theorem 21
From Theorem 18 (in particular 2.), we can write the differential equation satisfied by f in the form

$$\sum_{i=0}^N d_i \Gamma_0^{r_{i,0}} \Gamma_1^{r_{i,1}} \dots \Gamma_\mu^{r_{i,\mu}} = 0, \tag{6.53}$$

where $N \in \mathbf{N}$, $r_{i,j} \in \mathbf{R}$ $(i = 0, \dots, N,\ j = 0, \dots, \mu)$ and each d_i is a linear combination, with real coefficients, of the elements $c_1, \dots, c_J$. We may suppose without loss of generality that each $r_{i,k}$ is non-negative and that no coefficient d_i tends to zero.

Suppose that no $\gamma_1(d_i)$ is equal to any $\gamma_1(\Gamma_j)$, since otherwise we have the conclusion of the theorem. Consider a sum of the form

$$\sum_{j=0}^{M} d_{i_j} \Gamma_{l_{j,1}}^{r_{j,1}} \Gamma_{l_{j,2}}^{r_{j,2}} \cdots \Gamma_{l_{j,p_j}}^{r_{j,p_j}},$$

with $r_{j,1}, \ldots, r_{j,p_j}$ positive and $0 < l_{j,1} < l_{j.2} < \cdots < l_{j,p_j}$. We may assume that no two power products in the sum are identical. We call such a sum *cancellative* if there is an infinitely-increasing element, X, of a Hardy field and a positive constant, δ, such that

$$\left| \sum_{j=0}^{M} d_{i_j} \Gamma_{l_{j,1}}^{r_{j,1}} \Gamma_{l_{j,2}}^{r_{j,2}} \cdots \Gamma_{l_{j,p_j}}^{r_{j,p_j}} \right| < X^{-\delta},$$

with $\gamma_1(X) > \max\{\gamma_1(d_{i_j}),\ \gamma_1(\Gamma_{l_{j,1}})\}$, for all $j = 0, \ldots, M$. We define the *length* of the sum to be $\sum_{j=0}^{M}(p_j + 1)$. Because (6.53) holds, there exists a cancellative sum. Let

$$\sum_{j=0}^{M} d_{i_j} \Gamma_{l_{j,1}}^{r_{j,1}} \Gamma_{l_{j,2}}^{r_{j,2}} \cdots \Gamma_{l_{j,p_j}}^{r_{j,p_j}}, \tag{6.54}$$

be the cancellative sum of shortest length. Then no $\gamma_1(\Gamma_{l_{j,1}})$ is greater than all the $\gamma_1(d_{i_j})$. For otherwise, we may suppose $\gamma_1(\Gamma_{l_{0,1}})$ to be maximal. Let the highest power of $\Gamma_{l_{0,1}}$ occurring in the sum (6.54) be the r-th. Then (6.54) may be rewritten in the form

$$\Gamma_{l_{0,1}}^{r} \left(\sum_{j=0}^{M_1} d_{k_j} \Gamma_{m_{j,1}}^{s_{j,1}} \cdots \Gamma_{m_{j,q_j}}^{s_{j,q_j}} + S \right),$$

where the k_j are the indices of the terms of (6.54) which contain $\Gamma_{l_{0,1}}^{r}$ and S is formed from the remaining terms. It follows that there is a positive constant η, such that $|S| < \Gamma_{l_{0,1}}^{-\eta}$. Hence

$$\begin{aligned} \left| \sum_{j=0}^{M_1} d_{k_j} \Gamma_{m_{j,1}}^{s_{j,1}} \cdots \Gamma_{m_{j,q_j}}^{s_{j,q_j}} \right| &< |S| + \Gamma_{l_{0,1}}^{-r} X^{-\delta} \\ &< \Gamma_{l_{0,1}}^{-\eta/2}, \end{aligned}$$

and a cancellative sum of shorter length is obtained.

In view of our assumption that no $\gamma_1(d_i)$ is equal to any $\gamma_1(\Gamma_j)$, it is then clear that some $\gamma_1(d_{i_j})$ is greater than all the comparability classes of the $\Gamma_{l_{j,k}}$ in (6.54). We may suppose that d_{i_0} is the largest coefficient, and divide through by it, to obtain a cancellative sum of the form

$$\sum_{j=0}^{M} \left(\frac{d_{i_j}}{d_{i_0}} \right) \Gamma_{l_{j,1}}^{r_{j,1}} \Gamma_{l_{j,2}}^{r_{j,2}} \cdots \Gamma_{l_{j,p_j}}^{r_{j,p_j}}. \tag{6.55}$$

We may assume that (6.55) is the cancellative sum of that form with shortest length. As before, if some $\gamma_1(\Gamma_{l_{j,k}})$ is greater that all the $\gamma_1(d_{i_j}/d_{i_0})$, then a shorter cancellative sum would be obtained. On the other hand suppose some $\gamma_1(d_{i_j}/d_{i_0})$ were greater than the comparability classes of all the $\Gamma_{l_{j,k}}$. Any coefficients in (6.55) having that largest comparability class must tend to zero, since d_{i_0} was the largest coefficient in (6.54). If we separate off the terms containing a coefficient of this comparability class then the sum of the remaining terms, which certainly includes the first, must be cancellative. So again we obtain a cancellative sum of smaller length. The only possibility therefore is that some $\gamma_1(d_{i_j}/d_{i_0})$ is equal to some $\gamma_1(\Gamma_{l_{j,k}})$, and thus we have proved Theorem 21.

Theorem 21 is particularly suited to sparse differential equations and these often arise in applications. Unfortunately there seems to be no reason to believe that the equations which arise when nested forms are substituted back into the equation to obtain further terms will generally be sparse even if the original equation was.

To show how Theorem 21 can be used, consider a function $y(x)$ which satisfies an equation

$$(y')^2 + \log_2 x\, y' y^2 + y^3 + \log_2^2 x \log_n x = 0,$$

with $n > 2$. The Hardy field generated by $\mathbf{R}\langle \log_2 x, \log_2^2 x \log_n x\rangle$ has rank $n+1$ with comparability classes $\gamma_1(\log_n x),\ldots,\gamma_1(\log x)$ and $\gamma_1(x)$ (in addition to $\gamma_1(1)$). Hence possible asymptotic forms to be considered for the case when $y \to \infty$ are

$$y = \exp\{x^a \log^b x \ldots \log_n^p x(A+\varepsilon)\}$$

and

$$y = x^a \log^b x \ldots \log_n^p x \log_{n+1}^q x(A+\varepsilon).$$

In fact the only ones of these allowed by Theorem 21 are

$$y = \exp\{x \log_2^c x(A+\varepsilon)\} \quad \text{and} \quad y = x^a \log_2^c x(A+\varepsilon).$$

In all other cases, y and y' fail to generate one of the comparability classes $\gamma_1(\log_2 x)$, $\gamma_1(1)$, $\gamma_1(\log_n x)$ in the manner prescribed by the theorem.

7

Inverse Functions

Inverse functions have traditionally been regarded as problematic in asymptotics. For example, until recently it was unknown whether the inverse of an exp–log function was necessarily asymptotic to an exp–log function. Thus in [40], Hardy said '*whether or not it is true that, given an L-function* ϕ *and its inverse* $\bar{\phi}$, *there must be an L-function* ψ *such that* $\bar{\phi} \sim \psi$ *I cannot say; ... I am very doubtful whether this is so*'. Similarly [30] treats a number of examples and remarks on the difficulty of some of them. In fact once one has an understanding of scales and the mechanisms of nested forms and multiseries, inverse function are reasonably tractable. In particular, Hardy's conjecture was established in [98] by showing that the inverse of the function $\log\log x \cdot \log\log\log x$ is not asymptotic to any L-function. Subsequently it was shown in [113], and independently in [114], that the same is true of the inverse of $\log x \cdot \log\log x$, which was suggested as a possible counter-example by Hardy in [40].

One of the first concerns is whether the inverse functions to be considered belong to a Hardy field. The following is from [90], although similar results were proved earlier; see for example [9, 10].

THEOREM **22** *Let* f *be an element of a Hardy field which tends to infinity. Then* f *has a unique inverse function, which also tends to infinity and belongs to a Hardy field.*

Proof of Theorem 22
f' cannot have arbitrarily large zeros, and so must ultimately be positive. The usual Inverse Function Theorem then implies that f has a unique C^∞ inverse function, g. We show that $\mathbb{R}\langle g\rangle$ is a Hardy field. We shall make use of the following.

LEMMA **27** *Let* g *be the functional inverse of* f. *For every* $n \in \mathbb{N}^+$, *there is a polynomial* $Q_n \in \mathbb{R}[X_1, \ldots, X_n]$ *such that*

$$g^{(n)} = (f' \circ g)^{-2n} Q_n(f' \circ g, f'' \circ g, \ldots, f^{(n)} \circ g). \tag{7.1}$$

The coefficients of Q_n are independent of f and g.

Proof of Lemma 27
For $n = 1$, we have the well known relation $g' = (f' \circ g)^{-1}$. Suppose then that (7.1) holds for some value of $n \geq 1$. Differentiation then gives

$$\begin{aligned} g^{(n+1)} &= \frac{-2nQ_n\, g'\, f'' \circ g}{(f' \circ g)^{2n+1}} + \sum_{i=1}^{n} \frac{\partial_i Q_n \cdot g'\, f^{(i+1)} \circ g}{(f' \circ g)^{2n}} \\ &= \frac{-2nQ_n\, f'' \circ g}{(f' \circ g)^{2n+2}} + \sum_{i=1}^{n} \frac{\partial_i Q_n \cdot f^{(i+1)} \circ g}{(f' \circ g)^{2n+1}} \\ &= (f' \circ g)^{-2(n+1)} Q_{n+1}(f' \circ g, \ldots, f^{(n+1)} \circ g), \end{aligned}$$

where $\partial_i Q_n$ denotes the partial derivative of Q_n with respect to the i–th argument. This completes the proof of Lemma 27.

Continuation of the Proof of Theorem 22
Let P be a polynomial in $\mathbb{R}[X_1, \ldots, X_{n+1}]$ and suppose that $P(g, g', \ldots, g^{(n)})$ has arbitrarily large zeros. We prove that P must be the zero polynomial, which will establish that $\mathbb{R}\langle g \rangle$ is a Hardy field. On substituting the relations (7.1), we obtain

$$P(g, g', \ldots, g^{(n)}) = (f' \circ g)^{-N_n} R(g, f' \circ g, \ldots, f^{(n)} \circ g), \qquad (7.2)$$

for some polynomial $R \in \mathbb{R}[X_1, \ldots, X_{n+1}]$ and some $N_n \in \mathbb{N}$. If we make the change of variable $y = g(x)$, we see that $R(y, f'(y), \ldots, f^{(n)}(y))$ has arbitrarily large zeros. However $R(y, f'(y), \ldots, f^{(n)}(y))$ is an element of the Hardy field $\mathbb{R}\langle y, f \rangle$ and so it must be zero. But then P must be the zero polynomial, and Theorem 22 has been proved.

Next we may observe that the rank of $\mathbb{R}\langle g \rangle$ will be no larger than one plus the rank of $\mathbb{R}\langle f \rangle$. This is because

$$\mathbb{R}\langle g \rangle = \mathbb{R}(g, f' \circ g, \ldots, f^{(n)} \circ g, \ldots),$$

from (7.2). Note that the rank of $\mathbb{R}\langle \exp x \rangle$ is one, whereas that of $\mathbb{R}\langle \log x \rangle$ is two; so the extra one is sometimes needed. Similar remarks apply to $\mathbb{R}\langle\langle f \rangle\rangle$ and $\mathbb{R}\langle\langle g \rangle\rangle$, and if the rank of the latter is finite, g will have a nested expansion as in Theorem 18.

7.1 Inverting a Nested Expansion

This section is derived from [90].

We suppose first that f is an element of a Hardy–field, $\mathcal{F}$, with f tending to infinity and having a nested expansion which we can compute. We seek the nested expansion of $g(x)$ defined by

$$f(g(x)) = x. \tag{7.3}$$

As a first step, we show how to compute the nested form of $y(x)$. Let the first pnf of f be

$$f(y) = \exp_{s_1}(\log_{m_1}^{d_1} y \cdot f_1(y)), \tag{7.4}$$

with $\gamma_1(f_1(y)) < \gamma_1(\log_{m_1} y)$. By inverting (7.4), we get

$$y = g(x) = \exp_{m_1}(\log_{s_1}^{1/d_1} x \cdot g_1(y)), \tag{7.5}$$

where $g_1(y) = f_1^{-1/d_1}(y)$. It is not hard to see that $\gamma_1(g_1(y(x))) < \gamma_1(\log_{s_1} x)$. Indeed

$$\begin{aligned} \log g_1(y(x)) &= \frac{-1}{d_1} \log f_1(y(x)) = o(\log_{m_1+1} y(x)) \\ &= o(d_1^{-1} \log_{s_1+1} x + \log g_1(y(x))). \end{aligned}$$

Thus we have the first pnf for y, and if g_1 tends to a non–zero constant this is also the nested form of y.

Otherwise, we can compute the first pnf of $g_1(y) = f_1^{-1/d_1}(y)$. We use the following.

LEMMA **28** *If $|\phi| \to \infty$ and there are positive constants m and M such that $m < \psi < M$, then for $n > 0$,*

$$\psi \exp_n \phi = \exp_n(\phi(1+\delta)),$$

where δ tends to zero and can be computed.

Proof of Lemma 28
For $n = 1$, we have

$$\psi \exp \phi = \exp(\phi + \log \psi) = \exp(\phi(1 + o(1))).$$

Suppose then that the relation holds when n is replaced by $n - 1$. Then

$$\psi \exp_n \phi = \exp((1 + o(1)) \exp_{n-1} \phi) = \exp_n(\phi(1 + o(1)))$$

as required for the conclusion of Lemma 28.

Now let the first pnf of f_1 be

$$f_1(y) = \exp_{s_2}^{\epsilon_2}\{\log_{m_2}^{d_2} y \cdot f_2(y)\}.$$

If $s_2 = 0$ this gives

$$g_1(y) = \log_{m_2}^{-\epsilon_2 d_2/d_1} y \cdot f_2^{-1/d_1}(y),$$

while for $s_2 = 1$ we get

$$g_1(y) = \exp^{\epsilon_2}\{\log_{m_2}^{d_2} y \cdot (-f_2(y)/d_1)\}.$$

For $s_2 \geq 2$,

$$\begin{aligned} g_1(y) &= \exp^{-\epsilon_2}\{\exp_{s_2-1}\{\log_{m_2}^{d_2} y \cdot f_2(y)\}/d_1\} \\ &= \exp_{s_2}^{-\epsilon_2}\{\log_{m_2}^{d_2} \cdot f_2(y)(1+o(1))\} \end{aligned}$$

by Lemma 28. Thus in all case we get a pnf of the form

$$g_1(y) = \exp_{s_2}^{\epsilon_2'}\{\log_{m_2}^{d_2'} y \cdot g_2(y)\}. \tag{7.6}$$

We note that ϵ_2', d_2' and g_2 can be computed.

THEOREM **23** *Let $\mathcal{F}$ be a Hardy field which contains the variable y and is closed under the application of exponentials, logarithms and the taking of real powers of positive elements. Suppose that f belongs to $\mathcal{F}$ and has a nested form*

$$f(y) = \exp_{s_1}\{\log_{m_1}^{d_1} y \exp_{s_2}^{\epsilon_2}[\log_{m_2}^{d_2} y \cdots \exp_{s_n}^{\epsilon_n}(\log_{m_n}^{d_n} y \cdot (c+\nu(y))) \cdots]\}, \tag{7.7}$$

where c is a non-zero constant and $\nu \to 0$. Then the inverse function of f, $y = g(x)$, has nested form

$$\begin{aligned} g(x) = \exp_{m_1}\{&\log_{s_1}^{1/d_1} x \exp_{s_2}^{-\epsilon_2}[\log_{m_2-m_1+s_1}^{d_2'} x \exp_{s_3}^{\epsilon_3}[\cdots \\ &\exp_{m_n}^{\epsilon_n}(\log_{m_n-m_1+s_1}^{d_n'} x(C+\eta(x,g(x))) \cdots]]\}, \end{aligned} \tag{7.8}$$

where C is a non-zero constant and $\eta(x, g(x)) \to 0$. Moreover $\eta(f(y), y) \in \mathcal{F}$, and we can compute η and $d_2', \ldots, d_n'$.

Note that because of Theorem 10 and its second corollary, we are not really making any restriction in supposing that $\mathcal{F}$ contains exponentials, logarithms and real powers.

Proof of Theorem 23

When $n = 1$, the result is given by (7.5). For the general case, we use induction on the length, n, of the nested form of f. Essentially we substitute (7.5) into (7.7). We require the following lemma.

LEMMA **29** *Suppose that $f(y)$ has a nested form (7.7) and $y = y(x)$ satisfies $y = \exp_m(\log_s^{1/d} x \cdot \lambda(y))$ with $\lambda \in \mathcal{F}$, $m_1 > m$ and $\gamma_1(\lambda(y(x))) < \gamma_1(\log_s x)$. Then $f(y(x))$ has a nested form*

$$\begin{aligned} f(y(x)) = \exp_{s_1}\{&\log_{m_1-m+s}^{d_1} x \cdot \exp_{s_2}^{\epsilon_2}[\log_{m_2-m+s}^{d_2} x \cdot \exp_{s_3}^{\epsilon_3}[\cdots \\ &\cdots \exp_{s_n}^{\epsilon_n}[\log_{m_n-m+s}^{d_n} x(d+\theta(x,y(x)))]]]\}, \end{aligned}$$

where d is a non-zero constant, $\theta(f(y), y) \in \mathcal{F}$ and $\theta(x, y(x)) \to 0$ as $x \to \infty$. Moreover θ can be computed in terms of the parameters of (7.7).

Proof of Lemma 29
The case $n = 1$ is immediately clear from (7.5) and the remarks that follow it. Suppose then that $n \geq 2$ and that Lemma 29 holds when n is replaced by $n-1$. Let $f_1(y) = \log_{s_1}(f(y))/\log_{m_1}^{d_1} y$, so that

$$f_1(y) = \exp_{s_2}^{\epsilon_2}[\log_{m_2}^{d_2} y \cdots \exp_{s_n}^{\epsilon_n}(\log_{m_n}^{d_n} y \cdot (c + \nu(y))) \cdots].$$

By the induction hypothesis,

$$f_1(y(x)) = \exp_{s_2}^{\epsilon_2}[\log_{m_2-m+s}^{d_2} x \cdot \exp_{s_3}^{\epsilon_3}[\cdots \exp_{s_n}^{\epsilon_n}[\log_{m_n-m+s}^{d_n} x(d + \theta(x, y(x)))]]],$$

with θ as required. But then

$$f(y(x)) = \exp_{s_1}\{\log_{m_1}^{d_1} y \cdot f_1(y(x))\}$$

and substituting the nested form for $f_1(y(x))$ gives the result.

Continuation of the proof of Theorem 23
We may invert (7.7) as we did (7.4), obtaining

$$g(x) = \exp_{m_1}\{\log_{s_1}^{1/d_1} x \cdot g_1(y)\}$$

where g_1 is given by (7.6). Then we apply Lemma 29 to $g_1(y)$ with $\lambda = g_2$, yielding

$$g(x) = \exp_{m_1}\{\log_{s_1}^{1/d_1} x \exp_{s_2}^{-\epsilon_2}[\log_{m_2-m_1+s_1}^{d_2'} x \exp_{s_3}^{\epsilon_3}[\cdots \exp_{m_n}^{\epsilon_n}(\log_{m_n-m_1+s_1}^{d_n'} x(C + \eta(x, g(x))) \cdots]]\}.$$

This is of the required form, and it is clear that we can compute η and the d_i'. The proof of Theorem 23 is therefore complete.

Of course we would like to be able to compute the full nested expansion of $y(x)$, and it is natural to try substituting the nested form for $y(x)$ into $\eta_1(x, y(x))$. For the most part this works out as before, but there is one difference. Lemma 29 might not be applicable because we can no longer guarantee the condition $m_1 > m$. The modification is as follows.

Suppose that

$$y(x) = \exp_{m_1}(\log_{s_1}^{1/d_1} x\, g_1(y)) \qquad \text{and} \qquad g(y) = \exp_s^{\epsilon}\{\log_m^d y\, \lambda(y)\},$$

where $\gamma_1(\lambda(y)) < \gamma_1(\log_m y)$. If $m_1 = m$, substitution yields

$$g(y(x)) = \exp_s^{\epsilon}\{\log_{s_1}^{d/d_1} x \cdot g_1^d(y(x)) \cdot \lambda(y(x))\}.$$

The only new problem is that this pnf may need reduction if $d = d_1$, $g_1^d \lambda$ tends to a non-zero finite limit and s and s_1 are positive. The reduction is carried out as in Section 4.3.

Now suppose that $m_1 < m$. When we substitute we get

$$\begin{aligned} g(y(x)) &= \exp_s^\epsilon\{\exp^d_{m-m_1}(\log_{s_1}^{1/d_1} xg_1(y(x)))\cdot\lambda(y(x))\} \\ &= \exp_{s+1}^\epsilon\{d\exp_{m-m_1-1}(\log_{s_1}^{1/d_1} xg_1(y(x))) + \log\lambda(y(x))\} \\ &= \exp_{s+1}^\epsilon\{\exp_{m-m_1-1}(\log_{s_1}^{1/d_1} xg_1(y(x)))\Theta(x,y(x))\}, \end{aligned}$$

where

$$\Theta(x,y) = \left(d + \frac{\log\lambda(y)}{\exp_{m-m_1-1}(\log_{s_1}^{1/d_1} xg_1(y))}\right).$$

If $m_1 = m-1$, this immediately gives the pnf

$$g(y(x)) = \exp_{s+1}^\epsilon\{\log_{s_1}^{1/d_1} xg_1(y(x))\Theta(x,y(x))\}.$$

Otherwise we have

$$\begin{aligned} g(y(x)) &= \exp_{s+2}^\epsilon\{\exp_{m-m_1-2}(\log_{s_1}^{1/d_1} xg_1(y(x))) + \log\Theta(x,y(x))\} \\ &= \exp_{s+2}^\epsilon\{\exp_{m-m_1-2}(\log_{s_1}^{1/d_1} xg_1(y(x)))\Phi(x,y(x))\}, \end{aligned}$$

where

$$\Phi(x,y) = 1 + \frac{\log\Theta(x,y)}{\exp_{m-m_1-2}(\log_{s_1}^{1/d_1} xg_1(y(x)))}.$$

Now if $m_1 = m-2$, we have our pnf, and if not the last step needs to be iterated $m - m_1 - 2$ times until a pnf is reached.

Until now, we have only concerned ourselves with the case $y \to \infty$. However f might have finite singularities and we could try to find the asymptotics of inverses defined in the neighbourhood of each of these. The first problem is to find the singularities of f, and this can be very difficult. Progress has been made on the problem of locating and approximating zeros of exp–log functions, [68, 69, 57], and these can be used for finding singularities of such functions. Finite singularities of linear differential equations are either zeros of the leading coefficient or singularities of a coefficient. So there is a chance of handling this case when the coefficients are suitable exp–log functions. However if there is any non-linearity, all bets are off; witness for example (6.1).

If a finite singularity of f is discovered at $y = a$, say, we make substitutions $Y = \pm 1/(y-a)$ to form new functions $F_\pm(Y) = f(y)$. The two cases need to be considered separately. For example the function

$$f(y) = \frac{1}{y} + \frac{y}{y-1} + \exp\left(\frac{1}{y}\right)$$

has a singularity at $y = 0$. The functions $F_+(Y) = \exp(Y) + Y + Y/(1-Y)$ and $F_-(Y) = -Y - Y/(1+Y) + \exp(-Y)$, corresponding to $Y = y^{-1}$ and $Y = -y^{-1}$ respectively, have quite different asymptotics as $Y \to \infty$.

It may be that f, or one or both of F_+ and F_- may tend to $-\infty$. In this case we clearly have to substitute $X = -x$ in order to treat the asymptotics of the inverse function within our framework.

In addition, it is quite possible that $F_\pm$ may fail to lie in a Hardy field even if f does. An example is given by $f(y) = y + y^{-1}\sin(y^{-1})$, which yields $F_+(Y) = Y^{-1} + Y \sin Y$ on substituting $Y = y^{-1}$. The example $f(y) = y + y^{-1}\sin(\exp(y^{-1}))$ shows that the situation for F_+ can be different from that for F_-. Of course if f is an exp–log function the substitutions give exp–log functions which are guaranteed to be Hardy–field elements, and so these problems do not arise. In other cases it is necessary to check that F_+ and F_- are Hardy–field elements, and there is no general algorithm for doing this. If f is given in terms of exp, log, sin and cos, it may be possible to use Theorem 17 to establish that F_+ or F_- belongs to a Hardy field.

7.1.1 Summary of the Algorithm for Inversion

ALGORITHM **9** *Given a Hardy-field element $f(y)$, the principal steps of the algorithm are as follows:*

1. *Calculate the first pnf of f and check that $f \to \infty$.*
2. *If not, reject this singularity.*
3. *Invert the first pnf to give an expression of the form (7.5) and compute the nested form for $g_1(y)$.*
4. *In this, substitute the expression (7.5) for y. Then collect terms so as to determine the parameters $d'_2, \ldots, d'_n$ and c and the function η (of Theorem 23).*
5. *To compute further terms in the nested expansion, compute the nested form of $\eta(f(y), y)$ and again substitute for y using (7.5).*
6. *If inverses corresponding to other ranges of the variables are desired, determine the singularities of f.*
7. *For each singularity, a, make the substitutions $Y = \pm(y - a)^{-1}$ to obtain new functions $F_\pm$. For each of these, check (if possible) that the function belongs to a Hardy field; if not report fail for that branch. Check also that the function tends to infinity as $Y \to \infty$ and if not reject the singularity. Otherwise go to step 3.*

7.1.2 Example

Let

$$f(y) = \frac{y}{\log y - 1}. \tag{7.9}$$

Its first pnf is $f(y) = y g_1(y)$, where $g_1(y) = (\log y - 1)^{-1}$, which confirms that $f(y) \to \infty$. Inversion gives

$$y(x) = x(\log y - 1). \tag{7.10}$$

On substituting for y, we obtain

$$\begin{aligned} y(x) &= x[\log\{x(\log y - 1)\} - 1] = x[\log x + \log(\log y - 1)) - 1] \\ &= x[\log x + \log(\log\{x(\log y - 1)\} - 1) - 1] \\ &= x[\log x + \log\{\log x + \log(\log y - 1) - 1\} - 1] \\ &= x[\log x + \log_2 x - 1 + \log(1 + \{\log(\log y - 1) - 1\}\log^{-1} x)] \\ &= x\left[\log x + \log_2 x - 1 + \frac{\log_2 x}{\log x} + \cdots\right]. \end{aligned}$$

The function $f(y)$ also has a singularity at $y = e$. Let $Y = (y - e)^{-1}$, so that $y = e + Y^{-1}$. Then

$$f(y) = F_+(Y) = \frac{e + Y^{-1}}{\log(e + Y^{-1}) - 1} = \frac{e + Y^{-1}}{\log(1 + (eY)^{-1})}.$$

Hence the first pnf for F_+ is $F_+(Y) = Y g_1(Y)$, where

$$g_1(Y) = e(e + Y^{-1})(eY \log(1 + (eY)^{-1})^{-1}.$$

We invert this to give

$$\begin{aligned} Y(x) &= \frac{x}{g_1(Y)} = x\frac{eY\log(1 + (eY)^{-1})}{e(e + Y^{-1})} \\ &= \frac{x}{e(e + Y^{-1})}\sum_{n=1}^{\infty}\frac{(eY)^{1-n}}{n} \\ &= \frac{x}{e^2}\left(1 - \frac{1}{eY} + \frac{1}{(eY)^2} - \cdots\right)\left(1 - \frac{1}{2eY} + \frac{1}{3(eY)^2} - \cdots\right) \\ &= \frac{x}{e^2}\left(1 - \frac{3}{2eY} + \frac{11}{6(eY)^2} - \cdots\right). \end{aligned}$$

Substitution and iteration now give

$$Y(x)) = \frac{x}{e^2} - \frac{3}{2e} - \frac{5}{12x} + O(x^{-2}).$$

The results for F_- are very similar.

The example $f(y) = ye^y$ is similarly considered in [90]. A more complicated example is reported in [88]. If

$$f(y) = y\exp\{\log^2 y e^{\sqrt{\log_2 y}}\},$$

the algorithm produces

$$\begin{aligned} y(x) = \exp\Bigg\{ &\frac{\sqrt{\log x}}{e^{\sqrt{2\log_2 x/4}}} \cdot e^{1/8} \cdot \left(1 - \frac{\sqrt{2}}{64\log_2^{1/2} x} + \frac{1}{4096\log_2 x}\right. \\ &\left. + \frac{383\sqrt{2}}{786432\log_2^{3/2} x} - \frac{1535}{100663296\log_2^2 x} + \cdots\right)\Bigg\}. \end{aligned}$$

7.2 Multiseries of Inverse Functions

Our treatment here is based on [88], using a formula of Ecalle, [32]. See also [114].

Let $\mathcal{F}$ be an asymptotic field with scale $\{\mathbf{t}_1, \ldots, \mathbf{t}_n\}$ and shadow fields $\mathcal{S}_1(\mathcal{F}), \ldots, \mathcal{S}_n(\mathcal{F})$. Suppose that f is an element of $\mathcal{F}$ which tends to infinity. If possible we would like to construct an asymptotic field containing $\mathcal{F}$ and the functional inverse, f^{-1}, of f.

The function f has a computable shadow expansion of the form

$$\sum_{j=0}^{\infty} c_j(x)\mathbf{t}_k^{r_j}(x),$$

where $\mathbf{t}_k$ is element of largest comparability class which actually occurs in the expansion of f. Each c_j has a similar expansion in terms of the scale $\{\mathbf{t}_1, \ldots, \mathbf{t}_{k-1}\}$ and $c_0 \not\equiv 0$. We write $f_0(x) = c_0(x)t_k^{r_0}(x)$, and $g(x) = f(x) - f_0(x)$.

The first stage of the algorithm reduces the problem to the case when $r_0 = 0$ and $\gamma_1(c_0(x)) \le \gamma_1(x) < \gamma_1(\mathbf{t}_k)$. We have to show that we can do this in a finite number of steps. Then we show that we can obtain an expansion of $y(x) = f^{-1}(x)$ in the scale $\{\mathbf{t}_1(y_0), \ldots, \mathbf{t}_k(y_0)\}$, where y_0 is the inverse of f_0. Since the scale for f_0 is contained in $\{\mathbf{t}_1, \ldots, \mathbf{t}_{k-1}\}$, obtaining a suitable expansion for y_0 is a simpler problem, and we may call upon the principle of induction. However this will leave several remaining problems. We shall have to obtain expansions of $\{\mathbf{t}_1(y_0), \ldots, \mathbf{t}_k(y_0)\}$ in a scale containing $\{\mathbf{t}_1(x), \ldots, \mathbf{t}_n(x)\}$ and substitute into the expansion for y. We shall also have to consider the zero-equivalence testing of coefficients in this setting. When more than one inverse functions is involved it is not a trivial matter.

In practice our algorithm unwinds the recursion. So at the second stage, we produce an exact expression for $y(x)$ in terms of $y_0(x)$, then a similar expression for y_0 in terms of a function y_1, and so on, ending with an expression for some y_m in terms of x. The third stage produces a multiseries for y in terms of y_0, and similarly a multiseries for each y_i in terms of y_{i+1}, $i = 1, \ldots, m-1$. Finally at the fourth stage, we make the substitutions.

Stage 1

First note that since $f \to \infty$, we cannot have $r_0 > 0$. If $r_0 < 0$ and $\mathbf{t}_k$ is the only scale element occurring in the multiseries expansion of f, then of course the multiseries reduces to a simple power series in $\mathbf{t}_k$ and the usual power series algorithm for the inverse may be used. Otherwise if $r_0 < 0$ and more than one scale element occurs in the multiseries for f, we compute the inverse, Y, of $\log f$ and take $y(x) = Y(\log x)$. This is as required, since $y(f(x)) = Y(\log f(x)) = x$. Moreover

$$\log f = r_0 \log \mathbf{t}_k + \log c_0(x) + \log\left(1 + \frac{\mathbf{t}_k^{-r_0} g}{c_0(x)}\right). \tag{7.11}$$

So the expansion of $\log f$ still has $\mathbf{t}_k$ as its most significant monomial (since $\mathbf{t}_k$ occurs in the expansion of the final log) and there is a non–zero term in $\mathbf{t}_k^0$, namely $r_0 \log \mathbf{t}_k + \log c_0(x)$. Thus we arrive at the case $r_0 = 0$.

If $r_0 = 0$ but $\gamma_1(\mathbf{t}_k) \leq \gamma_1(x)$, we compute the inverse, Y, of $f(\exp x)$ and take $y(x) = \exp Y(x)$, noting that $f(y(x)) = f(\exp Y(x)) = x$. We see that r_0 remains equal to zero during this process and, after a finite number of iterations, we get $\gamma_1(\mathbf{t}_k) > \gamma_1(x)$ since $\gamma_1(\mathbf{t}_k) > \gamma_1(\log_N x)$ for some N.

If $r_0 = 0$ and $\gamma_1(x) < \gamma_1(c_0(x))$, then we replace f by $\log f$ as above. Equation (7.11) (with $r_0 = 0$) shows that this has the effect of replacing $c_0(x)$ by $\log c_0(x)$ while leaving $\mathbf{t}_k$ unchanged. After a finite number of iterations we will arrive at the case $\gamma_1(c_0(x)) \leq \gamma_1(x)$, and if we started with $\gamma_1(\mathbf{t}_k) > \gamma_1(x)$ that will remain true. Thus we can reduce to the case when $r_0 = 0$ and $\gamma_1(c_0(x)) \leq \gamma_1(x) < \gamma_1(\mathbf{t}_k)$ in a finite number of steps.

Stage 2

If y_0 denotes the inverse of f_0, then

$$y(x + g(y_0(x))) = y(f_0(y_0) + g(y_0)) = y(f(y_0)) = y_0(x). \tag{7.12}$$

In [32], Ecalle defined the operator $K : C^\infty(J_1) \longrightarrow C^\infty(J_2)$, for suitable intervals J_1 and J_2, by

$$K(h) = h(x + g(y_0(x))) - h(x); \tag{7.13}$$

Then (7.12) may be rewritten as

$$(I + K)y(x) = y_0(x), \tag{7.14}$$

which may be formally inverted to give

$$y(x) = \sum_{i=0}^{\infty} (-1)^i K^i y_0(x). \tag{7.15}$$

We prove below that that this formula stated by Ecalle gives an asymptotic expansion in our sense.

Next we want to obtain the inverse of $f_0 = c_0$ and we proceed in the same manner. So we let $f_{0,0}$ be the leading term in the expansion of f_0 with respect to its most important monomial. This will be $\mathbf{t}_{k-1}$ provided it actually occurs in the multiseries expansion of f_0. We reduce to the desired case as with f and consider the expansion of the leading term again. Eventually we arrive at the situation where only a simple power series expansion is involved.

In order to clarify stages 1 and 2, we look at an example. Let

$$f(x) = \frac{e^x}{\log x} + x. \tag{7.16}$$

The scale is $\{\mathbf{t}_1 = \log^{-1} x, \mathbf{t}_2 = x^{-1}, \mathbf{t}_3 = e^{-x}\}$. So $k = 3$ and $r_0 = -1$. Hence we take

$$f^{(1)} = \log f = x - \log_2 x + \log\left(1 + \frac{x \log x}{e^x}\right). \tag{7.17}$$

The new scale is $\{\mathbf{t}_1^{(1)} = \log_2^{-1} x,\ \mathbf{t}_2^{(1)} = \log^{-1} x,\ \mathbf{t}_3^{(1)} = x^{-1},\ \mathbf{t}_4^{(1)} = e^{-x}\}$. Now $c_0^{(1)} = x - \log_2 x$, $k = 4$ and we have the desired conditions $r_0^{(1)} = 0$ and $\gamma_1(c_0^{(1)}) \le \gamma_1(x) < \gamma_1(\mathbf{t}_4^{(1)})$.

We therefore consider the problem of finding the inverse of $f_0^{(1)} = x - \log_2 x$. Our new scale is a subset of the previous one, so we use the same notation. We note first that the leading term contains a negative power of $\mathbf{t}_k^{(1)} = x^{-1}$, and so we take logs. Thus

$$f^{(2)} = \log f_0^{(1)} = \log x + \log\left(1 - \frac{\log_2 x}{x}\right).$$

Now $r_0^{(2)} = 0$ and $\gamma_1(c_0^{(2)}) \le \gamma_1(x)$, but we still do not have $\gamma_1(x) < \gamma_1(\mathbf{t}_k^{(1)})$. So we replace x by e^x, giving

$$f^{(3)}(x) = f^{(2)}(e^x) = x + \log(1 - e^{-x}\log x). \tag{7.18}$$

Now we have the conditions we want and so we consider the inverse of $c_0^{(3)}(x) = x$, which is x of course.

Our results this far can be summarized as follows:

$$\begin{aligned}
&y(x) = y^{(1)}(\log x), && y^{(1)} = \operatorname{inv}\{x - \log_2 x + \log(1 + x\log x e^{-x})\},\\
&y^{(1)}(x + \log(1 + e^{-y_0} y_0 \log y_0)) && \\
&\quad = y_0^{(1)}(x), && y_0^{(1)} = \operatorname{inv}\{x - \log_2 x\},\\
&y^{(2)}(x) = y_0^{(1)}(\log x), && y^{(2)}(x) = \operatorname{inv}\{\log x + \log(1 - x^{-1}\log_2 x)\},\\
&y^{(3)} = \exp y^{(2)}, && y^{(3)} = \operatorname{inv}\{x + \log(1 - e^{-x}\log x)\},\\
&y^{(3)}(x + \log(1 - \log x e^{-x})) = x,\ y_0^{(3)}(x) = x. &&
\end{aligned}$$

Stage 3

We are going to use the formula (7.15) to compute y in terms of y_0. We want the result to be a multiseries in terms of y_0, but the iteration produces functions of $y_0(x+g(y_0))$ as well as functions given directly in terms of y_0. Moreover we shall at times obtain expansions in $\mathbf{t}_1(y_0(x+g(y_0))), \ldots, \mathbf{t}_k(y_0(x+g(y_0)))$, rather than directly in $\mathbf{t}_1(y_0), \ldots, \mathbf{t}_k(y_0)$. So to produce the first N terms of the $\mathbf{t}_k(y_0)$ expansion of y we proceed as follows.

1. Use the relation $y_0' = 1/f_0'(y_0)$ and the expansions of the derivatives of f_0 to successively compute, for $j = 1, \ldots, N$, the $\mathbf{t}_k(y_0)$–expansion of $y_0^{(j)}$ as far as the $(N-j)$–th term.
2. Hence obtain the first N terms of the $\mathbf{t}_k(y_0)$–expansion of $y_0(x + g(y_0))$ using the formula

$$y_0(x + g(y_0)) = \sum_{n=0}^{\infty} y_0^{(n)}(x)\frac{g^n(y_0(x))}{n!}.$$

3. If $\log x, \ldots, \log_J x$ are the iterated logarithms which are in the scale $\{\mathbf{t}_1, \ldots, \mathbf{t}_k\}$, compute successively the $\mathbf{t}_k(y_0)$–expansion of $\log[y_0(x + g(y_0))], \ldots, \log_J[y_0(x + g(y_0))]$ using the formula
$$\log\{\log_i(Y_0)\} = \log_{i+1}(y_0) + \operatorname{zlog}\left\{\frac{\log_i(Y_0) - \log_i(y_0)}{\log_i(y_0)}\right\}$$
with $Y_0 = y_0(x + g(y_0))$, and applying the standard expansion of zlog (recall that $\operatorname{zlog} t = \log(1+t)$).
4. For $\mathbf{t}_j$ of the form $\exp h$,
$$\mathbf{t}_j(Y_0) = \mathbf{t}_j(y_0)(1 + \operatorname{zexp}\{h(Y_0) - h(y_0)\}).$$
Compute the $\{\mathbf{t}_1(Y_0), \ldots, \mathbf{t}_{j-1}(Y_0)\}$–expansion of h, and then using the previous results, its $\{\mathbf{t}_1(y_0), \ldots, \mathbf{t}_{j-1}(y_0)\}$–expansion. Then expand the zexp; of course $\operatorname{zexp} t = \exp t - 1$.
5. The formula (7.15) is applied by first setting $\Phi_1 = y_0$. Then we set $\Phi_{i+1}(y_0) = \Phi_i(Y_0) - \Phi_i(y_0)$ and compute the expansion of $\Phi_i(Y_0)$ by substituting the expansions of the $\mathbf{t}_j(Y_0)$ obtained in the previous two steps into the expansion of Φ_i. Finally we return the $\mathbf{t}_k$–expansion of
$$\sum_{i=1}^{N}(-1)^{i+1}\Phi_i(y_0).$$

We illustrate this stage of the computation by continuing with our example of (7.16) with $N = 2$. From (7.17) (or the summary that follows it) we need to solve $(I + K)y = y_0$ with
$$K(h) = h(x + g(y_0(x))) - h(x) = h(x + \log(1 + e^{-y_0} y_0 \log y_0)) - h(x).$$
We have suppressed the superscript (1) to avoid cluttering the notation.

At step 1, we calculate
$$y_0' = \left(1 - \frac{1}{y_0 \log y_0}\right)^{-1} = \frac{y_0 \log y_0}{y_0 \log y_0 - 1}$$
and
$$y_0'' = -\frac{\log y_0 + 1}{(y_0 \log y_0 - 1)^2}.$$
The scale is $\mathbf{t}_1(y_0) = \log_2^{-1} y_0$, $\mathbf{t}_2(y_0) = \log^{-1} y_0$, $\mathbf{t}_3(y_0) = y_0^{-1}$, $\mathbf{t}_4(y_0) = \exp(-y_0)$, and we commence step 2 with the $\mathbf{t}_3^{-1}$. Writing $G = g(y_0)$, we have
$$\begin{aligned}\mathbf{t}_3^{-1}(y_0(x+G)) &= y_0(x+G)\\ &= y_0 + \frac{G y_0 \log y_0}{y_0 \log y_0 - 1} - \frac{G^2(\log y_0 + 1)}{2(y_0 \log y_0 - 1)^2} + O(e^{-3y_0})\\ &= y_0 + e^{-y_0}\frac{y_0^2 \log^2 y_0}{y_0 \log y_0 - 1}\\ &\quad - e^{-2y_0}\frac{y_0^2 \log^2 y_0 (y_0^2 \log^2 y_0 - y_0 \log y_0 + \log y_0 + 1)}{2(y_0 \log y_0 - 1)^2}\end{aligned}$$

plus terms of order e^{-3y_0}. Similarly working up to order e^{-2y_0} we have

$$\begin{aligned}\frac{1}{\mathbf{t}_2(y_0(x+G))} &= \log y_0(x+G) = \log y_0 + \operatorname{zlog}\left(\frac{y_0(x+G)-y_0}{y_0}\right) \\ &= \log y_0 + \frac{y_0(x+G)-y_0}{y_0} - \frac{(y_0(x+G)-y_0)^2}{2y_0^2} \\ &= \log y_0 + e^{-y_0}\frac{y_0\log^2 y_0}{y_0\log y_0 - 1} \\ -e^{-2y_0}&\frac{y_0\log^2 y_0(y_0^2\log^2 y_0 + y_0\log^2 y_0 - y_0\log y_0 + \log y_0 + 1)}{2(y_0\log y_0 - 1)^2}.\end{aligned}$$

Likewise

$$\begin{aligned}\frac{1}{\mathbf{t}_1(y_0(x+G))} &= \log_2 y_0(x+G) \\ &= \log\left\{\log y_0 + \operatorname{zlog}\left(\frac{y_0(x+G)-y_0}{y_0}\right)\right\} \\ &= \log_2 y_0 + e^{-y_0}\frac{y_0\log y_0}{y_0\log y_0 - 1} \\ -e^{-2y_0}&\frac{y_0\log y_0(y_0^2\log^2 y_0 + y_0\log^2 y_0 + \log y_0 + 1)}{2(y_0\log y_0 - 1)^2}.\end{aligned}$$

Then for $\mathbf{t}_4(y_0(x+G))$,

$$\begin{aligned}\mathbf{t}_4(y_0(x+G)) &= \exp(-y_0(x+G)) = e^{-y_0}\operatorname{zexp}(-y_0(x+G)+y_0) \\ &= e^{-y_0} - e^{-2y_0}\frac{y_0^2\log^2 y_0}{y_0\log y_0 - 1}\end{aligned}$$

from the calculation for $\mathbf{t}_3^{-1}(y_0(x+G))$.

Now we can start the iteration. We set $\Phi_1 = y_0$ and

$$\begin{aligned}\Phi_2(y_0) &= y_0(x+G) - y_0(x) \\ &= e^{-y_0}\frac{y_0^2\log^2 y_0}{y_0\log y_0 - 1} \\ &\quad -e^{-2y_0}\frac{y_0^2\log^2 y_0(y_0^2\log^2 y_0 - y_0\log y_0 + \log y_0 + 1)}{2(y_0\log y_0 - 1)^2}.\end{aligned}$$

Then $\Phi_3(y_0) = \Phi_2(y_0(x+G)) - \Phi_2(y_0(x))$ and substitution gives Φ_3 equal to

$$e^{-2y_0}\frac{y_0^3\log^3 y_0}{(y_0\log y_0 - 1)^3}\left\{y_0^2\log^2 y_0 - y_0\log^2 y_0 - 2y_0\log y_0 + 2\log y_0 + 2\right\}.$$

Now (7.15) yields $y(x) = \Phi_1 - \Phi_2 + \Phi_3 + O(e^{-3y_0})$. Thus

$$y^{(1)} = y_0 - e^{-y_0}\frac{y_0^2\log^2 y_0}{y_0\log y_0 - 1} + e^{-2y_0}\frac{y_0^2\log^2 y_0}{2(y_0\log y_0 - 1)^3}\times$$
$$\{5y_0^3\log^3 y_0 - 4y_0^2\log^3 y_0 - 10y_0^2\log^2 y_0 + 9y_0\log^2 y_0 + 10y_0\log y_0$$
$$-\log y_0 - 1\}. \quad (7.19)$$

We also need to invert (7.18), which is fortunately much simpler. Here we have $G(x) = \log(1 - e^{-x}\log x)$ and $y_0(x) = x$. At stages 3 and 4 we get $\log(x+G) = \log x + \log(1 + G/x)$ and $e^{-(x+G)} = e^{-x}e^{-G}$. Then $\Phi_1(x) = x$, $\Phi_2(x) = \Phi_1(x+G) - \Phi_1(x) = G$ and finally

$$\begin{aligned}\Phi_3(x) &= \Phi_2(x+G) - \Phi_2(x)\\ &= \log(1 - e^{-x}e^{-G}\log(x+G)) - \log(1 - e^{-x}\log x).\end{aligned}$$

A short calculation then reveals that

$$\Phi_3(x) = -e^{-2x}\left(\log^2 x - \frac{\log x}{x}\right) + \eta(x), \quad (7.20)$$

where we make the convention that $\eta(x)$ will stand for any quantity that tends to zero faster than $e^{-5x/2}$, and will not necessarily be the same quantity each time it is written.

Now we can obtain the expansion of $y^{(3)}$. We have

$$\begin{aligned}y^{(3)}(x) &= \Phi_1 - \Phi_2 + \Phi_3\\ &= x - \log(1 - e^{-x}\log x) - e^{-2x}\left(\log^2 x - \frac{\log x}{x}\right) + \eta(x)\\ &= x + e^{-x}\log x - e^{-2x}\left(\frac{3\log^2 x}{2} - \frac{\log x}{x}\right) + \eta(x).\end{aligned}$$

Stage 4

Having obtained y in terms of $y_0^{(1)}$, $y_0^{(1)}$ in terms of $y_0^{(2)}$, and so on, ending with some $y_0^{(j)}$ in terms of x, we have to make the substitutions to get y in terms of x. In principle this is now straightforward - we just use the method of Section 4.3.6. However the details are likely to be messy!

In our example, we start with

$$\begin{aligned}y^{(2)}(x) &= \log y^{(3)}(x)\\ &= \log x + \log\left\{1 + \frac{e^{-x}\log x}{x} - e^{-2x}\left(\frac{3\log^2 x}{2x} - \frac{\log x}{x^2}\right)\right\} + \eta(x)\\ &= \log x + \frac{e^{-x}\log x}{x} - e^{-2x}\left(\frac{3\log^2 x}{2x} - \frac{\log^2 x}{2x^2} - \frac{\log x}{x^2}\right) + \eta(x).\end{aligned}$$

Next

$$\begin{aligned}y_0^{(1)}(x) &= y^{(2)}(e^x)\\ &= x + e^{-e^x}e^{-x}x - \frac{e^{-2e^x}e^{-x}x}{2}(3x - e^{-x}(x+2)) + \eta(e^x). \quad (7.21)\end{aligned}$$

Now it is just(!) a matter of substituting (7.21) into (7.19) before replacing x by $\log x$ to give $y(x)$. The result is quite large, as may be checked with an algebra system. A slightly simpler example is given in [88].

7.2.1 Proof of the Iteration Formula

We begin with a lemma from [18]. The result proved there is slightly more general, but the following version is sufficient for our purpose.

LEMMA **30** *Let F and ϕ be two Hardy field elements such that $\phi F'/F \to 0$ and $\phi/x \to 0$. Then $F(x+\phi) \sim F(x)$.*

We note that F and ϕ do not have to belong to the *same* Hardy field.

Proof of Lemma 30
We use the notation F^{Δ} to mean F'/F, as in the previous chapter. It is sufficient to prove the result when F is positive and tends to zero. For if $|F| \to \infty$ we may consider $1/F$, while if F tends to a non–zero constant the result is trivial.

Now since $F \to 0$ we have $F' \to 0$ also, and so $|F'|$ decreases. The First Mean Value Theorem gives

$$F(x+\phi(x)) = F(x) + \phi(x)F'(x+\theta(x)\phi(x)), \tag{7.22}$$

where $0 < \theta(x) < 1$. If $\phi > 0$ then $|F'(x+\theta(x)\phi(x))| \leq |F'(x)|$ and since $\phi F^{\Delta} \to 0$, (7.22) gives the desired conclusion.

When $\phi < 0$, we look first at the case $|F^{\Delta}| \to \infty$. Then

$$|\phi(x)F^{\Delta}(x+\phi(x))| \leq |\phi(x)F^{\Delta}(x)|$$

and again (7.22) gives the conclusion. Similarly if F^{Δ} tends to a non–zero limit, $F^{\Delta}(x+\phi(x)) \sim F^{\Delta}(x)$. If $|xF^{\Delta}| \leq M$ for some real constant M then

$$|\phi(x)F^{\Delta}(x+\phi(x))| \leq \frac{M|\phi(x)|}{x+\phi(x)} = \frac{M|\phi(x)|}{x}\left(1+\frac{\phi(x)}{x}\right)^{-1} \to 0,$$

and we can argue as before. Finally if $F^{\Delta} \to 0$ but $|xF^{\Delta}| \to \infty$ then $1 > |F^{\Delta}| > x^{-1}$ and on taking logarithms and differentiating, we get that $x(F^{\Delta})^{\Delta}$ is bounded and by the previous case, $F^{\Delta}(x+\phi(x)) \sim F^{\Delta}(x)$. Since $\phi F^{\Delta} \to 0$, the remaining case now follows from (7.22). Thus the proof of Lemma 30 is complete.

Next we apply Lemma 30 to show that the terms in the series (7.15) decrease like powers of $G = g \circ y_0$.

LEMMA **31** *Let y_0, G and the operator K be as above. Then for $i \geq 1$*

$$\log|K^i(y_0)| \sim i \log|G|.$$

Proof of Lemma 31
We note that $\gamma_1(g) = \gamma_1(\mathbf{t}_k) > \gamma_1(x)$ and hence $\gamma_1(G) > \gamma_1(y_0)$.
For the case $i = 1$, we have

$$K(y_0) = y_0(x+G) - y_0(x) = Gy_0'(x+\theta G),$$

where $0 < \theta = \theta(x) < 1$. If we take $\phi = G$ and $F = y_0'$, then the hypotheses of Lemma 30 hold because $\gamma_1(G) > \gamma_1(y_0) \geq \gamma_1(x)$ (recall that $\gamma_1(c_0) \leq \gamma_1(x)$). Hence $K(y_0) \sim Gy_0'(x) = G^{1+o(1)}$ as required.
Suppose then that the result holds for i. We have

$$K^{i+1}(y_0) = K^i(y_0)(x+G) - K^i(y_0)(x) = G \cdot (K^i(y_0))'(x+\theta G), \quad (7.23)$$

with θ as before. Now by the induction hypothesis, $\log K^i(y_0) \sim i \log G$. By Lemma 8 of Chapter 3,

$$\gamma_1((K^i(y_0))^{\Delta}) = \gamma_1(G^{\Delta}) \leq \max\{\gamma_1(\log G), \gamma_1(x)\} < \gamma_1(G).$$

So if we take $\phi = G$ and $F = K^i(y_0)$ in Lemma 30, we get that

$$(K^i(y_0))'(x+\theta G) \sim (K^i(y_0))'(x)$$

and hence from (7.23), $K^{i+1}(y_0) = G^{i+1+o(1)}$. This completes the induction step and Lemma 31 is proved.

For $n \in \mathbb{N}$, let

$$S_n(x) = \sum_{i=0}^{n} (-1)^i K^i(y_0(x)).$$

In order to show that (7.15) is a valid asymptotic expansion, we must prove that for each n, $y - S_n = o(K^{n+1} y_0)$. In fact we prove a little more.

THEOREM **24** *For all $k, n \in \mathbb{N}$,*

$$\log |y^{(k)} - S_n^{(k)}| \sim (n+1) \log |G|, \quad (7.24)$$

where $G = g \circ y_0$.

Because we do not wish to make any unnecessary assumptions concerning Hardy–field membership at this stage, we write $\gamma_1(h_1) < \gamma_1(h_2)$ to mean $\log |h_1| = o(\log |h_2|)$, even when h_1 and h_2 might not be elements of the same Hardy field. This slight abuse of notation will not cause any difficulties.

Proof of Theorem 24
Our proof will use induction and we need the derivatives in (7.24) to make this go through. In fact it is the base case of the induction which needs the most work. We begin with two lemmas.

LEMMA **32** *Let h be a $\mathcal{C}^\infty$ function in a neighbourhood of $+\infty$, and let $k \in \mathbb{N}$,*

$$\frac{d^k h(x+G(x))}{dx^k} = h^{(k)}(x+G(x)) \cdot (1+G')^k + \Gamma_k, \tag{7.25}$$

where Γ_k is a polynomial in $G', \ldots, G^{(k)}, y'(x+G), \ldots, y^{(k-1)}(x+G)$ in which every monomial contains an element of the form $h^{(j)}(x+G(x))$, $j \in \mathbb{N}$, and a derivative of G.

Proof of Lemma 32
The statement of the lemma is tautological when $k = 0$, so suppose it holds for a given value of k. Then

$$\begin{aligned}\frac{d^{k+1}h(x+G(x))}{dx^{k+1}} &= (h^{(k)}(x+G(x))(1+G')^k + \Gamma_k)' \\ &= h^{(k+1)}(x+G(x))(1+G')^{k+1} + \\ &\qquad kh^{(k)}(x+G(x))(1+G')^{k-1}G'' + \Gamma_k'.\end{aligned}$$

We can therefore take $\Gamma_{k+1} = kh^{(k)}(x+G(x))(1+G')^{k-1}G'' + \Gamma_k'$, and so Lemma 32 holds by induction.

LEMMA **33** *For each $k \in \mathbb{N}$,*

$$\log|G^{(k)}| \sim \log|G|.$$

Proof of Lemma 33
For $k = 1$, we have $G' = g'(y_0)y_0'$, and a simple induction then shows that for $k \in \mathbb{N}$,

$$G^{(k)} = \sum_{j=1}^{k} P_j\langle y_0\rangle g^{(j)}(y_0),$$

where each P_j is a differential polynomial of order at most j. Hence

$$\gamma_1(G^{(k)}/G) \le \max\{\gamma_1(y_0^{(j)}), \gamma_1(g^{(j)}(y_0)/g(y_0));\ j = 1, \ldots, k\}. \tag{7.26}$$

By inductive hypothesis, y_0 has a multiseries expansion in base elements of comparability class less than g. Also Lemma 8 applied to $\log g$ gives $\gamma_1(g'/g) \le \max\{\gamma_1(\log g), \gamma_1(x)\}$. This is less than $\gamma_1(g)$ since $\gamma_1(g) = \gamma_1(\mathbf{t}_k) > \gamma_1(x)$. So $\log(g'/g) = o(\log g)$, and it follows that $\log g' \sim \log g$. Applying the same argument to g' gives $\log g'' \sim \log g' \sim \log g$. Hence $\log g^{(j)} \sim \log g$ for all $j \in \mathbb{N}$. Therefore $\gamma_1(g^{(k)}/g) < \gamma_1(g)$, and so $\log|G^{(k)}/G| = o(\log|G|)$, which is the conclusion of Lemma 33.

Continuation of the proof of Theorem 24
We have $y_0(x) = y(f(y_0(x))) = y(f_0(y_0(x)) + g(y_0(x))) = y(x+G)$. So if we apply Lemma 32 with $h = y$, we obtain

$$y_0^{(k)}(x) = y^{(k)}(x + G(x)) \cdot (1 + G')^k + \Gamma_k, \tag{7.27}$$

where Γ_k is a polynomial in $G', \dots, G^{(k)}, y'(x+G), \dots, y^{(k-1)}(x+G)$ in which every monomial contains a derivative of G. Now since we do not want to make any restrictive assumptions, we cannot be sure that G belongs to a Hardy field. However $\mathcal{F}$ is an asymptotic field which contains both f and its $\mathbf{t}_n$-shadow, f_0. Hence $g \in \mathcal{F}$ and $\gamma_1(g) > \gamma_1(x)$. On changing variable, we get

$$\gamma_1(G(x)) > \gamma_1(y_0(x)) \geq \gamma_1(x).$$

Because f_0 has a multiseries expansion in elements of comparability class less than $\gamma_1(g)$, Lemma 27 implies that $\gamma_1(y_0^{(k)}) < \gamma_1(G)$. However $\Gamma_k \to 0$ and $\gamma_1(\Gamma_k) = \gamma_1(G)$. So from (7.27) we have that

$$\gamma_1(y^{(k)}(x + G(x))) = \gamma_1(y_0^{(k)}) < \gamma_1(G).$$

Next,

$$\begin{aligned} y^{(k)}(x) - y_0^{(k)}(x) &= y^{(k)}(x) - y^{(k)}(x + G(x))(1 + G')^k - \Gamma_k \\ &= -y^{(k+1)}(x + \theta G)G + y^{(k)}(x + G)(1 - (1 + G')^k) - \Gamma_k, \\ &= -y^{(k+1)}(x + \theta G)G + W(x) \end{aligned} \tag{7.28}$$

say, where $0 < \theta(x) < 1$. Now Lemma 33 implies that $\gamma_1(W/G) < \gamma_1(G)$. Also for any $j \in \mathbb{N}$, $y^{(j)}(x + \theta G)$ lies between $y^{(j)}(x)$ and $y^{(j)}(x + G)$, and so $\gamma_1(y^{(j)}(x + \theta G)) < \gamma_1(G)$. Hence

$$y^{(k)} - y_0^{(k)} = G\psi_{0,k},$$

where $\gamma_1(\psi_{0,k}) < \gamma_1(G)$. Thus $\log|y^{(k)} - y_0^{(k)}| \sim \log|G|$, which is the conclusion of Theorem 24 for the case $n = 0$.

Now suppose that Theorem 24 holds for a particular value of n and all values of k. We write $h_n = y - S_n$. Then (7.14) gives $-Ky = y - y_0$ and hence

$$\begin{aligned} h_n(x) - h_n(x + G) = -Kh_n &= -Ky + K\sum_{i=0}^{n}(-1)^i K^i(y_0) \\ &= y - y_0 - \sum_{i=1}^{n+1}(-1)^{i+1}K^{i+1}(y_0) = y - S_{n+1}. \end{aligned}$$

Then differentiation yields

$$\begin{aligned} y^{(k)} - S_{n+1}^{(k)} &= h_n^{(k)}(x) - (h_n(x + G))^{(k)} \\ &= h_n^{(k)} - h_n^{(k)}(x + G) \cdot (1 + G')^k - \Gamma_{n,k}, \end{aligned} \tag{7.29}$$

where $\Gamma_{n,k}$ is as given by Lemma 32.

As in (7.28), this leads to the relation

$$y^{(k)}(x) - S_{n+1}^{(k)}(x) = -h_n^{(k+1)}(x + \theta G)G + W_n(x),$$

where $W_n = h_n^{(k)}(x+G)(1-(1+G')^k) - \Gamma_{n,k}$ and $0 < \theta(x) < 1$. Now Lemma 33 and our inductive assumption give the desired conclusion as above. This completes the proof of Theorem 24.

7.2.2 Asymptotic Fields and Inverse Functions

Ideally we would like to add functional inverses to asymptotic fields, obtaining new asymptotic fields. In principle this is now possible since we can obtain the shadows and ghosts from the multiseries. However there are some difficulties.

Because we give inverses in terms of other inverse functions, the closed forms of multiseries coefficients will generally be very complicated and hard to keep track of. Moreover zero equivalence becomes an issue, especially when more than one inverse function is involved. If we were to add the requirement to asymptotic fields that all elements can be specified using differential equations with initial conditions, then we could similarly specify functional inverses by using Lemma 27. The zero–equivalence methods of Section 2.2 would then be applicable. Most of the asymptotic fields we have been considering do satisfy the additional requirement, but this approach to zero equivalence is hardly elegant and is likely to be very slow in practice.

When we add only one inverse function, say $g = f^{-1}$, to an exp–log asymptotic field, we can do better by writing the variable x as $x = f(y)$ and substituting into any expressions under consideration. The resulting expressions are exp–log and can be tested in the usual way. This was exploited in [88], where the methods of the previous section were applied to some problems in combinatorics.

8

Implicit Functions

The exp–log functions of variables, x, $y_1, \ldots, y_n$ are those given by expressions with signature $\mathbb{K}$, x, $y_1, \ldots, y_n$, $+$, $-$, $\times$, $\div$, exp, log. Given n such functions, $F_1, \ldots, F_n$, the implicit function theorem guarantees the existence of roots $y_1 = y_1(x), \ldots, y_n = y_n(x)$ of the equations $F_1 = 0, \ldots, F_n = 0$ in a neighbourhood of any solution point where the rank of the Jacobian matrix is maximal. The root functions $y_1 = y_1(x), \ldots, y_n = y_n(x)$ are called *implicit functions*. We wish to be able to calculate the asymptotics of implicit functions in neighbourhoods of $+\infty$, that is to say on intervals of the form (a, ∞).

In the corresponding algebraic problem, where $F_1, \ldots, F_n$ are polynomials, elimination theory tells us that each $y_i(x)$ is the solution of a single algebraic equation $P(x, y_i) = 0$. So in the algebraic case all the new functions occur in the case $n = 1$. There is no corresponding result for exp–log functions known to this author, and so here we do need to consider general systems.

Our first concern is whether implicit functions are elements of a Hardy field. The answer to this lies within a very interesting development, with roots going back at least as far as the work of Descartes. The line includes a remarkable dissertation, by Khovanskii, [46], and the development of o–minimality, leading up to the work of Wilkie, and a great deal else besides. It is well worth several books in its own right. Here we take only the briefest look and refer the reader to the works of van den Dries, [112], Marker, Messmer and Pillay, [60], Wilkie, [124] and the references they contain.

Given a function $f \in \mathcal{C}^\infty(a, \infty)$ for some $a \in \mathbb{R}$, we can consider the field of functions generated by f and its derivatives, namely $\mathbb{Q}(f, f', f'', \ldots)$. If no non–zero element of this field has arbitrarily large zeros then it is (of course) a Hardy field containing f. For most of the cases we have been considering, it would make no difference if we were to require that non–zero elements of the field have only finitely many zeros. One way of looking at o–minimality is to say that it makes such a requirement, not just for polynomials in f and its derivatives but for any function *definable* from f using arithmetic and logical operations, and sometimes other operations besides. (In fact o–minimality is more general than this, but there is no need to pursue that here.). f is then said to be an *o–minimal function.* The derivative of f is definable using the normal ϵ–δ definition. Also inverse and implicit functions can be defined in the obvious way. So provided the defining functions are o–minimal, the implicit functions they specify will be Hardy–field elements.

The difficult thing, in the sort of cases we are interested in, is to prove that the defining functions are o–minimal. This was done for exp–log functions in Alex Wilkie's ground breaking paper, [124]. As a direct result, Angus Macintyre and Alex Wilkie were able to establish a conjecture of Tarski on the basis of the Schanuel conjecture, [55]. This was mentioned in Chapter 2. A more recent and equally remarkable result of Wilkie's has the consequence that functions defined by Pfaffian chains are o–minimal.

8.1 A Special Case

Our main interest is, of course, in algorithms to compute the asymptotics of implicit functions. In this section, which is based on previously unpublished joint work with Bruno Salvy, we look at a method which is only applicable to a special case and even there is not guaranteed to give the answer. However it is likely to be relatively fast when it does work, and it does represent an approach to the problem which is slightly different from that of our main algorithm.

We consider the problem of finding nested form solutions of single functional equations of the form

$$\sum_{i=1}^{N} F_i(x)G_i(y) = 0, \tag{8.1}$$

where the F_i and G_i are functions of a single variable with computable nested forms, for example exp–log functions.

For each solution $y(x)$ there exists a pair (i, j), $i \neq j$, such that

$$\gamma_0(F_i(x)G_i(y(x))) = \gamma_0(F_j(x)G_j(y(x))) \geq \gamma_0(F_k(x)G_k(y(x))) \tag{8.2}$$

for all k. The point is that if $\gamma_0(F_i(x)G_i(y(x)))$ is maximal it must be matched by at least one $\gamma_0(F_j(x)G_j(y(x)))$ since the left–hand side of (8.1) must cancel out when $y = y(x)$. The essence of our method is to consider each pair (i, j) and to compute a nested form solution of

$$F_i(x)G_i(y(x)) \sim KF_j(x)G_j(y(x)) \tag{8.3}$$

for K a constant which is undetermined at this stage. The nested form obtained, called a *candidate*, is checked for consistency with (8.2), and is discarded if it fails. Alternatively the nested form may be consistent for certain values of K only, or consistent for all values of K.

It should be emphasised that the nested forms found to be consistent do not necessarily correspond to actual solutions, nor even to formal solutions since an inconsistency may appear later on in the nested expansion. All that this method tells us is that any solution will have one of the nested forms we calculate. However it may be possible to show that solutions exist by using Sturm sequences [68, 69, 57].

8.1.1 Computation and Checking of Candidates

We may arrange the sum (8.1) so that $F_{i+1}/F_i \to \infty$ for each i. Then (8.3) may be cross multiplied to give a relation of the form

$$G(y) \sim KF(x), \tag{8.4}$$

with $F(x) \to \infty$. We make the change of variable $X = F(x)$ and compute a nested form for $y(X)$ as in Section 7.1. Then we replace X by the nested form for $F(x)$, and hence calculate a nested form for $y(x)$ using the techniques of Section 4.4.2.

Similar ideas may be used to check the conditions (8.2). For each $k \neq i, j$ we write G_k/G_i as a nested form in y and substitute our candidate for y. Then we compare the nested form obtained with that of F_i/F_k. Unless it is possible that for all k,

$$\gamma_0 \left(\frac{G_k(y(x))}{G_i(y(x))} \right) \leq \gamma_0 \left(\frac{F_i(x)}{F_k(x)} \right), \tag{8.5}$$

$y(x)$ must be rejected as a possible solution.

If we obtain strict inequality in (8.5) for all $k \neq i, j$, (8.1) allows only one value of K. Otherwise the matter is more complicated. Suppose that we obtain $F_i(x)G_i(y(x)) \sim K_p F_{j_p}(x) G_{j_p}(y(x))$, $p = 1, \ldots, q$. It is perfectly possible that one or more of these relations will lead to a nested form in which the constant K_p appears only in the error term, and then we have our nested form. For example $Ke^x = e^{x+o(1)}$. Otherwise each K_p appears in the body of the nested form. Since the different expressions for the nested form have to agree, we obtain a relation between K_p and K_1, $p = 2, \ldots, q$, and it is not hard to see that it must be of the form $K_p = B_p K_1^{\alpha_p}$ where B_p and α_p are constants which depend only on the original relation (8.1). On the other hand

$$F_i(x)G_i(y)+F_{j_1}(x)G_{j_1}(y)+\cdots+F_{j_q}(x)G_{j_q}(y) \sim (1-K_1-\cdots-K_p)F_i(x)G_i(y),$$

and since all other products are $o(F_i(x)G_i(y))$ this implies that

$$K_1 + \cdots + K_p = 1.$$

Thus we obtain a finite number of possible values for K in the nested form of our candidate.

8.1.2 Examples

First consider the equation

$$\frac{e^y \log x}{y} - \frac{xe^{\sqrt{y}}}{y} - e^{2x}\log y - e^{x^2} = 0. \tag{8.6}$$

There are two singularities to consider: $y = \infty$ and $y = 0$. We look first at the $y = \infty$ and the pair $(-e^{x^2}, e^y \log x/y)$. On applying the inversion algorithm of Section 7.1 to $e^y/y = X$, with $X = e^{x^2}/\log x$, we obtain

$$y \sim K \log X \left(1 + \frac{\log y}{\log X}\right),$$

and hence the nested form $y = Kx^2(1 + o(1))$. We check easily that when y is of this form, the pair $(-e^{x^2}, e^y \log x/y)$ does indeed dominate. So $K = 1$ and we have the nested form of a possible solution. Here and in many other cases, we can obtain a full nested expansion of $y(x)$ as follows.

We rewrite equation (8.6) as

$$\frac{e^y}{y} = \frac{e^{x^2}}{\log x} + \frac{e^{2x}\log y}{\log x} + \frac{xe^{\sqrt{y}}}{y \log x},$$

and then, as in Section 7.1, convert this to the iterative form

$$y = \log\left(\frac{e^{x^2}}{\log x} + \frac{e^{2x}\log y}{\log x} + \frac{xe^{\sqrt{y}}}{y\log x}\right) + \log y. \tag{8.7}$$

We use (8.7) obtain the following iteration.

$$\begin{aligned}
y_0 &= x^2 + \cdots,\\
y_1 &= x^2 + 2\log x - \log_2 x + \cdots,\\
y_2 &= x^2 + 2\log x - \log_2 x + \frac{2\log x - \log_2 x}{x^2} + \cdots,\\
y_3 &= x^2 + 2\log x - \log_2 x + \frac{2\log x - \log_2 x}{x^2} +\\
&\quad \frac{-2\log^2 x + 2\log x \log_2 x - \log_2^2 x/2}{x^4} + \cdots.
\end{aligned}$$

Next we consider the singularity at $y = 0$ for the same pair. On writing $X = e^{x^2}/\log x$ as before and $Y = y^{-1}$ we are led to the relation $Ye^{1/Y} \sim KX$. This yields the iterative form

$$Y = X(1 + (e^{1/Y} - 1)) = \frac{e^{x^2}}{\log x}(1 + (e^{1/Y} - 1)). \tag{8.8}$$

However the term $xe^{\sqrt{y}}/y = xYe^{\sqrt{1/Y}}$ dominates (8.6) when Y takes the form (8.8) and so this candidate is rejected. Examination of other pairs shows that these give no other possible nested forms and so the only possible solution is the one given above.

The iterative method of generating a nested expansion does not always work. Consider the equation

$$y - x^2 - \exp\left(y - x^2 + \frac{\log x \log_2 x}{\log_2 x - 1}\right) = 0. \tag{8.9}$$

Suppose we take the pair (y, x^2), so $y \sim Kx^2$. We see that for this pair to dominate we have to have $K = 1$. Now if we substitute $y = x^2$ into the exponential term of (8.9), we obtain $\exp(\log x \log_2 x(\log_2 x - 1)^{-1})$, which is easily seen to be $o(x^{1+\epsilon})$ for every $\epsilon \in \mathbb{R}^+$. We iterate using

$$y = x^2 + \exp\left(y - x^2 + \frac{\log x \log_2 x}{\log_2 x - 1}\right).$$

This yields

$$\begin{aligned}
y_0 &= x^2 \\
y_1 &= x^2 + \exp\left(\frac{\log x \log_2 x}{\log_2 x - 1}\right) \\
y_2 &= x^2 + \exp\left(\frac{\log x \log_2 x}{\log_2 x - 1} + \exp\left\{\frac{\log x \log_2 x}{\log_2 x - 1}\right\}\right).
\end{aligned}$$

The problem is seen more clearly if we note that

$$\exp\left(\frac{\log x \log_2 x}{\log_2 x - 1}\right) = x^{1+o(1)},$$

and so $y_2(x) = x^2 + x^{1+o(1)} \exp(x^{1+o(1)}) = \exp(x^{1+o(1)})$. But then

$$y_3 = x^2 + x^{1+o(1)} \exp[x^{1+o(1)} \exp(x^{1+o(1)})] = \exp_2(x^{1+o(1)}),$$

and the order of exponentiation increases with each iteration. The problem occurs because the 'error' in y_0 is about as big as x and this gets repeatedly exponentiated.

8.2 Systems of Exp–Log Equations

We consider systems of n equations. The case $n = 1$ was done in [87], and our method here is a generalization of that.

8.2.1 Zero Equivalence

Our main method of attack on implicit functions is to study the asymptotics of exp–log functions of several variables under various assumptions about the relative growths of the different variables. Zero equivalence does not pose any extra problems of a purely algorithmic nature here, although the extra variables do bring increased complexity. For one thing the method of Pélada-Germa, [63], already caters for an arbitrary number of variables. Alternatively, we may select a main variable, say x, and use one of the other zero–equivalence methods with $\mathbb{K}(y_1, \ldots, y_n)$ as constant field. Of course questions about the zero–equivalence of constants now take place in that field, but the number of variables is then reduced by one, and so by induction we have an algorithm modulo a method for $\mathbb{K}$.

8.2.2 Sketch of the Method

One way of summarizing our approach to the asymptotics of one–variable exp–log functions is to say that we rewrite given expressions in terms of base elements $\mathbf{t}_1, \ldots, \mathbf{t}_k$ using arithmetic operations and *z–functions*. Recall from Section 5.4 that the latter are of one of the forms $\mathrm{zexp}(t) = \exp t - 1$, $\mathrm{zlog}(t) = \log(1+t)$ and $\mathrm{zpow}(t, c) = (1+t)^c - 1, c \in \mathbb{R} \setminus \mathbb{N}$. In the one–variable case, the $\mathbf{t}_i$ are all of different comparability class, and so once the input expression has been rewritten in terms of them we only have to use the standard series expansions of the z–functions to obtain the required multiseries.

We use a similar idea with several variables, but with a number of differences. In particular we can no longer insist that the $\gamma_1(\mathbf{t}_i)$ are all different. Our main interest is in implicit functions, where the y_j will be functions of x, but we will not know a priori what their growth is. Indeed that is the problem to be solved!

For convenience with the examples given below we take the base functions $\mathbf{t}_i$ to tend to infinity, although one could obviously use their reciprocals instead. We allow $\mathbf{t}_i$ to be of one of the forms $\log_k x$, $\pm \log_k |y_j|$, $\exp w$, where $k \geq 0$, $1 \leq j \leq n$, and w tends to infinity and has an expansion in terms of previous $\mathbf{t}_\alpha$. We require that if $\log_k x$ belongs to the set of base elements with $k > 0$ then $\log_{k-1} x$ is also a base element, and similarly for $\log_k |y_j|$; c.f. Definition 10.

Although at this stage we do not know the relative growths of the y_j, we make assumptions about these growths and on the basis of each assumption give a nested form for the input expressions. In the case when each $y_i(x)$ tends to $\pm\infty$ or to 0, the assumptions will be conjunctions of one of the forms:

$$\gamma_1(\mathbf{t}_{i_1}) > \gamma_1(\mathbf{t}_{i_2})$$

$$\gamma_1(\mathbf{t}_{i_1}) = \gamma_1(\mathbf{t}_{i_2}) \ \& \ \mathrm{Pred}(\lim\{\log \mathbf{t}_{i_2} / \log \mathbf{t}_{i_1}\}, K_1)$$

$$\gamma_1(\mathbf{t}_{i_1}) = \gamma_1(\mathbf{t}_{i_2}) = \gamma_1(\mathbf{t}_{i_3}) \ \& \\ \mathrm{Pred}(\lim\{\log \mathbf{t}_{i_2} / \log \mathbf{t}_{i_1}\} + r_3 \lim\{\log \mathbf{t}_{i_3} / \log \mathbf{t}_{i_1}\}, K_1, K_2)$$

$$\begin{aligned} &\dots \quad \dots \quad \dots \qquad (8.10)\\ &\gamma_1(\mathbf{t}_{i_1}) = \gamma_1(\mathbf{t}_{i_2}) = \cdots = \gamma_1(\mathbf{t}_{i_{n+1}}) \;\&\; \mathrm{Pred}(\lim\{\log \mathbf{t}_{i_2}/\log \mathbf{t}_{i_1}\} + \\ &r_3 \lim\{\log \mathbf{t}_{i_3}/\log \mathbf{t}_{i_1}\} + \cdots + r_{n+1} \lim\{\log \mathbf{t}_{i_{n+1}}/\log \mathbf{t}_{i_1}\}, K_1, \ldots, K_l). \end{aligned}$$

Here Pred is one of '$\neq$','$=$' '$<$', '$>$' and $r_3, \ldots, r_{n+1}$ are given non-zero constants. In the case $\gamma_1(\mathbf{t}_{i_1}) = \cdots = \gamma_1(\mathbf{t}_{i_s})$, we shall have a linear equation for $K_1, \ldots, K_s$, but otherwise they might be undetermined.

In practice we shall also use similar, but more precise, conditions where γ_1 is replaced by γ_0 or $\lim\{\log \mathbf{t}_i/\log \mathbf{t}_j\}$ by $\lim\{\mathbf{t}_i/\mathbf{t}_j\}$. The various assumptions are to include all the possibilities, and so constitute a breakdown into cases. In fact we may need different lists of base variables in different cases, so each assumption has its associated list. We can give an asymptotic expression for the growth of the input expressions, except in cases where there is a possible cancellation. In that case we give the estimate '?'.

Solutions of the implicit equations can only occur when all the estimates for their left–hand sides are of this question–mark form. In other cases the explicit growth given for a left–hand side will imply that it cannot be zero. Moreover question–mark estimates can only occur when the conditions include an equality between comparability classes and the corresponding Pred is '$=$'. Question–mark estimates therefore lead to pnfs for solutions of the implicit equations and these can be refined to give nested forms. In the two–variable case the nested forms of all solutions will be among those calculated, but some of the calculated forms might not correspond to actual solutions. They might not even give formal solutions since a contradiction may emerge further on in the process of calculation. In general we cannot detect whether solutions with a particular form exist, but the intermediate value theorem can sometimes be used to good effect.

When there are three or more variables, it is possible that we may get the same asymptotic information from more than one equation, and so we may be unable to obtain nested forms for the unknown functions. Indeed it is possible that the system may degenerate or be contradictory, and we do not currently have algorithms to determine when these things occur.

In the case when one or more of the y_i tends to a non–zero finite limit (a priori undetermined), that limit appears as a parameter in the computation. So if $y_{i_j} \to K_j$, $j = 1, \ldots, l$ we work over the constant field $\mathbb{K}(K_1, \ldots, K_l)$. This is the reason for the appearance of the constants $K_1, \ldots, K_l$ in the conditions above.

A simple (indeed trivial) example from [87] which illustrates the case $y \to \infty$ is as follows

$$x^2 - y = \begin{cases} x^2(1 - y/x^2), & \text{if } \gamma_1(y) < \gamma_1(x) \text{ or } (\gamma_1(y) = \gamma_1(x) \\ & \quad \text{and } \lim\{\log y/\log x\} < 2), \\ ? & \text{if } \lim\{\log y/\log x\} = 2, \\ -y(1 - x^2/y), & \text{if } \gamma_1(y) > \gamma_1(x) \text{ or } (\gamma_1(x) = \gamma_1(y) \\ & \quad \text{and } \lim\{\log y/\log x\} > 2), \end{cases} \qquad (8.11)$$

We deduce that a solution of the implicit equation must satisfy the condition $\lim\{\log y/\log x\} = 2$, which leads to the pnf $y(x) = \exp\{(2 + o(1))\log x\} = x^{2+o(1)}$. The next step would be to substitute $y = x^2 z$ and obtain a multivariate estimate as above for the new function, but here of course the exact solution $z = 1$ drops out.

8.2.3 Building estimates from the Tower of Fields

To obtain the assumptions, lists and estimates, we build towards the given expression adding one new transcendental at a time. So given input expressions which are exp–log in $x, y_1, \ldots, y_n$, we build a tower of function fields

$$\mathbb{K}(x, y_1, \ldots, y_n) = \mathcal{F}_1 \subset \mathcal{F}_2 \subset \cdots \subset \mathcal{F}_N \tag{8.12}$$

with each $\mathcal{F}_i = \mathcal{F}_{i-1}(f_i)$ being a simple extension of $\mathcal{F}_{i-1}$ by an exponential or a logarithm, and our input expressions being elements of $\mathcal{F}_N$; c.f. (5.1). By working up the tower we build a set $\mathcal{D}_i = \{(\mathcal{T}_j^i, C_j^i),\ j = 1, \ldots, \beta_i\}$ for each $i = 0, \ldots, N$. Here each $\mathcal{T}_j^i$ is a list of the form $\{\mathbf{t}_1, \ldots, \mathbf{t}_k\}$ and its associated C_j^i is a conjunction of conditions as in (8.10). It will be convenient to write $\mathcal{G}(\mathcal{T}_j^i) = \mathcal{G}(\mathbf{t}_1, \ldots, \mathbf{t}_k)$ for the field of function germs generated over the exp–log constants by $\mathbf{t}_1, \ldots, \mathbf{t}_k$ and the application of z–functions (to arguments tending to zero of course).

We next look in more detail at the matter of building the assumptions and conditions from the tower (8.12).

8.2.4 Extension by a Logarithm

Let $f = f_i = \log g$ with $g \in \mathcal{F}_{i-1}$. We may suppose that we are extending the case $(\mathcal{T}_j^{i-1}, C_j^{i-1})$, $1 \le j \le \beta$, so that $g \in \mathcal{G}(\mathcal{T}_j^{i-1})$. Then we have a real–power $\mathcal{T}_j^{i-1}$ monomial asymptotic to g; i.e.

$$g = A\mathbf{t}_1^{r_1}\mathbf{t}_2^{r_2}\cdots\mathbf{t}_k^{r_k}(1 + h), \tag{8.13}$$

where A is a constant and h is an element of $\mathcal{G}(\mathcal{T}_j^{i-1})$ which tends to zero. Now $\log A$ and $\log(1 + h)$ belong to $\mathcal{G}(\mathcal{T}_j^{i-1})$ and so does the logarithm of any $\mathbf{t}_\alpha$ which is an exponential. Other $\log \mathbf{t}_\alpha$ are of the form $\log_{m+1} x$ or $\log_{m_\lambda+1} y_\lambda$, $\lambda = 1, \ldots, n$. Some of these may already be in $\mathcal{G}(\mathcal{T}_j^{i-1})$, but where this is not so the appropriate $\log_{m+1} x$ and $\log_{m_\lambda+1} y_\lambda$ will have to be added to $\mathcal{T}_j^{i-1}$ in order to form $\mathcal{T}_j^i$. It is clear from the conditions imposed on $\mathcal{T}_j^{i-1}$, that we shall have to add at most one $\log_{m+1} x$ and one $\log_{m_\lambda+1} y_\lambda$ for each $\lambda = 1, \ldots, n$.

Next we obtain the various C_j^i by adding the new logarithms one at a time into the list of variables. Since there will typically be different possible orderings between the comparability classes of the new logarithms and those of the existing $\mathbf{t}_\alpha$, we are likely to have to split C_j^{i-1} into subcases at this stage.

Suppose then that we are adding $\log_{m+1} x$; the addition of a $\log_{m_\lambda+1} y_\lambda$ is similar. Two things need to be taken into account. Firstly $\log_m x$ will already be present in the C_j^{i-1} list and obviously we insert $\log_{m+1} x$ before $\log_m x$. Secondly existing order relations involving $\gamma_1(\log_m x)$ may have implications for $\gamma_1(\log_{m+1} x)$. So for example if $\gamma_1(\log_m x) \leq \gamma_1(\log_\mu y_i)$ then $\gamma_1(\log_{m+1} x) \leq \gamma_1(\log_{\mu+1} y_i)$. Subject to these restrictions we create a new C_j^i for each possible position of $\log_{m+1} x$, adjusting the index j as necessary.

8.2.5 Extension by an Exponential

Suppose now that $f = f_i = \exp g$ where $g \in \mathcal{F}_{i-1}$. As before we may assume that we are extending the case $(\mathcal{T}_j^{i-1}, C_j^{i-1})$ and that we have a $\mathcal{T}_j^{i-1}$ monomial asymptotic to g given by (8.13). Often this will be enough for us to determine that the limit of g is infinite, or perhaps zero. Otherwise, when for example there two different $\mathbf{t}_i$ of maximal comparability class in the monomial, we will have to split into the subcases that give $g \to \infty$, $g \to -\infty$, $g \to 0$ and $g \to K$, $K \in \mathbb{R} \setminus \{0\}$. When the limit of g is finite, we can express f as an element of $\mathcal{G}(\mathcal{T}_j^{i-1})$ using the z-function $\mathrm{zexp}(g - \lim g)$. Otherwise $g \to \pm\infty$. We treat the case $g \to \infty$, that of $g \to -\infty$ being closely similar.

We consider the limit of $g/\log \mathbf{t}_\alpha$ for each $\mathbf{t}_\alpha \in \mathcal{T}_j^{i-1}$. If $\log \mathbf{t}_\alpha \in \mathcal{F}_{i-1}$ this limit can be calculated in $\mathcal{G}(\mathcal{T}_j^{i-1})$ using the condition C_j^{i-1}, again splitting into cases as above if necessary. This determines the order relation between $\gamma_1(\exp g)$ and $\gamma_1(\mathbf{t}_\alpha)$. If we are in a case where $g \sim K \log \mathbf{t}_\alpha$, with K a non-zero constant, we write $g_1 = g - K \log \mathbf{t}_\alpha$. Then $\exp g = \mathbf{t}_\alpha^K \exp g_1$ and we continue by applying our method to $\exp g_1$. Clearly $g_1/\log \mathbf{t}_\alpha \to 0$ and so $\gamma_1(\exp g_1) < \gamma_1(\mathbf{t}_\alpha)$. Moreover g_1, like g, belongs to $\mathcal{G}(\mathcal{T}_j^{i-1})$ which contains only finitely many comparability classes. Hence the process of replacing $\exp g$ by $\exp g_1$ can only happen a finite number of times.

In cases where $\gamma_0(g)$ is not equal to any $\gamma_0(\log \mathbf{t}_\alpha)$ with $\log \mathbf{t}_\alpha \in \mathcal{G}(\mathcal{T}_j^{i-1})$, we add $f = \exp(g)$ to $\mathcal{G}(\mathcal{T}_j^{i-1})$ and split C_j^{i-1} according to the possible positions of f, taking into account the order relations already obtained form the calculations of the limits of $g/\log \mathbf{t}_\alpha$ for the various values of α.

If we find that we have a condition of the form $\gamma_1(\exp g) = \gamma_1(\exp h)$, then g/h tends to some finite limit, K. We substitute g in our expression by $Kh + h_0$, where $K \in \mathbb{R} \setminus \{0\}$ and $\gamma_0(h_0) < \gamma_0(h)$. The condition $\gamma_1(\exp g) = \gamma_1(\exp h)$ is replaced by $\gamma_1(\exp(h_0)) < \gamma_1(\exp h)$. This ensures that at any stage at most $n+2$ of the $\mathbf{t}_\alpha$ can have the same comparability class, namely $\log_m x$, $\log_{m_\lambda} y_\lambda$, $1 \leq \lambda \leq n$ and one $\exp g$ as above. Thus the requirements on the new $\mathcal{D}_i$ will be met by our construction. Note however that in some cases the monomial for the given expression may contain some undetermined constants and it may later prove necessary to split such cases according to the values of these constants.

8.2.6 Obtaining the Nested Forms

We commence by showing how to obtain a first pnf in the case when there are only two variables, x and y. Here there must be at least one question–mark estimate for the input function F, or else there are no solutions to the equation $F = 0$. Any question–mark estimate corresponds to an equality between comparability classes. If we have an equality of the form $\gamma_1(\log_n y) = \gamma_1(\log_m x)$, we can deduce the pnf

$$y = \exp_{n+1}(\log_{m+1} x \cdot \phi(x)).$$

In the case when three comparability classes are equal we must have such an equality, so the only remaining cases are $\gamma_1(\log_n y) = \gamma_1(\exp w)$ and $\gamma_1(\log_n x) = \gamma_1(\exp w)$. We treat the first of these, the second being almost identical. Thus in the case considered we must have $\log_{n+1} y \sim K_1 w$ and the value of K_1 will be known. Now we can apply our method to the equation

$$\log_{n+1} y = K_1 w. \tag{8.14}$$

It may be necessary to add $\log_{n+1} y$ as a new scale element. Nonetheless there is no danger of non-termination since (8.14) has at least one fewer exponential than our original equation, $F_1(x, y) = 0$. Moreover, because of the way our method compares asymptotics, any solution of $\log_{n+1} y \sim K_1 w$ will be found as a possible solution of (8.14).

Once we have a pnf we can refine our estimates to obtain either a nested form, or indeed a nested expansion, or else the information that no solution of this type exists. Suppose that our pnf is $y(x) = \exp_n(\log_m x\phi(x))$. we substitute for y in our original equation, $F_1(x, y) = 0$, to obtain a new equation in x and ϕ. We compute a first pnf for this, which gives us a second pnf for y. Then we can continue the process. We must obtain a nested form after a finite number of steps because of the following result from [88].

LEMMA **34** *Let $F(x, y)$ be an exp–log function of two variables and let $y = y(x)$ be a solution of $F(x, y) = 0$. Then $y(x)$ has a nested form whose length can be bounded in terms of the structure of F.*

Proof of Lemma 34
We already know that y will be an element of a Hardy field. In fact it also satisfies an algebraic differential equation over $\mathbb{R}$. For we may assume that F is irreducible and we can build a tower of function fields

$$\mathbb{R} \subset \mathbb{R}(y) \subset \mathcal{F}_2 = \mathbb{R}(x, y) \subset \mathcal{F}_3 \subset \cdots \subset \mathcal{F}_k$$

where for $i = 3, \ldots, k$, $\mathcal{F}_i = \mathcal{F}_{i-1}(h_i)$ with h_i an exponential or a logarithm, and $F \in \mathcal{F}_k$. Then we can eliminate $h_k, h_{k-1}, \ldots, h_3$ between the equations

$$F(x, y(x)) = 0, \quad \frac{d}{dx}(F(x, y(x))) = 0, \quad \cdots, \quad \frac{d^{k-2}}{dx^{k-2}}(F(x, y(x))) = 0$$

to obtain an algebraic differential equation for y of order no more than k over $\mathbb{R}(x)$. Then Theorem 18 implies that y has a nested form $\{(\epsilon_i, s_i, m_i, d_i, \phi_i), i = 1, 2, \ldots, R\}$ such that

$$\sum_{j=1}^{R} s_j + \delta_k + m_k \leq k - 1. \tag{8.15}$$

This suffices to prove Lemma 34.

In theory there is still the possibility that we might generate an infinite sequence of pnfs which do not correspond to any solution, but we will know k and can reject any sequence of pnfs which violates (8.15) as not corresponding to any solution.

Another potential source of difficulty is that we might obtain a pnf of the form $\exp_n\{\log_m x(K+z)\}$ with K a non-zero constant and z tending to zero; see Section 4.4. We then substitute $\exp_{n-1}(\log_{m-1}^K x \cdot \exp(z \log_m x))$ for y and take $\phi = \exp(z \log_m x))$ as the new variable. We can then continue as above. If $K = 1$ it is possible that the same situation may recur at the next stage. However n and m will have been reduced by one, and so this can only happen a finite number of times.

At any stage in the calculation of the pnfs, we might arrive at a situation where there is no question–mark estimate. In that case there is no solution corresponding to that branch of the calculation. Otherwise we arrive at a nested form in a finite number of steps. If we want more terms in the nested expansion, we just continue substituting and computing pnfs.

8.2.7 Example 1

We start with the single implicit equation

$$f(x, y) = \log|xe^{-y} - y| - y + x + 1 = 0, \tag{8.16}$$

and look for solutions with $y \to \infty$. Our function tower is

$$\mathcal{F}_1 = \mathbb{R}(x, y) \subset \mathcal{F}_2 = \mathbb{R}(x, y, e^y) \subset \mathcal{F}_3 = \mathbb{R}(x, y, e^y, \log|xe^{-y} - y|). \tag{8.17}$$

We have $\mathcal{D}_1 = \{(\mathcal{T}_1^1, C_1^1), (\mathcal{T}_2^1, C_2^1), (\mathcal{T}_3^1, C_3^1)\}$ with $\mathcal{T}_1 = \mathcal{T}_2 = \mathcal{T}_3 = \{x^{-1}, y^{-1}\}$ and $C_1^1 = (\gamma_1(x) < \gamma_1(y))$, $C_2^1 = (\gamma_1(x) = \gamma_1(y))$, $C_3^1 = (\gamma_1(y) < \gamma_1(x))$.

To obtain $\mathcal{D}_2$ we must add e^{-y} to the lists $\mathcal{T}_i^1$, and give the new conditions bearing in mind that $\gamma_1(e^y) > \gamma(y)$ (since $y \to \infty$). So $\mathcal{D}_2 = \{(\mathcal{T}_i^2, C_i^2), i = 1, \ldots, 5\}$ where $\mathcal{T}_i^2 = \{x^{-1}, y^{-1}, e^{-y}\}$ for all i, and

$$C_1^2 = (\gamma_1(x) < \gamma_1(y) < \gamma_1(e^y)), \qquad C_2^2 = (\gamma_1(x) = \gamma_1(y) < \gamma_1(e^y)),$$
$$C_3^2 = (\gamma_1(y) < \gamma_1(x) < \gamma_1(e^y)), \qquad C_4^2 = (\gamma_1(y) < \gamma_1(x) = \gamma_1(e^y)),$$
$$C_5^2 = (\gamma_1(y) < \gamma_1(e^y) < \gamma_1(x)).$$

Before adding the logarithm we must compute a monomial asymptotic to $xe^{-y} - y$ in each of the five cases. We have $xe^{-y} - y \sim -y$ in cases C_1^2, C_2^2 and C_3^2 and $xe^{-y} - y \sim xe^{-y}$ in case C_5^2. We see that we need to add $\log y$ to $\mathcal{T}_i^2$, $i = 1, 2, 3$ and $\log x$ to $\mathcal{T}_5^2$. In case C_4^2 we obtain a question–mark estimate for $xe^{-y} - y$. We make no addition to $\mathcal{T}_4^2$ at this stage. The new conditions are

$$\begin{aligned}
C_1^3 &= (\gamma_1(x) < \gamma_1(\log y) < \gamma_1(y) < \gamma_1(e^y)),\\
C_2^3 &= (\gamma_1(x) = \gamma_1(\log y) < \gamma_1(y) < \gamma_1(e^y)),\\
C_3^3 &= (\gamma_1(\log y) < \gamma_1(x) < \gamma_1(y) < \gamma_1(e^y)),\\
C_4^3 &= (\gamma_1(\log y) < \gamma_1(x) = \gamma_1(y) < \gamma_1(e^y)),\\
C_5^3 &= (\gamma_1(\log y) < \gamma_1(y) < \gamma_1(x) < \gamma_1(e^y)),\\
C_6^3 &= (\gamma_1(y) < \gamma_1(x) = \gamma_1(e^y)),\\
C_7^3 &= (\gamma_1(y) < \gamma_1(e^y) < \gamma_1(\log x) < \gamma_1(x)),\\
C_8^3 &= (\gamma_1(y) < \gamma_1(e^y) = \gamma_1(\log x) < \gamma_1(x)),\\
C_9^3 &= (\gamma_1(y) < \gamma_1(\log x) < \gamma_1(e^y) < \gamma_1(x)),\\
C_{10}^3 &= (\gamma_1(y) = \gamma_1(\log x) < \gamma_1(e^y) < \gamma_1(x)).
\end{aligned}$$

These yield the two–variable estimate for $f(x, y)$

$$f(x,y) \sim \begin{cases} -y\,, & \text{if } C_1^3 \text{ or } C_2^3 \text{ or } C_3^3 \text{ or } C_5^3 \\ x\,, & \text{if } C_7^3 \text{ or } C_8^3 \text{ or } C_9^3 \text{ or } C_{10}^3 \\ ?\,, & \text{if } C_4^3 \text{ or } C_6^3. \end{cases} \tag{8.18}$$

The question–mark estimate in case C_6^3 gives $y \sim K \log x$ for some non–zero constant K, while in case C_4^3 we get $y \sim x$. We can refine the case C_6^3 into the sub-cases $K > 1$, $K = 1$ and $K < 1$. The question mark persists only when $K = 1$ since $f \sim -y$ when $K > 1$ and $f \sim x$ when $K < 1$. Thus there are possible solutions asymptotic to $\log x$ and to x, and here we are able to make use of the intermediate value theorem to show that such solutions actually exist.

Let $b(x)$ denote the inverse function of ye^y. It is known that

$$b(x) \sim \log x - \log\log x + \frac{\log\log x}{\log x}(1 + o(1)); \tag{8.19}$$

see for example [30], pp 25-26, or [90]. We may rewrite (8.16) as $f(x, y) = \log|x - ye^y| - 2y + x + 1$ and we then see that $f(x, b(x+1)) = x - 2b(x+1) + 1 \sim x$. On the other hand

$$f(x, b(x + e^{-x})) = -2b(x + e^{-x}) + 1 \sim -2b(x + e^{-x}).$$

So for a large value of x, $f(x, b(x + 1)) > 0$ and $f(x, b(x + e^{-x})) < 0$. By the intermediate value theorem there is then a $y_1(x)$ with $b(x + e^{-x}) < y_1(x) <$

$b(x+1)$ such that $f(x, y_1(x)) = 0$. So there is indeed a solution of (8.16) which is asymptotic to $\log x$.

Similarly $f(x, x + \sqrt{x}) < 0$ and $f(x, x - \sqrt{x}) > 0$ for large x, and so there is also a solution of (8.16) asymptotic to x. Other examples are given in [88].

Now suppose that we have a system of $n \geq 2$ equations, $F_1 = F_2 = \cdots = F_n = 0$. We obtain $(n+1)$–variable estimates for $F_1, \ldots, F_n$ as above. For a solution set to exist we must have at least one question–mark estimate for each equation and this will have associated one of the 'equality' conditions of the type (8.10), with at most one exponential $\mathbf{t}_i$. Single equalities of the $\gamma_1(\log_n y) = \gamma_1(\exp w)$ or $\gamma_1(\log_n x) = \gamma_1(\exp w)$ can be dealt with as in the case $n = 1$. Hence we may assume that we have n or more equalities of the one of the forms $\gamma_1(\log_s y_i) = \gamma_1(\log_t y_j)$ or $\gamma_1(\log_s y_i) = \gamma_1(\log_t x)$, $1 \leq i < j \leq n$. These will be linked to linear equations for the limits of the various $\log_{s+1} y_i / \log_{t+1} y_j$ as in (8.10). In many cases we will then able to obtain a set of equalities of the form $\gamma_1(\log_{s_i} y_i) = \gamma_1(\log_{t_i} x)$ for each $i = 1, \ldots, n$, and from these we can get pnfs for $y_1, \ldots, y_n$ as in the case $n = 1$. Then we can make substitutions as before and repeat the whole process with the new variables, thus arriving at a nested form for each y_i (or the message 'no solution') for each set of question–mark estimates.

However we might get a degenerate set of equalities between the γ_1s even if the system $F_1 = F_2 = \cdots = F_n = 0$ itself is not degenerate. We could try substituting in any pnfs that we can get, or in some way manipulating the system of equations, but there is no guarantee that such techniques would work. Of course the problem is asymptotic cancellation between different equations. Therefore all that can be said is that in the generic case we do get sets of nested forms which must contain the nested forms of actual solutions. In other cases the algorithm may have to report failure, but at least it will not produce nested forms of non–solutions.

8.2.8 Example 2

Consider the system of equations

$$z \log y - xy - \log x \log z = 0, \tag{8.20}$$

$$yz \log x - 2x \log y + \log z = 0. \tag{8.21}$$

If we were to strictly follow the given algorithm, we would add the variables and the logarithms one at a time, splitting into cases as we go, eventually arriving at 97 cases where y and z each have infinite limits. The other cases would require further analysis. However working by hand it is not too hard to see that only the cases where $\gamma_1(y) < \gamma_1(z) = \gamma_1(x)$ lead to a solution. We must then have $\gamma_1(\log z) = \gamma_1(\log x)$. Thus we suppose that we are about to add $\log y$ and have the following existing cases.

$$
\begin{aligned}
C_5^1 &: \gamma_1(\log x) = \gamma_1(\log z) < \gamma_1(y) < \gamma_1(x) = \gamma_1(z),\\
C_5^2 &: \gamma_1(\log x) = \gamma_1(\log z) = \gamma_1(y) < \gamma_1(x) = \gamma_1(z),\\
C_5^3 &: \gamma_1(y) < \gamma_1(\log x) = \gamma_1(\log z) < \gamma_1(x) = \gamma_1(z).
\end{aligned}
$$

This does not take account of the possibility of y tending to a non–zero finite limit, but in fact there are no solutions of that form, as can been seen by inspection.

Since $\gamma_1(y) < \gamma_1(x)$ we must have $\gamma_1(\log y) \le \gamma_1(\log x)$. So C_5^1 splits into

$$
\begin{aligned}
C_6^1 &: \gamma_1(\log y) < \gamma_1(\log x) = \gamma_1(\log z) < \gamma_1(y) < \gamma_1(x) = \gamma_1(z),\\
C_6^2 &: \gamma_1(\log y) = \gamma_1(\log x) = \gamma_1(\log z) = \gamma_1(y) < \gamma_1(x) = \gamma_1(z).
\end{aligned}
$$

Similarly C_5^2 and C_5^3 lead to

$$
\begin{aligned}
C_6^3 &: \gamma_1(\log y) < \gamma_1(\log x) = \gamma_1(\log z) = \gamma_1(y) < \gamma_1(x) = \gamma_1(z),\\
C_6^4 &: \gamma_1(\log y) < \gamma_1(y) < \gamma_1(\log x) = \gamma_1(\log z) < \gamma_1(x) = \gamma_1(z).
\end{aligned}
$$

In each of these we have $\gamma_1(z) = \gamma_1(x)$ and so we write $z = x^\lambda$. On substituting into (8.20) we see that λ must tend to one, and so we put $z = x\alpha$ with $\gamma_1(\alpha) < \gamma_1(x)$. Substitution then gives the equations

$$x(\alpha \log y - y) - \log x(\log x - \log \alpha) = 0, \tag{8.22}$$

$$x(\alpha y \log x - 2\log y) + \log x + \log \alpha = 0. \tag{8.23}$$

From here we get $\alpha \log y \sim y$ and $\alpha y \log x \sim 2 \log y$. Hence

$$\frac{y}{\log y} \sim \pm\sqrt{\frac{2}{\log x}}. \tag{8.24}$$

We can then apply the methods of 7.1 to (8.24), giving

$$y \sim \pm\sqrt{\frac{2}{\log x}} \log y \sim \pm\sqrt{\frac{2}{\log x}} \log\left\{\pm\sqrt{\frac{2}{\log x}} \log y\right\}.$$

Hence

$$y = \frac{\mp \log\log x}{\sqrt{2\log x}}(1+\epsilon_1), \tag{8.25}$$

where $\epsilon_1 \to 0$. Then since $\alpha \sim y/\log y$, we have

$$z = x\alpha = \frac{xy}{\log y}(1+o(1)) = \pm x\sqrt{\frac{2}{\log x}}(1+\epsilon_2), \tag{8.26}$$

where $\epsilon_2 \to 0$. These relations come from the case C_6^3, with C_6^1, C_6^2 and C_6^4 yielding contradictions.

Further terms in the expansions of y and z may be obtained by using (8.25) and (8.26) to substitute into (8.20) and (8.21) and solving for ϵ_1 and ϵ_2.

8.2.9 Example 3

We consider the pair of equations

$$z \exp\left(\frac{\log^2 z}{\log y}\right) - yx^2 - \frac{1}{x} = 0 \tag{8.27}$$

$$x^2 - \exp\left(\frac{\log^2 y \log z}{\log^2 x}\right) + \frac{1}{y} = 0. \tag{8.28}$$

We suppose that we are in the case $\gamma_1(z) = \gamma_1(y) = \gamma_1(x)$. Of course it is a direct consequence that $\gamma_0(\log z) = \gamma_0(\log y) = \gamma_0(\log x)$. We take $\log y = \alpha \log x$ and $\log z = \beta \log x$ with $a = \lim \alpha$ and $b = \lim \beta$. Then a and b are finite and non–zero.

Our next step is to add the exponential of $\log^2 z/\log y$. We see that $\log^2 z/\log y = (\beta^2/\alpha)\log x$ and hence

$$\exp\left(\frac{\log^2 z}{\log y}\right) = x^{\beta^2/\alpha}.$$

Then after similarly adding the exponential of $\log^2 y \log z/\log^2 x$, we may rewrite (8.27) and (8.28) respectively as

$$x^{\beta+\beta^2/\alpha} - x^{2+\alpha} - x^{-1} = 0, \tag{8.29}$$

$$x^2 - x^{\alpha^2\beta} + x^{-\alpha}. \tag{8.30}$$

The algorithm will now obtain estimates for the left–hand side under the various assumptions concerning a and b. However it is easy for us to see that we only get a question–mark estimate for the left–hand side of (8.29) when one of the conditions

[A1] $2 + a = -1 = b + b^2/a,$

[A2] $2 + a = -1 > b + b^2/a,$

[A3] $b + b^2/a = -1 > a + 2,$

[A4] $b + b^2/a = a + 2 > -1$

holds. Similarly for(8.30) we must have one of

[B1] $a^2 b = -a = 2,$

[B2] $a^2 b = -a > 2,$

[B3] $a^2 b = 2 > -a,$

[B4] $-a = 2 > a^2 b.$

We obtain one solution ($(a, b) \approx (-1.08, 1.71)$) from A3 and B3 and two ($(a, b) \approx (1.18, 1.44)$ and $(a, b) \approx (-1.60, 0.78)$) from A4 and B3. The pair A4, B4 yield the real solution $a = -2$, $b = 0$. This value of b is not compatible with the assumptions of the case under consideration, but a possible solution in the case $\gamma_1(z) < \gamma_1(y) = \gamma_1(x)$ is indicated. All other combinations Ai, Bj give contradictions.

9

Star–Product Expansions

In this chapter we look at expansions which use star products. We generalize star products and apply the corresponding expansions to some functions which grow faster than any iterated exponential. This section continues Section 4.5 and provides a development of the second part of [104].

9.1 Exp–Log Expansions

Let N and K be fixed natural numbers. We wish to consider expansions using a field of real constants, base functions $\exp_n(x)$, $|n| \leq N$, and the operations $\star_k$ and $/_k$, $0 \leq k \leq K$. We are seeking standard forms of expansions for expressing asymptotic growth. There is room for considerable debate as to what these should be.

One clear requirement is that the limit of a function should be immediately recognisable from its expansion in standard form. The same should be true after the removal of a finite number of leading terms. This leads us to our first principle.

> **[P1]** The most important information should be given first. In particular, terms should generally decrease from left to right.

Cancellation is one thing that might prevent us recognising the limit from the expansion, and we certainly do not want it to be present within our standard expansions. So we should avoid writing $a \star_k b$ when this composition is significantly smaller than either of the components, a and b. In other words, we at least want $\log_k(a \star_k b) \sim C \max\{\log_k(a), \log_k(b)\}$ for some non-zero constant C. A stronger condition, and indeed a more natural one, is $\gamma_k(b) < \gamma_k(a)$. For example when $k = 0$, this says that when we write $a + b$ we should have $b = o(a)$, which is precisely the requirement of asymptotic series. We now have our second principle.

> **[P2]** When we write $a \star_k b$ we must have $\gamma_k(b) < \gamma_k(a)$.

We have seen with multiseries and nested forms that exponentials, for example, have to be treated differently according to whether their arguments have finite or infinite limits. We may expect the same to be true for star expansions. Indeed infinite expansions for exp–log functions can only come from the z–functions. Thus we have the following.

> **[P3]** Exponentials of terms tending to finite limits, and logarithms and powers of terms tending to non–zero finite limits need to be written as z–functions and be regarded as standing for their series expansions.

Our fourth principle also embodies something that we have seen to be necessary with multiseries.

> **[P4]** Series expansion should always be carried out first with respect to the most rapidly varying term tending to zero.

Experience with multiseries might also suggest adding the following.

> **[P5]** An expansion should be presented as a $\star_k$–expansion for the smallest k present, with coefficients which are $\star_k$–expansions for the next smallest value of k, and so on.

Notice that if only $\star_0$ and $\star_1$ occur, the fifth principle asserts that we should write our expansion as a sum of products, which is of course the most usual fashion. However it is not the only possibility, and we look at others later in this section.

In order to comply with **[P1]** - **[P5]**, we need to rewrite a given expression into the form

$$E = \prod_{m_0}^{\star_0} \prod_{m_1}^{\star_1} \cdots \prod_{m_k}^{\star_k} a_{\mathbf{m}}, \tag{9.1}$$

where $\mathbf{m} = \{m_0, \ldots, m_k\}$ and each $a_{\mathbf{m}}$ is either a power of a base element $\exp_n(x)$, $n \in \mathbb{Z}$ or a real constant. The multiproduct in (9.1) needs further comment.

Firstly any of the products may in theory be infinite, but in practice one would truncate after a certain stage. A point arises with

$$c \exp\left(\frac{x \log x}{\log x - 1}\right) = \prod_{i=0}^{\infty} \exp(x \log^{-i} x) c = \text{sign}(c) \prod_{i=0}^{\infty} \exp(x \log^{-i} x)\, |c|,$$

where c is a constant. Although by our rules the factor c comes after all the others, we need to keep at least its sign when we truncate because that determines the limit.

Secondly for $0 \le i < k$ and each $p_1, \ldots, p_i$ in range, **[P2]** requires that

$$\gamma_{m_i}\left(\prod_{m_{i+1}}^{\star j_{i+1}} \cdots \prod_{m_k}^{\star j_k} \hat{a}_{p_i}\right) > \gamma_{m_i}\left(\prod_{m_{i+1}}^{\star j_{i+1}} \cdots \prod_{m_k}^{\star j_k} \hat{a}_{p_i+1}\right),$$

where $\hat{a}_q = a_{p_0,\dots,p_{i-1},q,m_{i+1},\dots,m_k}$. That is to say the γ_{m_i}s of the terms in the $\star_{m_i}$-product decrease.

Finally we must comply with **[P4]**. So for each α, $0 \le \alpha \le k-1$, and each $i_0, \dots, i_{\alpha+1}$ there must be only a finite number of values of

$$\gamma_{\alpha+1}\left(\prod_{m_{\alpha+2}}^{\star_{\alpha_2}} a_{i_0,\dots,\, i_{\alpha+1},m_{\alpha+2},\dots,\, m_k}\right),$$

and for $\alpha \le k-2$ the expansion

$$\prod_{m_\alpha}^{\star_\alpha}\prod_{m_{\alpha+1}}^{\star_{\alpha+1}}\left(\prod_{m_{\alpha+2}}^{\star_{\alpha_2}} a_{i_0,\dots,\, i_{\alpha+1},m_{\alpha+2},\dots,\, m_k}\right)$$

must be grouped as an expansion in decreasing powers of the highest $\gamma_{\alpha+1}$ with coefficients which are expansions in the next highest $\gamma_{\alpha+1}$, and so on. For the case $\alpha = 0$ this is our basic multiseries requirement, while for other values of α it is equivalent to the requirement that the same is true of expansions which take place inside $\exp_\alpha(\;)$.

Once we have an expression for an exp–log function f in the form (9.1), we can obtain the limit of f as follows. If $\prod_{m_1}^{\star_1}\cdots\prod_{m_k}^{\star_k} a_{0,m_1,\dots,m_k}$ has only one factor, which is a constant, then that is the limit of f. Otherwise if there is a last factor which is a constant, let $s = \pm 1$ be its sign; if not let $s = +1$. Now $a_{0,\dots,0}$ is either a constant or a constant power of $\exp_n x$ for some $n \in \mathbb{Z}$. If $a_{0,\dots,0} \to 0$ then $f \to 0$. If $a_{0,\dots,0} \to \infty$ then $f \to \infty$ if $s = 1$ and $f \to -\infty$ if $s = -1$.

In 9.1.1 we look at how to rewrite an exp–log expression as a standard star product expansion. Then in 9.1.2 we examine the problem of comparing the γ_j values of two star products.

9.1.1 Rewriting Exp–Log Expressions Into Standard Star Expansion Form

Our input will be a tree where each leaf is either the variable x or a real constant and each node is one of the operators $+$, $-$, $\times$, $\div$, ^, exp or log. We show how to obtain the standard form of a op b when a and b are in standard form and op is any of the above operators.

Case $a + b$.

If $a = a_1 + \cdots + a_n$ and $b = b_1 + \cdots + b_m$ with $n, m \ge 1$ and the a_i and b_j given as products, then we need to order the combined sum so that the summands are of strictly decreasing γ_0. For this we need to determine the order relations between each $\gamma_0(a_i)$ and $\gamma_0(b_j)$ and to rewrite $a_i + b_j$ if these are equal. We make use of the fact that $\gamma_0(a_1) > \gamma_0(a_2) > \cdots > \gamma_0(a_n)$ and similarly for the b_j.

So, making a slight change of notation, it is sufficient to deal with the case when a and b are products; say $a = a_1 a_2 \cdots a_n$ and $b = b_1 b_2 \cdots b_m$. This matter is treated in §9.1.2. We note that by induction we can determine the order relations between the $\gamma_1(a_i)$ and the $\gamma_1(b_j)$ since this is equivalent to comparing the $\gamma_0(\log a_i)$ and $\gamma_0(\log b_j)$, which is similar to the previous paragraph but with the largest $\star_k$ replaced by $\star_{k-1}$.

Thus we can reduce to the case when at least one of a and b, say a, is not presented as a star product. Hence it is a real constant or a constant power of a base function. If b is also one of these, then it is easy to compute the limit of a/b. If b is a product, $\prod b_i$, we compare $\gamma_1(a)$ and $\gamma_1(b_1)$ as above, and if these are equal, we rewrite and if necessary compare the residue with b_2 etc.

If it turns out that $b/a \to l \in \mathbb{R} \setminus \{0\}$, we rewrite $a+b$ as $(l+1)a + (b - la)$ and partially expand $b - la$ to remove the cancelling leading terms. Note that we are able to do this for the same reason that we can do it when dealing with multiseries, namely because **[P4]** holds.

Since the expansion is likely to involve z–functions, it is useful to have some notation for the tails. We therefore define

$$\operatorname{zexp}_i t = \exp t - \sum_{j=0}^{i} \frac{t^j}{j!}, \tag{9.2}$$

$$\operatorname{zlog}_i t = \log(1+t) + \sum_{j=1}^{i} \frac{(-t)^j}{j}, \tag{9.3}$$

$$\operatorname{zpow}_i(t,c) = (1+t)^c - 1 - \sum_{j=1}^{i} \frac{\Gamma(c+1)t^j}{\Gamma(c+1-j)j!}, \tag{9.4}$$

where $c \in \mathbb{R}$ and $i \geq 1$. The problem that arises when t is a sum of elements of different comparability class was described in Section 5.1. It is handled here in similar fashion.

Cases $a - b$ and $a \times b$.

$a - b$ is dealt with in the same way as $a + b$. As for the product,

$$(a_1 + a_2 + \cdots + a_n) \times (b_1 + b_2 + \cdots + b_m)$$

is expanded by the usual distributive law and the resulting factors ordered as in the case $a + b$.

Case $a \div b$

First we check that b is not equivalent to zero. Then if $a = a_1 + a_2 + \cdots + a_n$ and $b = b_1 + b_2 + \cdots + b_m$, we write

$$a \div b = \frac{a_1}{b_1} \operatorname{zpow}\left(\frac{b_2 + \cdots + b_m}{b_1}, -1\right) + \cdots + \frac{a_n}{b_1} \operatorname{zpow}\left(\frac{b_2 + \cdots + b_m}{b_1}, -1\right).$$

If desired, the zpows may be expanded using (9.4). Then the products in the summands can be handled as above.

Case a^b.
If b is a positive integer less than some threshold value, we expand

$$(a_1 + \cdots + a_n)^b$$

using the distributive law and order the resulting summands as above. If b is otherwise given explicitly as a real constant, we make the appropriate zero–equivalence checks and then write

$$a^b = a_{11}^b \cdots a_{1p}^b \operatorname{zpow}\left(\frac{a_2 + \cdots + a_n}{a_1}, b\right),$$

where $a_1 = a_{11} \cdots a_{1p}$. Again we can expand using (9.4) if we want.

If b is not given explicitly as a constant we can determine if it is equivalent to one by differentiating with respect to x and seeing whether the result is equivalent to zero. If so we replace b by its limit, which by induction we can calculate, and continue as above.

If b is not equivalent to a constant, we rewrite a^b as $\exp b \star_2 a$.

Case $\exp a$.
With $a = a_1 + \cdots + a_n$, we may assume that we can find the limit, l_i, of each a_i. If this is finite and non zero for some value of i, we replace a_i by $l_i + (a_i - l_i)$. Then $a_{i+1}, \ldots, a_n$ tend to zero, and we write

$$\exp a = \exp a_1 \exp a_2 \cdots \exp a_{i-1} e^{l_i}(1 + \operatorname{zexp}((a_i - l_i) + a_{i+1} + \cdots + a_n)).$$

Exponentials of products are rewritten as $\star_2$ products of exponentials, and in general exponentials of $\star_j$ products as $\star_{j+1}$ products of exponentials, using Proposition 6.7.

Case $\log a$.
We write

$$\log(a_1 + \cdots + a_n) = \log a_1 + \operatorname{zlog}\left(\frac{a_2 + \cdots + a_n}{a_1}\right), \tag{9.5}$$

possible expanding the zlog using (9.3). Then for $j = 1, 2, \ldots$ the logarithm of a $\star_j$ product, $j \geq 1$, is written as the $\star_{j-1}$ product of the logarithms of the $\star_j$ factors, by Proposition 6.7 again.

Thus we can convert an exp–log expression to the form (9.1). If we wish to further expand our star product, we may use the relations (9.2)–(9.4) and the Generalized Distributive Law. In practice it would probably be more efficient to carry out this expansion while building the product from the exp–log expression as we have indicated.

It is clear that there are some underlying similarities between the algorithm we have just given and the exp–log algorithm for multiseries of Section 5.1. To some extent at least that is a reflection of the choice we made with [**P5**]. Even so, there may be some advantages in the new notation. For, from a star point of view, the usual multiseries notation stresses addition and multiplication while sweeping the higher order operations into the definitions of the monomials. By contrast, the star notation gives all the operations equal weight, making the algebraic structure clearer. Star products are also potentially more general than multiseries but in some situations, for example in the paragraphs preceding this section, they do seem to be more difficult to handle, perhaps because of unfamiliarity.

One way of looking at **[P5]** is to regard it as a strategy of expanding by the lowest possible star operation. At the other end of the scale, if $\star_k$ is the highest order star product occurring in the expression, then apart from an initial constant and perhaps a couple of signs, one can get an asymptotic expansion (say of an exp-log function) using just $\star_k$ and $\star_{k-1}$. Indeed such expansions correspond to nested expansions. In this situation, there is something to be said for using $/_k$ and $\star_{k-1}$ rather than $\star_k$ and $\star_{k-1}$ because later terms, which typically tend to zero, usually involve negative powers. In the case $k = 1$, the expansion is then of the form

$$a_0 \star_0 a_1 /_1 (a_2 \star_0 a_3 /_1 (a_4 \star_0 \cdots,$$

which is otherwise written

$$a_0 + a_1/(a_2 + a_3/(a_4 + \cdots.$$

Thus nested expansions may be seen as generalized continued fractions rather than generalized series.

9.1.2 Order Comparisons

In this section we look at the problems of comparing γ_ks of star polynomials in standard form and rewriting $a \star_k b$ if we discover that $\gamma_k(a) = \gamma_k(b)$. The rewriting is fairly simple, at least in theory.

LEMMA **35** *Suppose that $k > 0$ and $\gamma_k(a) = \gamma_k(b)$. Then we can compute $r \in \mathbb{R}$ and c with $\gamma_k(c) < \gamma_k(a)$ such that*

$$a \star_k b = a^{\star_k^r} \star_k c = (a \star_{k+1} \exp_k r) \star_k c.$$

Proof of Lemma 35
The hypotheses imply that $\log_k b \sim K \log_k a$, for some non-zero constant K. We write $r = K + 1$ and take

$$c = \exp_k(\log_k b - (r-1)\log_k a).$$

We have

$$\begin{aligned} a^{\star_k^r} \star_k c &= \exp_k(r \log_k a + \log_k c) = \exp_k(r \log_k a + \log_k b - (r-1)\log_k a) \\ &= \exp_k(\log_k a + \log_k b) = a \star_k b. \end{aligned}$$

Also

$$\frac{\log_k c}{\log_k a} = \frac{\log_k b}{\log_k a} - (r-1) = o(1).$$

This completes the proof of the lemma.

In many cases the expression for c will simplify.

The more difficult question is to decide if $\gamma_k(a) = \gamma_k(b)$. When there is cancellation, we must expect this decision to involve full expansions. We begin with two results concerning powers of base functions. In this context, it is also worth noting Lemmas 42 and 43, given in the next section within a more general setting.

LEMMA **36** *If $k, j \in \mathbb{R}$ with $k > 0$ and $m, n, s, t \in \mathbb{Z}$, then $s > t$ implies that*

$$\gamma_N\left((\exp_s x)^{\star^k_n}\right) > \gamma_N\left((\exp_t x)^{\star^j_m}\right),$$

for all $N \in \mathbb{Z}$.

In other words, taking star powers doesn't alter the order relations between the basic functions $\exp_i x$, $i \in \mathbb{Z}$.

Proof of Lemma 36

Clearly it is enough to prove this for N sufficiently large. We first observe that $\log_n((\exp_s x)^{\star^k_n}) = k \exp_{s-n} x$. Hence if we choose N greater than both m and n, we have

$$\log_N((\exp_t x)^{\star^j_m}) = \exp_{t-N}\{x(1+o(1))\} = o\left(\log_N((\exp_s x)^{\star^k_n})\right),$$

since $\log_N((\exp_s x)^{\star^k_n}) = \exp_{s-N}\{x(1+o(1))\}$. The result then follows.

LEMMA **37** *Suppose that $m, n, s \in \mathbb{Z}$ and $k, j \in \mathbb{R}$.*

(i) *If $n > m$ then*

$$\gamma_{n-1}\left((\exp_s x)^{\star^k_n}\right) > \gamma_{n-1}\left((\exp_s x)^{\star^j_m}\right)$$

if and only if $k > 1$; the reverse inequality holds if and only if $k < 1$. For the case $k = 1$, we have

$$\gamma_{n-1}\left((\exp_s x)^{\star^1_n}\right) = \gamma_{n-1}\left((\exp_s x)^{\star^j_m}\right).$$

Moreover

$$\gamma_{m-1}\left((\exp_s x)^{\star^1_n}\right) > \gamma_{m-1}\left((\exp_s x)^{\star^j_m}\right)$$

if and only if $j < 1$, and the reverse inequality holds if and only if $j > 1$.

(ii) *If $n = m$, then*

$$\gamma_{n-1}\left((\exp_s x)^{\star_n^k}\right) > \gamma_{n-1}\left((\exp_s x)^{\star_m^j}\right)$$

if and only if $k > j$.

Proof of Lemma 37
By making the change of variable $x \mapsto \log_s x$, we may take s to be zero. Then if $n > m$, we have $\log_{n-1}(x^{\star_n^k}) = \log_{n-1}^k x$ and

$$\begin{aligned}\log_{n-1}(x^{\star_m^j}) &= \log_{n-1}(\exp_m(j \log_m x)) = \log_{n-m-1}(j \log_m x)\\ &\sim \log_{n-1} x\\ &= o(\log_{n-1}^k x) \text{ if and only if } k > 1.\end{aligned}$$

If $k = 1$, we have $\log_{m-1}(x^{\star_n^k}) = \log_{m-1} x$ and $\log_{m-1}(x^{\star_m^j}) = \log_{m-1}^j x$. Thus $\gamma_{m-1}(x^{\star_m^j}) < \gamma_{m-1}(x^{\star_n^k})$ if and only if $j < 1$.

Finally if $n = m$,

$$\log_{n-1}(x^{\star_n^j}) = \log_{n-1}^j x = o(\log_{n-1}(x^{\star_n^k}))$$

if and only if $j < k$. The proof of Lemma 37 is therefore complete.

As a simple example of the application of Lemma 37, we obtain immediately that $x^{\star_2^2} > x^n$ for any constant n.

It is sometimes useful to be able to rewrite $\star_j$ products as $\star_k$ products. This is possible when $j < k$, as the following result shows, but not generally when $j > k$.

LEMMA **38** *Suppose $c/a \to 0$ and $k > 0$. Then*
(i)

$$a + c = a \star_k \exp_{k-1}(c_k),$$

where $c_1 = 1 + c/a$ and for $j = 1, \dots, k-1$,

$$c_{j+1} = 1 + \frac{\log c_j}{\log_j(a)}.$$

(ii) *More generally for $i > 0$,*

$$a \star_i c = a \star_{k+i} \exp_{k+i-1}(C_k),$$

where $C_1 = 1 + \log_i(c)/\log_i(a)$ and for $j = 1, \dots, k-1$,

$$C_{j+1} = 1 + \frac{\log C_j}{\log_{i+j}(a)}.$$

(iii) *Similarly for $i > 0$, $a \star_{-i} c = a + B_{i-1}$, where we take $B_{-1} = c$ and for $0 \le j \le i-1$,*

$$B_j = \log_{i-j}\left(1 + \frac{\exp_{i-j}(B_{j-1})}{\exp_{i-j}(a)}\right).$$

The condition that $c/a \to 0$ is inessential, but it indicates the intended use of the result. Of course we would usually write the exponentials with arguments tending to finite limits in terms of zexp.

Proof of Lemma 38
For $k = 1$ part (i) is clear, so suppose it holds for $k = j$. Then

$$\begin{aligned} a + c &= a_j \star_j \exp_{j-1}(c_j) = \exp_j(\log_j(a) + \log c_j) \\ &= \exp_j\left(\log_j(a) \cdot \left(1 + \frac{\log c_j}{\log_j(a)}\right)\right) = a \star_{j+1} \exp_j(c_{j+1}), \end{aligned}$$

which establishes (i).

Now, using (i), we have

$$\begin{aligned} a \star_i c &= \exp_i(\log_i(a) + \log_i(c)) = \exp_i(\log_i(a) \star_k \exp_{k-1}(C_k)) \\ &= \exp_i(\exp_{k-1}(\log_{k+i-1}(a) \cdot \log_{k-1}(\exp_{k-1}(C_k))) \\ &= \exp_{k+i-1}(\log_{k+i-1}(a) \cdot \log_{k+i-1}(\exp_{k+i-1}(C_k))) \\ &= a \star_{k+i} \exp_{k+i-1}(C_k). \end{aligned}$$

This gives (ii), and for (iii) we observe that

$$\begin{aligned} a \star_{-i} c &= \log_i(\exp_i(a) + \exp_i(c)) = \log_i\left(\exp_i(a)\left(1 + \frac{\exp_i(c)}{\exp_i(a)}\right)\right) \\ &= \log_{i-1}\left(\exp_{i-1}(a) + \exp_{i-1}\left(\log_i\left(1 + \frac{\exp_i(c)}{\exp_i(a)}\right)\right)\right) \\ &= a \star_{-i+1} \log_i\left(1 + \frac{\exp_i(c)}{\exp_i(a)}\right). \end{aligned}$$

Part (iii) follows on iterating, and this completes the proof of the lemma.

Suppose that we wish to know the order relation between $\gamma_j(A_1 \star_k A_2)$ and $\gamma_j(B_1 \star_n B_2)$, where $k, n \geq 0$, $\gamma_k(A_2) < \gamma_k(A_1)$ and $\gamma_n(B_2) < \gamma_n(B_1)$. We treat here only the case when $B_1 \to \infty$. For $k \geq 2$ there is no essential difference if $B_1 \to 0$ since $\log|\log|B_1^{-1}|| = \log|\log|B_1||$. However cases where $B_1 \to \exp_i(0)$, $1 \leq i < k$ would need further investigation.

By Lemma 38 we may take n to be equal to k, and by taking logs we may reduce to the case $j = 0$. So we want to calculate

$$l = \lim\left(\frac{A_1 \star_k A_2}{B_1 \star_k B_2}\right).$$

Lemma **39**

(i) *If* $\gamma_k(A_1) < \gamma_k(B_1)$, *then* $l = 0$.

(ii) *If $k \geq 2$ and*

$$\frac{\log_k A_1}{\log_k B_1} \to r \in \mathbb{R}^+ \setminus \{1\}$$

then $l = 0$ if $r < 1$ and $l = \infty$ if $r > 1$.

Proof of Lemma 39
(i) Suppose that $\gamma_k(A_1) < \gamma_k(B_1)$. Then

$$\frac{\log_k(A_1 \star_k A_2)}{\log_k(B_1 \star_k B_2)} = \frac{\log_k A_1 + \log_k A_2}{\log_k B_1 + \log_k B_2} \to 0,$$

since $\log_k B_1$ dominates the other three terms.

Now if $D \to \infty$ and $\log C = o(\log D)$ then for every $\epsilon \in \mathbb{R}^+$, $|C| < |D|^\epsilon$. Hence $C/D < D^{\epsilon-1} \to 0$. On applying this with $C = \log_i(A_1 \star_k A_2)$ and $D = \log_i(B_1 \star_k B_2)$ for $i = k-1, k-2, \ldots, 1$, we see that $l = 0$. So we have established (i).

The proof of (ii) follows roughly the same lines. Suppose that $r < 1$ and choose R such that $r < R < 1$. Then $\log_k(A_1 \star_k A_2) < R \log_k(B_1 \star_k B_2)$ and hence

$$\frac{\log_{k-1}(A_1 \star_k A_2)}{\log_{k-1}(B_1 \star_k B_2)} < (\log_{k-1}(B_1 \star_k B_2))^{R-1} \to 0.$$

Hence $l = 0$ as in (i). Similarly if $r > 1$, we take R such that $r > R > 1$ to give the conclusion that $l = \infty$. So Lemma 39 is proved.

So now we assume that $r = 1$. The proof of Lemma 35 shows that we may write $A_1 = (B_1 \star_k B_2) \star_k C$ with $\gamma_k(C) < \gamma_k(B_1)$. On using the associativity of $\star_k$, we get $A_1 \star_k A_2 = (B_1 \star_k B_2) \star_k (C \star_k A_2)$. Then with a slight change of notation, l is the limit of $(B_1 \star_k C_1)/B_1$.

The proof of Lemma 42 below requires nothing from this section and so we may use the lemma here. When C_1 tends to infinity, part 2 of the lemma with $j = 0$ implies that $l = \infty$. For the case when $C_1 \to 0$, we take D_1 to be the $\star_k$ inverse of C_1 and Lemma 42 gives $\gamma_0(B_1 \star_k C_1 \star_k D_1) > \gamma_0(B_1 \star_k C_1)$. Hence $l = 0$. Similarly for $k \geq 2$, if $C_1 \to L$ with $L > \exp_k(0)$ then $l = \infty$, while if $L < \exp_k(0)$ then $l = 0$.

For the case $L = \exp_k(0)$, we write $C_1 = L + c_1$. Then the Generalized Distributive Law yields

$$\frac{B_1 \star_k (L + c_1)}{B_1} = \frac{(B_1 \star_k \exp_k(0)) \star_{k-1} (B_1 \star_k \lambda)}{B_1} = \frac{B_1 \star_{k-1} (B_1 \star_k \lambda)}{B_1},$$

since $\exp_k(0)$ is the $\star_k$ identity. Here $\lambda = \exp_{k-2}(1 + d_{k-2})$ where $d_0 = c_1/B_1$ and $d_i = \log(1 + d_{i+1})/\log_i B_1$. So

$$\frac{\log_{k-1}(B_1 \star_k \lambda)}{\log_{k-1} B_1} = \log_{k-1} \lambda = \log(1 + d_{k-2}).$$

An easy induction shows that $d_{k-2} \to 0$, and so $\gamma_{k-1}(B_1 \star_k \lambda) < \gamma_{k-1}(B_1)$. Thus we now are looking at a problem similar to the one we started with but with k reduced by 1.

It is not claimed that the above constitutes an algorithm for determining l since other cases need to be investigated and termination needs to be proved more rigorously. However it is hoped that it may serve as a framework for such an algorithm.

The following example shows how a cancellation may sometimes be handled by using $\star_k$ for negative k and applying the distributive law.

Example

Let $E = \log(x+\log(1+x^{-1})) - \log x$. A cancellation is involved here, although not of the type which might present a danger of non–termination. In terms of star products, we have

$$E = \log(x \star_0 \log(1 \star_0 x^{-1})) \star_0 ((\log x)^{\star_0^{-1}})$$

(where $(\log x)^{\star_0^{-1}} = -\log x$). Here the inner log has been left unchanged since its argument tends to 1. Applying Proposition 6.7 from Chapter 4 gives

$$E = (\log x \star_{-1} \log_2(1 \star_0 x^{-1})) \star_0 ((\log x)^{\star_0^{-1}}),$$

from which the distributive law yields

$$\begin{aligned} E &= (\log x \star_0 ((\log x)^{\star_0^{-1}}) \star_{-1} (\log_2(1 \star_0 x^{-1}) \star_0 ((\log x)^{\star_0^{-1}}) \\ &= 0 \star_{-1} (\log_2(1 \star_0 x^{-1}) \star_0 ((\log x)^{\star_0^{-1}}) \\ &= \log(1 + x^{-1} \log(1 + x^{-1})). \end{aligned}$$

Now to obtain the expansion of E we use the log series. Note how the process has handled the cancellation.

9.2 Growth Classes in Hardy Fields

Star products are very obviously based on the iterated exponential and logarithmic functions, and this makes them a good vehicle for generalizing the theory to cover transexponential functions and the like. We merely replace the exponential function by another sufficiently rapidly increasing function.

9.2.1 Generalizing the γ_ns

Let $\mathcal{F}$ be a Hardy field. By the first corollary to Theorem 10 in Chapter 3, we may assume that the identity function $x = \int 1$ belongs to $\mathcal{F}$. We write $\mathcal{F}^\infty$ for the set of elements of $\mathcal{F}$ which tend to infinity.

Suppose that E is an element of $\mathcal{F}$ which tends to infinity faster than any power of x, and that $\mathcal{F}^\infty$ is closed under composition with E and its compositional inverse, E^{-1}.

For $n \in \mathbb{Z}$, we write E_n for the n-times iterated composition of E. As before, we use the convention that E_0 is the identity function. Thus

$$E_n(x) = \begin{cases} x & \text{if n} = 0 \\ E(E_{n-1}(x)) & \text{if n} > 0 \\ E^{-1}(x) & \text{if n} = -1 \\ E_{-1}(E_{n+1}(x)) & \text{if n} < -1. \end{cases}$$

DEFINITION **23**
Let f and g be any two elements of $\mathcal{F}^\infty$, and let $n \in \mathbb{N}$. We define $f \approx_n g$ to mean that $E_{-n}(f)/E_{-n}(g)$ tends to a non-zero constant, and $f \prec_n g$ to mean that $E_{-n}(f)/E_{-n}(g) \to 0$.

The relations $\approx_n$ are easily seen to be equivalence relations, and moreover the truth of $f \prec_n g$ depends only on the equivalence classes of f and g. We write $\Upsilon_n(\mathcal{F})$ for the set of equivalence classes of the relation $\approx_n$, and denote the map which takes an element of $\mathcal{F}^\infty$ to its $\approx_n$–equivalence class by γ_n^E. We put $\gamma_n^E(f) < \gamma_n^E(g)$ whenever $f \prec_n g$. It is not hard to see that $\gamma_0^E(\ )$ is equivalent to $\gamma_0(\ )$.

Our next goal is to show that $\gamma_n^E(f) = \gamma_n^E(g)$ implies that $\gamma_{n+1}^E(f) = \gamma_{n+1}^E(g)$.

LEMMA **40**
Let $f \in \mathcal{F}^\infty$ and let K be a strictly positive constant. Then

$$E_{-1}(Kf) \sim E_{-1}(f).$$

Proof of Lemma 40
We may assume without loss of generality that $K > 1$. From the fact that $E(x) > x^N$ for every $N \in \mathbb{N}$, it follows easily that $E_{-1}(x) < x^{1/N}$ for $N \in \mathbb{N}^+$. Now by the first corollary to Theorem 10, $\log E$ belongs to a Hardy field containing $\mathcal{F}$, and we have

$$\frac{E'_{-1}(x)}{E_{-1}(x)} = (\log E_{-1}(x))' < \frac{1}{Nx}.$$

Hence

$$\log\left(\frac{E_{-1}(Kf)}{E_{-1}(f)}\right) = \int_f^{Kf} \frac{E'_{-1}(t)}{E_{-1}(t)}dt < \int_f^{Kf} \frac{1}{Nt}dt = \log(K^{1/N}).$$

Thus $E_{-1}(f) < E_{-1}(Kf) < K^{1/N}E_{-1}(f)$ for every $N \in \mathbb{N}^+$, and the conclusion of Lemma 40 follows immediately.

LEMMA **41**
Let $n \in \mathbb{N}^+$, let f and g be elements of $\mathcal{F}^\infty$, and suppose that $E_{-n}(f) \sim KE_{-n}(g)$ for some strictly positive constant K. Then $E_{-n-1}(f) \sim E_{-n-1}(g)$.

Proof of Lemma 41
We may take $K > 1$. Then for ε a positive constant, we have

$$E_{-n}(f) < (K+\varepsilon)E_{-n}(g).$$

Hence

$$E_{-n-1}(f) = E_{-1}(E_{-n}(f)) < E_{-1}((K+\varepsilon)E_n(g)) < (1+\varepsilon)E_{-n-1}(g),$$

by Lemma 40. Similarly

$$E_{-n-1}(g) < E_{-1}((K^{-1}+\varepsilon)E_{-n}(f)) < (1+\varepsilon)E_{-n-1}(f).$$

Since these relations hold for every $\varepsilon > 0$, the lemma follows.

COROLLARY
$f \approx_n g \Rightarrow f \approx_{n+1} g$; that is to say

$$\gamma_n^E(f) = \gamma_n^E(g) \qquad \Rightarrow \qquad \gamma_{n+1}^E(f) = \gamma_{n+1}^E(g).$$

This conclusion allows us to define a map η_n from Υ_n to Υ_{n+1} as follows: Given $\upsilon \in \Upsilon_n$, we choose $f \in \mathcal{F}^\infty$ such that $\gamma_n^E(f) = \upsilon$ and define $\eta_n(\upsilon) = \gamma_{n+1}^E(f)$. By the corollary, this does not depend on the choice of f. We therefore have a directed system of sets and maps

$$\cdots \to_{\eta_{n-2}} \Upsilon_{n-1} \to_{\eta_{n-1}} \Upsilon_n \to_{\eta_n} \Upsilon_{n+1} \to_{\eta_{n+1}} \cdots.$$

We let Υ_∞ be the direct limit, and denote the canonical map from $\mathcal{F}$ to Υ_∞ by γ_∞^E. Thus $\gamma_\infty^E(f) = \gamma_\infty^E(g)$ if and only if $\gamma_n^E(f) = \gamma_n^E(g)$ for some $n \in \mathbb{Z}$. Similarly we define $f \prec_\infty g$ to mean that $f \prec_n g$ for every $n \geq 0$. Since this relation depends only on the classes of f and g in Υ_∞, we may also write $\gamma_\infty^E(f) < \gamma_\infty^E(g)$ under these circumstances. One might also consider the inverse limit, $\Upsilon_{-\infty}$, of the above sequence, but we do not do so here.

Since $E(f)/f \to \infty$ whenever $f \to \infty$, it is immediate that $\gamma_\infty^E(E(f)) > \gamma_\infty^E(f)$ for every $f \in \mathcal{F}^\infty$. So for any such f, the elements $E_n(f)$, $n \in \mathbb{Z}$, generate a chain in Υ_∞ order-isomorphic to $\mathbb{Z}$, and Υ_∞ is the union of these chains and $\gamma_\infty^E(1)$. Note that the identity function, x, belongs to every chain.

Let us suppose that we have somehow chosen a representative (not equivalent to x) in each chain, and let $\mathcal{R} = \mathcal{R}(\mathcal{F})$ be the set of these. Then

$$\Upsilon_\infty = \gamma_\infty^E(1) \cup \bigcup_{\theta \in \mathcal{R}} \{\gamma_\infty^E(E_n(\theta)) : n \in \mathbb{Z}\}.$$

Let $f \in \mathcal{F}^\infty$. Then $\gamma_\infty^E(f) = \gamma_\infty^E(E_r(\theta))$ for some $r \in \mathbb{Z}$ and $\theta \in \mathcal{R}$, which implies that $\gamma_n^E(f) = \gamma_n^E(E_r(\theta))$ for some $n \in \mathbb{N}^+$. We take the smallest such n. Then $E_{-n}(f)/E_{-n}(E_r(\theta)) \to K$ for some positive constant K. If we take

$$\phi = E_n(E_{-n}(f) - KE_{-n}(E_r(\theta))),$$

we obtain

$$f = E_n(KE_{-n}(E_r(\theta)) + E_{-n}(\phi)). \tag{9.6}$$

A comparison of this formula with that of Proposition 6(3) suggests that (9.6) might be recast in terms of a generalized star product. This we now define.

9.3 Generalized Star Products

Let f and g be elements of $\mathcal{F}$, and let $n \in \mathbb{Z}$. By analogy with Definitions 15 and 16 (see also [104]), we define

$$f *_n g = E_n(E_{-n}(f) + E_{-n}(g)),$$

and

$$f \nearrow_n g = E_n(E_{-n}(f) - E_{-n}(g)),$$

provided that the relevant compositions with E_{-1} are defined. We shall implicitly assume this to be so whenever we write $f *_n g$ or $f \nearrow_n g$. Note that $f *_n g$ is not generally equal to $E_{n-1}(E_{1-n}(f)E_{1-n}(g))$.

The following result generalizes the equivalent parts of Proposition 6.

PROPOSITION **8**

1. *$a *_0 b = a + b$.*
2. *For each $n \in \mathbb{Z}$, $*_n$ is commutative and associative.*
3. *$E_n(0)$ is the identity for the operation $*_n$. Moreover the $*_n$-inverse of a is $E_n(-E_{-n}(a))$.*
4. *We have*

$$E(a *_n b) = Ea *_{n+1} Eb \quad \textit{and} \quad E_{-1}(a *_n b) = E_{-1}a *_{n-1} E_{-1}b.$$

The other parts of Proposition 6, do not generalize. For example if we take $E(x) = \exp(x^3)$, then $E_{-1}(x) = \log^{1/3} x$ and

$$\begin{aligned} a *_1 b &= \exp\{(\log^{1/3} a + \log^{1/3} b)^3\} \\ &= \exp\{\log a + 3\log^{2/3} a \log^{1/3} b + 3\log^{1/3} a \log^{2/3} b + \log b\} \\ &= ab \exp\{3\log^{2/3} a \log^{1/3} b + 3\log^{1/3} a \log^{2/3} b\} \neq ab \end{aligned}$$

for general values of a and b. Similarly

$$\begin{aligned} a *_1 (b *_0 c) &= E_1(E_{-1}a + E_{-1}(b + c)) = \exp\{(\log^{1/3} a + \log^{1/3}(b + c))^3\} \\ &= a(b + c) \exp\{3\log^{2/3} a \log^{1/3}(b + c) + 3\log^{1/3} a \log^{2/3}(b + c)\}, \end{aligned}$$

whereas $(a *_1 b) *_0 (a *_1 c)$ is equal to

$$\begin{aligned} &ab \exp\{3\log^{2/3} a \log^{1/3} b + 3\log^{1/3} a \log^{2/3} b\} + \\ &ac \exp\{3\log^{2/3} a \log^{1/3} c + 3\log^{1/3} a \log^{2/3} c\}. \end{aligned}$$

Our next two results link the $*_n$s with the γ^Es.

LEMMA **42**
Let a and b belong to $\mathcal{F}^\infty$.

1. *If $j \geq k$, then*
$$\gamma_j^E(a *_k b) = \max\{\gamma_j^E(a), \gamma_j^E(b)\},$$
and provided that $\gamma_j^E(a) > \gamma_j^E(b)$,
$$\gamma_j^E(a \nearrow_k b) = \gamma_j^E(a).$$

2. *Suppose that $E(x) > \exp cx$ for some constant $c > 0$, and that $j < k$. Then*
$$\gamma_j^E(a *_k b) > \max\{\gamma_j^E(a), \gamma_j^E(b)\},$$
and if $a \nearrow_k b \to \infty$, then
$$\gamma_j^E(a \nearrow_k b) < \gamma_j^E(a).$$

Proof of Lemma 42
We have

$$\begin{aligned} \frac{E_{-j}\{E_k(E_{-k}a + E_{-k}b)\}}{E_{-j}a} &= \frac{E_{k-j}(E_{-k}a + E_{-k}b)}{E_{k-j}(E_{-k}a)} \\ &= \exp\left\{\int_{E_{-k}a}^{E_{-k}a+E_{-k}b} (\log E_{k-j})'(t)\, dt\right\}. \end{aligned} \tag{9.7}$$

Suppose that $a \geq b$ and $j \geq k$. Then the right–hand side of (9.7) is greater than 1 and, since $E_{k-j}(x) \leq x$, it is less than

$$\exp\left\{\int_{E_{-k}a}^{E_{-k}a+E_{-k}b} (\log t)'\, dt\right\} = \frac{E_{-k}a + E_{-k}b}{E_{-k}a} \leq 2. \tag{9.8}$$

Thus $a *_k b \approx_k a$, which gives the first half of part 1. The second half follows on replacing a by $a \nearrow_k b$.

Now suppose that $E(x) > \exp cx$ for some constant $c > 0$, and that $j < k$. Then the right–hand side of (9.7) is greater than

$$\exp\left\{\int_{E_{-k}a}^{E_{-k}a+E_{-k}b} c\, dt\right\} = \exp(cE_{-k}b).$$

Hence $E_{-j}(a \star_k b)/E_{-j}a \to \infty$, which gives the first half of part 2, and again the second half follows from this on replacing a by $a \nearrow_k b$. This completes the proof of Lemma 42.

Note that part 2 of Lemma 42 may fail without the assumption that $E(x) > \exp cx$. For example, take $E(x) = \exp \log^2 x$, $a(x) = e^x$ and $b(x) = x$. Then $E_{-1}(x) = \exp\sqrt{\log x}$, $E_2(x) = \exp \log^4 x$ and $E_{-2}(x) = \exp \log^{1/4} x$. Hence

$$\begin{aligned}\frac{E_{-1}(e^x *_2 x)}{E_{-1}(e^x)} &= \frac{E_{-1}(\exp\{(x^{1/4}+\log(1+x^{1/4}\exp(-x^{1/4})))^4\})}{E_{-1}(e^x)}\\ &= \exp\{(x^{1/4}+\log(1+x^{1/4}\exp(-x^{1/4})))^2-\sqrt{x}\}\\ &= \exp\{2x^{1/4}\log(1+x^{1/4}\exp(-x^{1/4}))\\ &\qquad +\log^2(1+x^{1/4}\exp(-x^{1/4}))\}\\ &\sim 1.\end{aligned}$$

So here $\gamma_1^E(a *_2 b) = \gamma_1^E(a) = \max\{\gamma_1^E(a), \gamma_1^E(b)\}$.

Our next result gives comparison between different star products with the same left–hand side.

LEMMA **43**
Suppose that $E(x) > \exp cx$ for some constant $c > 0$. Suppose also that $a \to \infty$ and that $\gamma_{m_2}^E(b_2) < \gamma_{m_2}^E(a)$. Then

$$a \star_{m_1} b_1 > a \star_{m_2} b_2 \tag{9.9}$$

in the Hardy–field ordering, if any of the following hold:

1. $m_1 > m_2$ *and* $\lim b_1 > E_{m_1}(0)$.
2. $m_1 < m_2$ *and* $\lim b_2 < E_{m_2}(0)$
3. $m_1 = m_2$ *and* $b_1 > b_2$.

The condition $\gamma_{m_2}^E(b_2) < \gamma_{m_2}^E(a)$ ensures that any $\star_{m_2}$ power of b_2 is less than a. The cases when $m_1 > m_2$ and $b_1 \to E_{m_1}(0)$, and when $m_1 < m_2$ and $b_2 \to E_{m_2}(0)$ may involve cancellation between dominant terms, which as we have seen in Chapter 5, can cause significant difficulties. See also Section 9.1.

Proof of Lemma 43
Suppose that $m_1 > m_2$ and write $r = m_1 - m_2$. Then

$$\begin{aligned}a \star_{m_1} b_1 &= E_{m_1}\{E_{-m_1}a + E_{-m_1}b\} = E_{m_2}\{E_r[E_{-r}(E_{m_2}a) + E_{-r}(E_{m_2}b_1)]\}\\ &= E_{m_2}\{E_{m_2}a \star_r E_{m_2}b_1\}.\end{aligned}$$

On the other hand,

$$a \star_{m_2} b_2 = E_{m_2}\{E_{m_2}a + E_{m_2}b_2\},$$

and hence $a \star_{m_1} b_1 > a \star_{m_2} b_2$ if and only if $E_{m_2}a \star_r E_{m_2}b_1 > E_{m_2}a + E_{m_2}b_2$. Thus we may restrict attention to the case when $m_2 = 0$.

Now if $\lim b_1 > E_{m_1}(0)$, let $l = \lim E_{-m_1}(b_1)$. Since $l > 0$, we can choose $l_1 \in \mathbb{R}$ such that $0 < l_1 < l$; then $E_{-m_1}(b_1) > l_1$. Also

$$\begin{aligned}\log(a *_{m_1} b_1) - \log a &= \log(E_{m_1}(E_{-m_1}a + E_{-m_1}b_1)) - \log(E_{m_1}(E_{-m_1}a))\\ &> \int_{E_{-m_1}a}^{E_{-m_1}a+l_1} (\log E_{m_1})'(t)\, dt,\end{aligned}$$

and the last expression can be seen to be greater than $\exp cl_1$ as in the proof of Lemma 42. Hence $a *_{m_1} b_1 > e^{cl_1} a > a + b_2$, since $\exp cl_1 > 1$ and $b_2/a \to 0$. This establishes (i).

A similar analysis shows that if $\lim b_1 < E_{m_1}(0)$ then $a \star_{m_1} b_1 < a \star_{m_2} b_2$ which gives (ii) on interchanging the subscripts 1 and 2. Part (iii) is immediate on applying the operator E_{-m_1} to each side. We have therefore completed the proof of Lemma 43.

9.3.1 Expansions

Continuing the analogy with star products, we use the notation $a^{*_n^c}$ for $E_n(cE_{-n}a)$. Then we may rewrite (9.6) in the form

$$f = E_r(\theta)^{*_n^K} *_n \phi. \tag{9.10}$$

Moreover, it is clear that $\gamma_n^E(\phi) < \gamma_n^E(E_r(\theta))$, and so we can regard (9.10) as the beginning of a generalized star expansion.

DEFINITION **24** *Let u and u_i, $i = 1, \ldots$ be elements of $\mathcal{X}$, and let $n \in \mathbb{N}$. We write*

$$\begin{aligned} u &\sim \prod_i^{*_n} u_i \\ &\sim u_1 *_n u_2 *_n \cdots *_n u_i *_n \cdots \end{aligned}$$

if for each $i \in \mathbb{N}$

$$\gamma_n(u \nearrow_n (u_1 *_n u_2 *_n \cdots *_n u_i)) < \gamma_n(u_i).$$

Note that the case $n = 0$ is equivalent to the classical definition of an asymptotic series.

If $f \to -\infty$ we can apply the above to $-f$, and if $f \to A \in \mathbb{R}$ we may similarly consider $\pm(f - A)^{-1}$. Hence we can obtain partial expansions in these cases. Thus the ϕ in (9.10) may be similarly expanded, and the process continued indefinitely (unless it terminates in an exact representation of f after a finite number of stages). We have therefore proved the following.

THEOREM **25**
Let $\mathcal{F}$, E and $\mathcal{R}$ be as above. Then any element of $\mathcal{F}$ has a generalized star-product expansion in terms of the elements of $\{E_r(\theta) : r \in \mathbb{Z}, \theta \in \mathcal{R}\}$.

Note how general this result is. $\mathcal{F}$ can have infinite rank and can contain elements of transexponential growth (i.e. elements which grow faster than any $\exp_n x$). In the latter case, with $E(x) = \exp(x)$, the set of representatives, $\mathcal{R}$, will have to contain transexponential functions. Alternatively one can take E to be transexponential, as is shown by the examples in the next section.

In a specific Hardy field, Theorem 25 will not give us all that we might desire. Ideally we would like full multiple expansions, analogous to multiseries expansions with closed forms for coefficients, as we obtained for exp–log functions in Section 9.1. However one cannot possibly expect to be able to do this at the level of generality of Theorem 25.

9.3.2 Non-integral Iterates and Multiple Scales

We use the framework developed above to investigate the solutions of the equation $h(x+1) = \exp(h(x))$ and those of the related equation $g(g(x)) = e^x$. It is easy to see that h must grow faster than every $\exp_n(x)$ as $x \to \infty$. This result and the others we obtain in this subsection were already present, implicitly or explicitly, in [13]. Our purpose here is to show how these examples illustrate the general picture. In particular, we use the functions h and g themselves as representatives of a chain in the expansions of their respective derivatives.

In the Hardy field generated by h and the functions $\exp_n(x)$, $n \in \mathbb{Z}$, take E to be the exponential function and choose the set of representatives to be $\mathcal{R} = \{x, h\}$. Differentiation of the relation $h(x+1) = \exp(h(x))$ yields $h'(x+1) = h(x+1)h'(x)$. Using iteration and noting that $h(x-n) = \log_n h(x)$, we have

$$\begin{aligned} h'(x) &= h(x)h'(x-1) \\ &= h(x)h(x-1)h'(x-2) \\ &= h(x)h(x-1)\cdots h(x-n)h'(x-n-1) \\ &\sim h(x)\log h(x)\log_2 h(x)\cdots\log_n h(x)\cdots. \end{aligned} \tag{9.11}$$

In order to justify the last step, we need to know that $\log h'(x-n-1) = o(\log h(x-n))$. This can be proved as follows (see [13], Lemma 3.5).

The relation $h'(x+1) = h(x+1)h'(x)$ yields $\log h'(x+1) - \log h'(x) = \log h(x+1)$. Summation then gives

$$\begin{aligned} \log h'(x) &= \log h'(x-[x]) + \log h(x) + \log h(x-1) + \cdots + \log h(x-[x]+1) \\ &< 2\log h(x), \end{aligned}$$

since $\log h(x-1) = o(\log h(x))$ and $\log h'(x-[x])$ is bounded. So

$$\log h'(x-n-1) < 2\log h(x-n-1) = o(\log h(x-n)),$$

as required.

We have assumed that h belongs to a Hardy field, but in fact we could prove it from the expansion (9.11), which can be obtained without that assumption. If we write

$$h^{\Delta} = h'/h \quad \text{and} \quad h^{\Delta^i} = (h^{\Delta^{i-1}})^{\Delta}, \quad \text{for } i > 1,$$

then a simple induction shows that

$$h^{\Delta^i} \sim \prod_{n=i}^{\infty} \log_n h.$$

Now any polynomial in h and its derivatives can be rewritten as a polynomial in the h^{Δ^i}s. However for each i, h^{Δ^i} dominates any power of $h^{\Delta^{i+1}}$. Hence $P(h, h^{\Delta}, \ldots, h^{\Delta^i})$ is asymptotic to its largest monomial under the ordering

$$\ldots > h^n > h^{n-1} > \ldots > h > \ldots > (h^{\Delta})^n > \ldots > h^{\Delta} > \ldots (h^{\Delta^2})^n > \ldots .$$

So if P is not constant, it will tend to $\pm\infty$. Hence non-zero no polynomial in h and its derivatives can have arbitrarily large zeros; i.e. h generates a Hardy field. A slightly stronger result is proved in a similar way in Theorem 5.2 of [13].

The function g is defined by $g(x) = h(h^{-1}(x) + 1/2)$. It may be verified that $g \circ g = \exp$. We take $\mathcal{F}$ to be the Hardy field generated by the functions $\exp_n x$ and $\exp_n g(x)$ for $n \in \mathbb{Z}$. This exists because the exponentials can always be added to the Hardy field generated by g. Let $E = g$ and $\mathcal{R} = \{g\}$. The function E satisfies the required conditions since if n is positive and even the n–times iterated composite, g_n, is equal to $\exp_{n/2} x$, while for n positive and odd, $g_n = \exp_{(n-1)/2} g$. Also the compositional inverse g_{-1} is equal to $\log g$, since $(\log g) \circ g = \log(g \circ g) = x$. So g_{-n} is equal to $\log_{n/2} x$ for n positive and even, and to $\log_{(n-1)/2} g$ for n positive and odd.

Then $g \circ g = \exp x = (\exp x)' = g'(g' \circ g)$. So $g' \circ g = (g \circ g)/g'$ and hence by repeated substitution, we obtain

$$\begin{aligned} g' = \frac{g}{g' \circ g^{-1}} &= \frac{g}{g \circ g^{-1}} \cdot g' \circ g^{-1} \circ g^{-1} = \frac{g}{x} \cdot \frac{g^{-1}}{g' \circ g_{-3}} \\ &\sim \frac{g}{x} \cdot \frac{g_{-1}}{g_{-2}} \cdot \frac{g_{-3}}{g_{-4}} \cdots = \frac{1}{x} \prod_{n=0}^{\infty} (\log_n g)^{(-1)^n}, \end{aligned}$$

since $g^{-1} = \log g$.

9.4 Real Iterates of Increasing Functions

The construction of g from h can be generalized. If for $q \in \mathbb{N}^+$ we define g_q to be $h(h^{-1} + 1/q)$, then it is not hard to see that g_q composed with itself q times is the exponential function. g_q is then a natural candidate for $\exp_{1/q}$. Similarly for real α, the function $h(h^{-1} + \alpha)$ is a candidate for $\exp_\alpha$. However if we are to pursue such a line, we need to ask what meaning can be attached to the statement that ϕ is an α–th iterate of a function f when α is irrational. It may well be that this notion is nothing more than a mathematical curiosity, but it is nonetheless rather fun to finish this chapter by looking more closely at it.

Let $\mathcal{F}$ be a Hardy field and let $\phi, \lambda \in \mathcal{F}^\infty$. We suppose that the compositional iterates ϕ_n and λ_n are also elements of $\mathcal{F}^\infty$ for all $n \in \mathbb{Z}$.

Definition 25
If there is a sequence of integers $\{n_k\}$ such that for all k, $\lambda_{n_k - 1} \le \phi_k \le \lambda_{n_k}$ and $n_k/k \to \alpha$, we say that ϕ is an asymptotic α–th iterate *of λ.*

LEMMA **44**
Suppose that there are positive integers, n and m such that $\phi \leq \lambda_n$ and $\lambda < \phi_m$. Then ϕ is an asymptotic α–th iterate of λ for some α.

We remark that we may take one of n, m to be 1.

Proof of Lemma 44
Suppose first that $\phi \geq \lambda$. For each $k \in \mathbb{N}^+$ we choose n_k to be the least integer such that $\phi_k \leq \lambda_{n_k}$. Then $n_{j!} \geq j!$ and for $j \geq 2$, $\phi_{(j-1)!} \leq \lambda_{n_{(j-1)!}}$. This implies that

$$\phi_{j!} = \phi_{(j-1)!} \circ \phi_{(j-1)!} \circ \cdots \circ \phi_{(j-1)!} \leq \lambda_{n_{j-1}} \circ \lambda_{n_{j-1}} \circ \cdots \circ \lambda_{n_{j-1}} = \lambda_{jn_{j-1}}.$$

Hence $n_{j!} \leq jn_{(j-1)!}$, and so $n_j/j!$ decreases; it is bounded below by 1. We take α to be the limit of $n_{j!}/j!$ as $j \to \infty$.

Similarly if $\phi < \lambda$. We choose n_k to be the greatest integer such that $\phi_k > \lambda_{n_k}$. Then a similar calculation shows that $n_{j!}/j!$ increases and is bounded above by 1. We again write α for the limit.

Now any $k \geq 1$, may be written in the form

$$k = r_j j! + r_{j-1}(j-1)! + \cdots + r_1,$$

with $0 \leq r_i \leq i$ for each i and $r_j \geq 1$. Then

$$\begin{aligned}\phi_k &= (\phi_{j!})_{r_j} \circ (\phi_{(j-1)!})_{r_{j-1}} \circ \cdots \circ \phi_{r_1} \\ &\leq (\lambda_{n_{j!}})_{r_j} \circ (\lambda_{n_{(j-1)!}})_{r_{j-1}} \circ \cdots \circ (\lambda_{n_1})_{r_1} = \lambda_{r_j n_{j!} + r_{j-1} n_{(j-1)!} + \cdots + r_1 n_1}.\end{aligned}$$

Likewise

$$\begin{aligned}\phi_k &\geq (\lambda_{n_{j!}-1})_{r_j} \circ (\lambda_{n_{(j-1)!}-1})_{r_{j-1}} \circ \cdots \circ (\lambda_{n_1-1})_{r_1} \\ &= \lambda_{r_j(n_{j!}-1) + r_{j-1}(n_{(j-1)!}-1) + \cdots + r_1(n_1-1)},\end{aligned}$$

and so

$$\begin{aligned}r_j(n_{j!} - 1) + r_{j-1}(n_{(j-1)!} - 1) + \cdots + r_1(n_1 - 1) &\leq n_k \\ &\leq r_j n_{j!} + r_{j-1} n_{(j-1)!} + \cdots + r_1 n_1.\end{aligned}$$

Now $r_{j-2}(j-2)! < (j-1)! = o(r_j j!)$ as $j \to \infty$. Hence

$$r_j j! + r_{j-1}(j-1)! + \cdots + r_1 = (r_j j! + r_{j-1}(j-1)!)(1 + o(1)).$$

Similarly $r_{j-2} n_{(j-2)!} < (j-1) n_{(j-2)!} = o(j!)$, since $n_{(j-2)!}/(j-2)! \to \alpha$, and so

$$r_j n_{j!} + r_{j-1} n_{(j-1)!} + \cdots + r_1 n_1 = r_j n_{j!} + r_{j-1} n_{(j-1)!} + o(j!).$$

Thus

$$\frac{r_j n_{j!} + r_{j-1} n_{(j-1)!} + \cdots + r_1 n_1}{r_j j! + r_{j-1}(j-1)! + \cdots + r_1} = \frac{r_j n_{j!} + r_{j-1} n_{(j-1)!} + o(j!)}{(r_j j! + r_{j-1}(j-1)!)(1 + o(1))}$$
$$\sim \frac{\frac{n_{j!}}{j!} + \frac{r_{j-1}}{j r_j}\frac{n_{(j-1)!}}{(j-1)!}}{1 + \frac{r_{j-1}}{j r_j}},$$

and the last expression is easily seen to tend to α. It then follows quite easily that

$$\frac{r_j(n_{j!} - 1) + r_{j-1}(n_{(j-1)!} - 1) + \cdots + r_1(n_1 - 1)}{r_j j! + r_{j-1}(j-1)! + \cdots + r_1} \to \alpha.$$

So $n_k/k \to \alpha$ as required, and this completes the proof of Lemma 44.

LEMMA **45**
If ϕ and ψ are two α–th iterates of λ, suppose that $\phi < \psi$. Then $\phi^{-1} \circ \psi$ tends to infinity faster than $x = \lambda_0$ but slower than every positive iterate of λ.

Proof of Lemma 45
Suppose that $\lambda_{n_k - 1} \le \phi_{k!} \le \lambda_{n_k}$, $\lambda_{m_k - 1} \le \psi_{k!} \le \lambda_{m_k}$ and that $n_k/k!$ and $m_k/k!$ both tend to α. We write $\nu = \phi^{-1} \circ \psi$. Then

$$(\phi \circ \nu)_{k!} = \psi_{k!} \le \lambda_{m_k}. \tag{9.12}$$

If $\nu \circ \phi \le \phi \circ \nu$, then

$$\begin{aligned}(\phi \circ \nu)_{k!} &= \phi \circ \nu \circ \phi \circ \cdots \circ \nu \\ &\ge \nu \circ \phi \circ \phi \circ \nu \circ \cdots \circ \nu \\ &\cdots\cdots \quad \cdots \quad \cdots \\ &\ge \nu \circ \nu \circ \nu \circ \cdots \circ \nu \circ \phi \circ \phi \circ \cdots \circ \phi \\ &= \nu_{k!} \circ \phi_{k!}.\end{aligned}$$

So $\nu_{k!} \circ \phi_{k!} \le \lambda_{m_k}$. On applying $\phi_{k!}^{-1}$ on the right, and using the fact that $\phi_{k!}^{-1} \le \lambda_{n_k - 1}^{-1}$, we obtain that $\nu_{k!} \le \lambda_{n_k - m_k - 1}$.

Similarly if $\nu \circ \phi > \phi \circ \nu$, we obtain

$$(\phi \circ \nu)_{k!} \ge \phi_{k!} \circ \nu_{k!}.$$

Hence $\phi_{k!} \circ \nu_{k!} \le \lambda_{m_k}$ and composing on the left with $\phi_{k!}^{-1}$ gives the same conclusion as before, namely that $\nu_{k!} < \lambda_{n_k - m_k - 1}$. But $(n_k - m_k - 1)/k! \to 0$, and hence $\nu_n < \lambda$ for every $n \in \mathbb{N}$. It follows that ν is smaller than every positive iterate of λ. However $\nu > x = \lambda_0$. This completes the proof of Lemma 45.

Examples
As our first example we take $\phi(x) = x^2$ and $\lambda(x) = x^3$. Then $\phi_n(x) = x^{2^n}$ and $\lambda_m(x) = x^{3^m}$, and for ϕ to be an α–th iterate of λ, we must have $\alpha = \lim n_k/k$ where $3^{n_k - 1} \le 2^k \le 3^{n_k}$. Hence $\alpha = \log 2/\log 3$. Another α–th iterate of λ is $\psi(x) = x^2 \log x$, and then $\nu(x) = x\sqrt{\log x}$.

More interestingly we can obtain an example where $\gamma_\infty^E(\nu)$ is greater than $\gamma_\infty^E(E_0) = \gamma_\infty^E(x)$ but smaller than every $\gamma_\infty^E(E_\alpha)$ with $\alpha > 0$. We take E to be the exponential function, and with h as above, we define

$$f(x) = h \circ (h^{-1}(x) + 1/h^{-1}(x)).$$

Then

$$\begin{aligned} f \circ f &= h \circ \left(h^{-1} + \frac{1}{h^{-1}}\right) \circ h \circ \left(h^{-1} + \frac{1}{h^{-1}}\right) = h\left(x + \frac{1}{x}\right) \circ \left(h^{-1} + \frac{1}{h^{-1}}\right) \\ &= h \circ \left(h^{-1} + \frac{1}{h^{-1}} + \frac{1}{h^{-1} + \frac{1}{h^{-1}}}\right). \end{aligned}$$

Since $h^{-1} \to \infty$, it is then easy to see that for every $n \in \mathbb{N}$,

$$f_n = h \circ (h^{-1} + o(1)). \tag{9.13}$$

We now show that for all $n, N \in \mathbb{N}$, $\log_n f > \log_n^N x$. We have

$$\begin{aligned} \log_3(\log_n^N x) &= \log_2(N \log_{n+1} x) = \log(\log_{n+2} x + \log N) \\ &= \log_{n+3} x + \log\left(1 + \frac{\log N}{\log_{n+2} x}\right) < \log_{n+3} x + \frac{1}{h^{-1}(x)}, \end{aligned}$$

since $h^{-1}(x)$ tends to infinity more slowly than every $\log_n x$. So

$$\log_{n+3}(\exp_n(\log_n^N x)) - \log_{n+3} x < 1/h^{-1}(x),$$

and it follows that

$$\begin{aligned} h^{-1}(\exp_n(\log_n^N x)) - h^{-1}(x) &= \int_{t=x}^{\exp_n(\log_n^N x)} (h^{-1}(t))' dt \\ &< \int_{t=x}^{\exp_n(\log_n^N x)} (\log_{n+3} t)' \, dt < \frac{1}{h^{-1}(x)}. \end{aligned}$$

So, on applying h, we get

$$\exp_n(\log_n^N x) < h\{h^{-1}(x) + 1/h^{-1}(x)\} = f(x),$$

which proves that $\log_n^N x < \log_n f$ as required.

Our next task is to show that f belongs to a Hardy field. Using (9.11), we have

$$\begin{aligned} f' &= \frac{h'(h^{-1} + 1/h^{-1})}{h'(h^{-1})} \left\{1 - \frac{1}{(h^{-1})^2}\right\} \\ &\sim \frac{h(h^{-1}(x) + 1/h^{-1}(x)) \cdot \log h(h^{-1}(x) + 1/h^{-1}(x)) \cdots}{x \cdot \log x \cdots} \\ &= \prod_{n=0}^{\infty} \frac{\log_n f}{\log_n x}. \end{aligned} \tag{9.14}$$

We are going to define elements $F(i)$ which are rational functions of the derivatives of f, such that $F(0) = f$ and for $i \geq 1$,

$$F(2i-1) \sim \frac{1}{\log_{i-1} x} \prod_{n=i}^{\infty} \frac{\log_n f}{\log_n x} \quad \text{and} \quad F(2i) \sim \prod_{n=i}^{\infty} \frac{\log_n f}{\log_n x}; \tag{9.15}$$

c.f. Section 3.4 and [99]. First we note that (9.15) implies that

$$\left(\frac{1}{F(2i-1)}\right)^{\Delta} \sim \frac{1}{\prod_{j=0}^{i-1} \log_j x}. \tag{9.16}$$

For the case $i = 1$, we take $F(1) = f^{\Delta}$ and $F(2) = -F(1)/F(1)^{\Delta}$. Now

$$F(1)^{\Delta} = (\log F(1))' \sim (-\log x + \log_2 f - \log_2 x + \cdots)',$$

and so

$$F(2) \sim \prod_{n=1}^{\infty} \frac{\log_n f}{\log_n x}$$

as required. Now supposing that $F(1), \ldots, F(2i)$ have been defined satisfying (9.15), we take

$$F(2i+1) = \frac{F(2i)^{\Delta}}{(1/F(2i-1))^{\Delta}}, \quad F(2i+2) = \frac{F(2i)^{\Delta}}{F(2i+1)^{\Delta}}.$$

Now

$$F(2i)^{\Delta} = (\log F(2i))' \sim (\log_{i+1} f)' = \frac{f'}{\prod_{j=0}^{i} \log_j f}$$

$$\sim \prod_{j=0}^{i} \frac{1}{\log_j x} \prod_{n=i+1}^{\infty} \frac{\log_n f}{\log_n x}, \tag{9.17}$$

by (9.14). So (9.16) implies that

$$F(2i+1) \sim \frac{1}{\log_i x} \prod_{n=i+1}^{\infty} \frac{\log_n f}{\log_n x}$$

as required. Similarly, if we replace i by $i+1$ in (9.16) and use (9.17), we obtain

$$F(2i+2) \sim \prod_{n=i+1}^{\infty} \frac{\log_n f}{\log_n x}.$$

Now any polynomial in f and its derivatives can be written as a rational function in the $F(i)$s. However for all $i \geq 0$, $F(i)$ dominates every integer power of $F(i+1)$. Hence, as with the function h, a polynomial in the $F(i)$s is asymptotic to its largest monomial in the analogous ordering and so cannot have arbitrarily large zeros. It follows that rational functions in f and its derivatives form a Hardy field. However we have shown that $\log_n^N x < \log_n f$ for every $n, N \in \mathbb{N}$ and so $\gamma_\infty(f) > \gamma_\infty(x)$.

Moreover $h^{-1}(x)$ goes to infinity slower than any $\log_n x$, and from (9.11)

$$(h^{-1})' = \frac{1}{h'(h^{-1})} \sim \prod_{n=0}^{\infty} \frac{1}{\log_n x}.$$

So we can similarly define a sequence $\{h_i\}$ with h_i a rational function of the first i derivatives of h^{-1}, such that

$$h_i \sim \prod_{n=i}^{\infty} \frac{1}{\log_n x}.$$

It is then clear that $\mathbb{R}(f, h^{-1})$ is a Hardy field.

Using this fact we can show that $\gamma_\infty(f) < \gamma_\infty(\exp_\alpha)$ for all $\alpha > 0$. We shall need the following.

Lemma 46

Let $\mathcal{F}$ be a Hardy field and let $a, b, f, g \in \mathcal{F}^\infty$. Suppose that $f(a), f(b), g(a)$ and $g(b)$ belong to $\mathcal{F}$ and that $a < b$ and $f < g$. Then

$$\frac{g(a)}{g(b)} > \frac{f(a)}{f(b)}.$$

Proof of Lemma 46

$$\frac{g(a)}{g(b)} = \exp\left(-\int_a^b (\log g)'(t)\, dt\right) > \exp\left(-\int_a^b (\log f)'(t)\, dt\right) = \frac{f(a)}{f(b)},$$

which is the desired conclusion.

Next we recall that from (9.13), $f_n = h \circ (h^{-1} + o(1))$. Since $h(x+1) = \exp(h(x))$, we have $\exp x = h \circ (h^{-1} + 1)$. Moreover $\log_N h > \exp_2 x$ for $n, N \in \mathbb{N}$. So we can apply Lemma 46 with $a = f_n$, $b = \exp x$, $f = \log_N h$ and $g = \exp_2 x$. We obtain

$$\begin{aligned}\frac{\log_N f_n}{\log_N(\exp x)} &= \frac{\log_N\{h(h^{-1} + o(1))\}}{\log_N\{h(h^{-1}+1)\}} \\ &< \frac{\exp_2(h^{-1} + o(1))}{\exp_2(h^{-1}+1)} = \exp\{e^{h^{-1}+o(1)} - e^{h^{-1}+1}\} \\ &= \exp\{-e^{h^{-1}}(e - e^{o(1)})\} \\ &< \exp\{-e^{h^{-1}}\} \to 0.\end{aligned}$$

Hence $\gamma_\infty(f) < \gamma_\infty(\exp_\alpha x)$ for all $\alpha > 0$. However we have already proved that $\gamma_\infty(f) > \gamma_\infty(x)$, and so the function f has the properties we required.

10

Oscillating Functions

When we compare the algorithms given in the previous chapters with the expansions typically used by researchers in differential equations and in mathematical physics, one thing stands out - the absence of trigonometric functions from our theory. Of course, we have seen in Section 5.3 how trigonometric functions composed with functions tending to finite limits can be handled, but these composed functions do not exhibit oscillation. This, the last chapter, is concerned with how to deal with oscillating functions, including trigonometric ones, in the context we have built up.

The difficulties are not hard to find. The whole theory we have described has been founded on the basis of Hardy fields, but oscillating functions can never be elements of Hardy fields. As soon as we divide by a function which has arbitrarily large zeros, we get a sequence of singularities at points tending to infinity. How can we make sense of the asymptotics of such functions in neighbourhoods of $+\infty$?

The presence of trigonometric functions in denominators can cause difficulties even when no singularity results. The following result appears on pages 97-99 of [5].

THEOREM **26** *Let $\phi(t)$ be an arbitrary monotone increasing function of t. There exists an irrational number α such that the function*

$$u(t) = \frac{1}{2 - \cos t - \cos \alpha t}, \tag{10.1}$$

which is real and continuous for all t and satisfies an equation of the form $P(t, u, u', u'') = 0$, satisfies the inequality

$$\operatorname{limsup}_{t\to\infty} \left\{ \frac{u(t)}{\phi(t)} \right\} \geq 1.$$

Furthermore difficult, and in many case unsolved, questions of transcendental number theory can become involved. For example, consider the function $(\sin^2 \pi x + \sin^2 x)^{-1}$. For this to be greater than M we must have both

$\sin^2 \pi x < M^{-1}$ and $\sin^2 x < M^{-1}$. For M greater than 4, say, we may use the inequality $\sin x \geq 2x/\pi$ for $0 < x < \pi/2$ to deduce that there must then exist integers j, k such that $|x - k\pi| < \pi/(2\sqrt{M})$ and $|x - j| < 1/(2\sqrt{M})$. This in turn implies that

$$\left|\pi - \frac{j}{k}\right| < \frac{\pi + 1}{2k\sqrt{M}}. \tag{10.2}$$

So if we want to know how rapidly $\sup_{[0,x]}\{(\sin^2 \pi x + \sin^2 x)^{-1}\}$ grows as $x \to \infty$ we need to know how closely π can be approximated by rational functions with a given maximum size of denominator. Such matters are far from easy, but in this particular case some results are known, see [56] for example. However π could be replaced by some other constant, or we might look at the function $(\sin^2 \pi x + \sin^2 ex + \sin^2 x)^{-1}$, which will similarly involve the simultaneous approximation of e and π with the same denominator.

Finally the undecidability results of Richardson show that if unrestricted composition of the sine function is allowed in the signature, there may be no algorithm to decide whether a function tends to a limit. Let $\mathcal{A}$ denote the class of functions generated from the constants 1 and π and the variable x, by addition, multiplication, functional composition and the application of the sine function. Then it is proved in [65] that no algorithm exists to decide whether a given function in $\mathcal{A}$ tends to a limit; see also [101].

Despite all these difficulties, a surprising amount can be done when trigonometric functions with arguments tending to infinity are present.

10.1 An Interval Calculus Algorithm

Our first algorithm concerning oscillating functions is relatively crude. Essentially it gives intervals in which the limit points of the functions must lie. In some cases the interval is $[-\infty, \infty]$ and so no information is obtained. However in many cases something useful may be learned. This section is based on [95].

The main idea is to replace functions such as sine and cosine by the intervals in which their values oscillate and then to push these intervals down into the coefficients. Functions such as the tangent have singularities and this has to be reflected in the intervals that replace them. As we shall see, some operations result in single intervals being replaced by two intervals. So actually we need finite unions of intervals for coefficients.

Thus our expansions are as in Section 4.2 except that the c_m in Definition 11 become sets of intervals instead of real numbers. The methods are basically those of Section 5.1.

We use the techniques of interval analysis to handle the intervals, but we need a version of that theory which is non–standard in a number of ways. In particular, we adjoin points $-\infty$ and ∞ as in [44] and [61].

10.1.1 Our Calculus of Intervals

Let $\mathcal{K}$ be a totally ordered field. We have in mind the case when $\mathcal{K}$ is a field of real constants, but we do not have to restrict to that; c.f. [2]. We choose two new symbols, which we take here to be '∞' and '$-\infty$', and write $\overline{\mathcal{K}}$ for the set $\mathcal{K} \cup \{\infty, -\infty\}$.

We make $\overline{\mathcal{K}}$ into a totally ordered partial field with the following definitions:

1. $\infty > 0$ and $-\infty < 0$.
2. For unary minus, $-(\infty) = -\infty$ and $-(-\infty) = \infty$.
3. If $a \neq -\infty$ then $a + \infty = \infty + a = \infty$.
4. $a - b = a + (-b)$ provided that the latter is defined.
5. $a > b$ if and only if $a - b > 0$.
6. If $a > 0$ then $a \times \infty = \infty$ and $a \times -\infty = -\infty$.
7. If $a < 0$ then $a \times \infty = -\infty$ and $a \times -\infty = \infty$.
8. $\infty^{-1} = (-\infty)^{-1} = 0$ and if $b \neq 0$, $a/b = a \times b^{-1}$.

Of course we require $+$ and $\times$ to be commutative. The operations $\infty + (-\infty)$, $\infty - \infty$, $\infty \times 0$, $-\infty \times 0$ are undefined as are the operations obtained by applying commutation to these. Note that $\infty^{-1} = 0$ does not imply that $0^{-1} = \infty$ since $\infty \times 0$ is undefined, and in fact 0^{-1} is undefined also.

Suppose that $a, b \in \overline{\mathcal{K}}$ with $a \leq b$. Then, as usual, we define

$$[a, b] = \{x \in \overline{\mathcal{K}};\ a \leq x \leq b\}.$$

We write $I(\mathcal{K})$ for the set of all such intervals, and $O(\mathcal{K})$ for the set of all finite sets of non-overlapping intervals from $I(\mathcal{K})$. Of course 'non-overlapping' means that all the pairwise intersections are empty. Thus

$$O(\mathcal{K}) = \{\{[a_1, b_1], \ldots, [a_n, b_n]\} \in \mathcal{P}(I(\mathcal{K}));\ n \in \mathbb{N}^+, a_{i+1} > b_i \text{ for } 1 \leq i < n\},$$

where $\mathcal{P}(I)$ denotes the power set of I. The elements of $O(\mathcal{K})$ will provide the coefficients in our expansions. A real number r may at times have to stand duty for the element $\{[r, r]\}$, and likewise elements of $I(\mathcal{K})$ will sometimes be used in place of the singleton set containing them.

We want to specify various operations on $O(\mathcal{K})$, and it would be convenient to do this by defining the operation on single intervals and then using the *map* function (see below) to apply it to other finite sets of intervals. However this may sometimes result in a set of overlapping intervals. We therefore define a join operation, jn, which takes any finite sequence of elements of $I(\mathcal{K})$ and produces the corresponding element of $O(\mathcal{K})$. Let $\{[a_1, b_1], \ldots, [a_n, b_n]\}$ be a sequence in $I(\mathcal{K})$, and suppose that we have ordered the intervals so that $a_1 \leq a_2 \leq \cdots \leq a_n$. We define $jn(\emptyset) = \emptyset$, $jn(\{[a_1, b_1]\}) = \{[a_1, b_1]\}$, and for $n \geq 2$,

$$jn(\{[a_1, b_1], \ldots, [a_n, b_n]\}) = \begin{cases} \{[a_1, b_1]\} \cup jn(\{[a_2, b_2], \ldots, [a_n, b_n]\}), & b_1 < a_2 \\ jn(\{[a_1, \max(b_1, b_2)], \ldots, [a_n, b_n]\}), & b_1 \geq a_2 \end{cases}.$$

Then if Λ is a function from $I(\mathcal{K})$ to $O(\mathcal{K})$, our standard way of extending Λ to a function from $O(\mathcal{K})$ to $O(\mathcal{K})$ is to set

$$\Lambda(\{I_1,\ldots,I_n\}) = jn(mp(\Lambda)(\{I_1,\ldots,I_n\})). \tag{10.3}$$

Here $mp = sort \circ map$, where $sort$ orders intervals by increasing lower end point. map is the higher–order function, rightly beloved by functional programmers, defined by

$$map(f)(\{X_1, X_2,\ldots,X_n\}) = \{f(X_1), f(X_2),\ldots,f(X_n)\},$$

for f a function defined on $X_1, X_2,\ldots,X_n$. We shall also make the obvious extension of (10.3) when some $\Lambda(X_i)$ is a list of elements in $I(\mathcal{K})$ rather than a single element. Thus we define addition and subtraction on $I(\mathcal{K})$ by

$$[a_1, b_1] \pm [a_2, b_2] = [a_1 \pm a_2, b_1 \pm b_2].$$

Then for $\{I_1,\ldots,I_n\}$ and $\{J_1,\ldots,J_m\}$ two elements of $O(\mathcal{K})$, we set

$$\begin{aligned}\{I_1,\ldots,I_n\} \pm \{J_1\} &= jn(mp(\pm J_1)(\{I_1,\ldots,I_n\})) \\ &= jn(\{I_1 \pm J_1,\ldots,I_n \pm J_1\})\end{aligned} \tag{10.4}$$

and for $m \geq 2$,

$$\begin{aligned}\{I_1,\ldots,I_n\} \pm \{J_1, J_2,\ldots,J_m\} = (&\{I_1,\ldots,I_n\} \pm \{J_1\}) \\ &\pm(\{I_1,\ldots,I_n\} \pm \{J_2,\ldots,J_m\}).\end{aligned} \tag{10.5}$$

Multiplication in $I(\mathcal{K})$ is more complicated. Firstly it is required to be commutative. Then provided all the products in $\mathcal{K}$ are defined, we set

$$[a_1, b_1] \times [a_2, b_2] = [m, M],$$

where m and M are respectively the minimum and the maximum of $\{a_1 \times a_2, a_1 \times b_2, b_1 \times a_2, b_1 \times b_2\}$. We further define

$$\begin{aligned}
[a, 0] \times [b, \infty] &= [-\infty, \max\{0, a \times b\}] && \text{if } a < 0,\\
[0, a] \times [b, \infty] &= [\min\{0, a \times b\}, \infty] && \text{if } a > 0,\\
[-\infty, a] \times [0, b] &= [-\infty, \max\{0, a \times b\}] && \text{if } b > 0,\\
[-\infty, a] \times [b, 0] &= [\min\{0, a \times b\}, \infty] && \text{if } b < 0.
\end{aligned}$$

It remains to consider the multiplication of infinite intervals by the zero interval, $[0, 0]$. One point here is that unless we test all denominators in subexpressions for zero equivalence before starting interval calculations, there is the possibility that an infinite interval might arise as the inverse of an expression which is functionally equivalent to zero. If we are not going make these tests, we need to treat the appearance of an infinity in an interval as a warning that the function might be undefined and use a "safe" set of rules for multiplication by $[0,0]$ such as the following.

$$\begin{aligned} [0,0] \times [a,\infty] &= [0,\infty], \qquad &&\text{for } a \neq -\infty \\ [0,0] \times [-\infty,b] &= [-\infty,0], \qquad &&\text{for } b \neq \infty \\ [0,0] \times [-\infty,\infty] &= [-\infty,\infty]. \end{aligned}$$

If on the other hand we agree to check that our function is well defined almost everywhere, there is still a decision to be taken. For example is $0 \cdot \tan x$ the zero function on (∞,∞)? Most mathematicians would probably answer "yes". We follow that line here and define

$$[0,0] \times [a,b] = [0,0] \quad \text{for all } a,b \in \overline{\mathcal{K}}. \tag{10.6}$$

Now multiplication can be extended to $O(\mathcal{K})$ in an analogous way to addition by making the obvious changes in (10.4) and (10.5).

Next we define a partial map $^{-1} : I(\mathcal{K}) \to O(\mathcal{K})$ as follows: $[0,0]^{-1}$ is undefined and $[-\infty,\infty]^{-1} = [-\infty,\infty]$. Now if $a,b \in \overline{\mathcal{K}}$ with $a \leq b$ and $[a,b] \neq [-\infty,\infty]$, we define

$$[a,b]^{-1} = \begin{cases} \{[b^{-1},a^{-1}]\} & \text{if } a > 0 \text{ or } b < 0, \\ \{[b^{-1},\infty]\} & \text{if } a = 0 < b, \\ \{[-\infty,a^{-1}]\} & \text{if } a < 0 = b, \\ \{[-\infty,a^{-1}],[b^{-1},\infty]\} & \text{if } a < 0 < b. \end{cases}$$

For $\{I_1,\ldots,I_n\} \in O(\mathcal{K})$, we define

$$\{I_1,\ldots,I_n\}^{-1} = jn(sort(I_1^{-1} \cup I_2^{-1} \cup \cdots \cup I_n^{-1})).$$

We next suppose that $\mathcal{K}$ has operations $\exp(\ )$ and $\log(\ ) = \log|\ |$ obeying the usual rules. We extend to $\overline{\mathcal{K}}$ by

$$\exp(-\infty) = 0, \quad \exp(\infty) = \infty, \quad \log|\infty| = \infty, \quad \log|0| = -\infty.$$

Then for $[a,b] \in I(\mathcal{K})$ we put $\exp([a,b]) = [\exp a, \exp b]$ and if $[a,b] \neq [0,0]$

$$\log|[a,b]| = \begin{cases} [\log|a|, \log|b|], & \text{if } a \geq 0, b > 0, \\ [\log|b|, \log|a|], & \text{if } a < 0, b \leq 0, \\ [-\infty, \max\{\log|a|, \log|b|\}] & \text{if } a < 0 < b. \end{cases}$$

The logarithm of $[0,0]$ is undefined.

The functions exp and log are then extended to $O(\mathcal{K})$ in the standard way.

Let $\mathcal{K}$ be a field of real constants which contains the rationals and the number π and is closed under the maps $a \mapsto \exp a$, $a \mapsto \sin a$ and $a \mapsto \log|a|$, $a \neq 0$. Let $\mathcal{E}$ be the set of functions defined by expressions with signature $\mathcal{K}, x, +, -, \times, \div, \exp(\ ), \log|\ |, \sin(\ )$. Note that we can regard $\mathcal{E}$ as having a power operation by treating a^b as an abbreviation for $\exp(b \log a)$, $a > 0$. We sketch an algorithm to give a multiseries expansion of an element of $\mathcal{E}$ with coefficients in $O\mathcal{K})$.

ALGORITHM **10** *Given an expression E representing a function in $\mathcal{E}$, perform the following steps:*

1. *Check all relevant sub–expressions of E for zero equivalence to ensure that E defines a function.*
2. *Replace every element k of $\mathcal{K}$ in E by $\{[k,k]\} \in O(\mathcal{K})$.*
3. *Proceed with the multiseries algorithm of Section 4.2 using coefficients in $O(\mathcal{K})$, except that when we meet an expressions of the form $\sin F(x)$ we see whether $F(x)$ must tend to a finite limit. If so $\sin F(x)$ is treated as in Section 5.3. If the values of $F(x)$ can be seen to be confined to a finite interval $[a,b]$ for large values of x we replace $\sin F(x)$ by $\{[\min(\sin y; y \in [a,b]), \max(\sin y; y \in [a,b])]\}$. In all other case we replace $\sin F(x)$ by $\{[-1,1]\}$.*

One can ask how useful Algorithm 10 is likely to be in practice. Its only real virtue is that it is applicable to cases where no other algorithm is possible. In some of these cases it will doubtless produce useful information, but one suspects that far too often the interval $[-\infty,\infty]$ will be the final result. Moreover if the coefficient of some term contains non–trivial intervals, the meaning of any term that follows is unclear unless closed forms are maintained alongside the sets of intervals which estimate them.

A simple example where the above algorithm works reasonably well is given by

$$f(x) = \frac{e^x \cos x^2}{x - \sin x}.$$

This converts to

$$\begin{aligned}
\frac{e^x[-1,1]}{x-[-1,1]} &= \frac{e^x}{x}\left(1 - \frac{[-1,1]}{x}\right)^{-1} \\
&= e^x\left\{\frac{[-1,1]}{x} + \frac{[-1,1]^2}{x^2} + \frac{[-1,1]^3}{x^3} + \cdots\right\} \\
&= [-1,1]e^x\left\{\frac{1}{x} + \frac{1}{x^2} + \frac{1}{x^3} + \cdots\right\}.
\end{aligned}$$

10.2 Lim-sups and Lim-infs

Most mathematicians are aware that if λ is an irrational number then every point in $[-1,1]$ is a limit point of the sequence $\sin(\lambda\pi n)$. Much more is true. Let T^n denote the n–dimensional torus, $\mathbb{R}^n/\mathbb{Z}^n$. For $\lambda \in \mathbb{R}^n$, X a measurable subset of T^n and I an interval of $\mathbb{R}$ write

$$\rho(I, X, \lambda) = \frac{\mu(\{x \in I; p_n(x\lambda) \in X\})}{\mu(I)},$$

where μ denotes Lebesgue measure on $\mathbb{R}$ and p_n is the projection from $\mathbb{R}^n$ to T^n. The following result is classical.

THEOREM **27**
Let $\lambda = (\lambda_1, \ldots, \lambda_n)$ with $\lambda_1, \ldots, \lambda_n$ being $\mathbb{Q}$–linearly independent real numbers. For $i = 1, \ldots, n$, let $a_i, b_i \in \mathbb{R}$ with $0 \leq a_i \leq b_1 \leq 1$, and let

$$X = [a_1, b_1) \times \cdots \times [a_n, b_n) \subseteq T^n.$$

Then

$$\rho(I, X, \lambda) \to \mu_n(X),$$

as $\mu(I) \to \infty$, uniformly in I.

Here $\times$ denotes the Cartesian product, μ_n is Lebesgue measure on T^n, and to say that the property $\mathcal{P}$ holds uniformly in I means that $\mathcal{P}$ holds uniformly in a where a is the left–hand end point of I.

In [115], Joris Van Der Hoeven exploits this circle of ideas to give an algorithm for calculating lim-sups and lim-infs in a class of elementary functions which includes trigonometric functions with exp–log arguments. This section is based on [115]. The paper contains, among other things, a proof of Theorem 27 and also a generalization. Before stating that, we need some more definitions.

For any $a \in \mathbb{R}$ and any $f \in \mathcal{C}^2(a, \infty)$ we say that f is *steadily dominated* by x if the following hold:

(i) $f \to \infty$ as $x \to \infty$;
(ii) $f' \to 0$;
(iii) $f''/f' \to 0$.

It is worth pointing out that if f belongs to a Hardy field, (i) and (ii) imply (iii). For we must have $\gamma_1(f') < \gamma_1(e^x)$ and Lemma 7 in Chapter 3 then implies that $\gamma_0(f''/f') < \gamma_0(1)$; i.e. $f''/f' \to 0$.

Then if g is also an element of $\mathcal{C}^2(a, \infty)$ which strictly increases to infinity, we say that f is steadily dominated by g if $f \circ g^{-1}$ is steadily dominated by x. In that case we write $f \lll_s g$.

Now let $f_1, \ldots, f_p$ be elements of $\mathcal{C}^2(a, \infty)$, $a \in \mathbb{R}$, which strictly increase to infinity such that $f_1 \lll_s f_2 \lll_s \cdots \lll_s f_p$. Let $n_1, \ldots, n_p \in \mathbb{N}^+$ with $n = n_1 + \cdots + n_p$, and for each $i = 1, \ldots, p$ let $\lambda_{i,1}, \ldots, \lambda_{i,n_i}$ be $\mathbb{Q}$–linearly independent real numbers. Put

$$g(x) = (f_1(\lambda_{1,1}x), \ldots, f_1(\lambda_{1,n_1}x), \ldots, f_p(\lambda_{p,1}x), \ldots, f_p(\lambda_{p,n_p}x)),$$

so that $x \mapsto p_n(g(x))$ is a curve in T^n. Now for X a measurable subset of T^n and I an interval of $\mathbb{R}$ with sufficiently large left–hand end point, we define

$$\rho_{f,g}(I, X) = \frac{\mu(\{x \in I; p_n(g(f_1^{-1}(x)) \in X\})}{\mu(I)}.$$

Van Der Hoeven's generalization is as follows.

THEOREM **28**
Let $f_1, \dots, f_p, g$ be as above and let $X = [a_1, b_1) \times \cdots \times [a_n, b_n) \subseteq T^n$ as in Theorem 27. Then for intervals, I, with sufficiently large left–hand end point

$$\rho_{f,g}(I, X) \to \mu_n(X),$$

as $\mu(I) \to \infty$, uniformly in I.

For the proof see [115].

To apply this we need another result from [115].

THEOREM **29**
Let $f_1, \dots, f_p$ be exp–log functions which tend to infinity and satisfy

$$\gamma_0(f_1) < \gamma_0(f_2) < \cdots < \gamma_0(f_p).$$

Let $\lambda_{i,j}$, $1 \le j \le n_i$, $i = 1, \dots, p$ be positive real numbers such that $\lambda_{i,1}, \dots, \lambda_{i,n_i}$ are $\mathbb{Q}$–linearly independent for each i. Put $n = n_1 + \cdots + n_p$, let ϕ be a continuous function from $[0,1]^n$ into $\mathbb{R}$, and write

$$\psi(x) = \phi(e^{\sqrt{-1}\lambda_{1,1} f_1(x)}, \dots, e^{\sqrt{-1}\lambda_{1,n_1} f_1(x)}, \dots, e^{\sqrt{-1}\lambda_{p,n_p} f_p(x)}).$$

Then

$$\text{lim-sup}_{x \to \infty} \psi(x) = \sup_{\mathbf{t} \in [0,1]^n} \phi(\mathbf{t}).$$

Proof of Theorem 29
We start by noting that we do have $f_1 \lll_s f_2 \lll_s \cdots \lll_s f_p$ in this situation, and so Theorem 28 can be applied.

In a suitable neighbourhood of $+\infty$, the map $\psi : V \to \mathbb{R}$ can be factored as $\psi = \psi_1 \circ \psi_2$ where $\psi_1 : V \to T^n$ and $\psi_2 : T^n \to \mathbb{R}$ are given by

$$\psi_1(x) = \left(\frac{\lambda_{1,1} f_1(x)}{2\pi}, \dots, \frac{\lambda_{p,n_p} f_p(x)}{2\pi} \right)$$

and

$$\psi_2(t_1, \dots, t_n) = \phi(e^{2\pi\sqrt{-1}t_1}, \dots, e^{2\pi\sqrt{-1}t_n}).$$

Both ψ_1 and ψ_2 are continuous, and since T^n is compact, ψ_2 attains its maximum at some point $\mathbf{Y} = (Y_1, \dots, Y_n)$. Suppose that $\varepsilon \in \mathbb{R}^+$, and let Ω be a neighbourhood of $\mathbf{Y}$ such that $|\psi_2(\mathbf{y}) - \psi_2(\mathbf{Y})| < \varepsilon$ for $\mathbf{y} \in \Omega$. By Theorem 28, there exist arbitrarily large x such that $\psi_1(x) \in \Omega$ and for such x $|\psi(x) - \sup_{\mathbf{t} \in [0,1]^n} \phi(\mathbf{t})| < \varepsilon$. This proves Theorem 29.

THEOREM **30**
Let $F_1, \dots, F_q$ be exp–log functions and let $\Phi : [0,1]^n \to \mathbb{R}$ be an algebraic function on the real algebraic variety $[0,1]^n$. Assuming that we can test the $\mathbb{Q}$–linear dependence of exp–log constants, then there is an algorithm to compute the lim-sup of

$$\Psi(x) = \Phi(e^{\sqrt{-1}F_1(x)}, \dots, e^{\sqrt{-1}F_q(x)}).$$

It is worth noting that by the result of Macintyre and Wilkie, [55], the $\mathbb{Q}$–linear dependence of exp–log constants is decidable on the basis of the Schanuel conjecture.

The essence of the method is as follows. Rewrite Φ and $F_1, \ldots, F_q$ in terms of ϕ, $f_1, \ldots, f_p$ and $\lambda_{1,1}, \ldots, \lambda_{1,n_1}, \ldots, \lambda_{p,n_p}$ so that Theorem 29 applies. Then calculate the sup of this ϕ on $[0,1]^n$. The details for limsup are given in the following algorithm, which requires only minor changes for the liminf.

Algorithm **11**

By replacing $e^{\sqrt{-1}F_j(x)}$ by $e^{-\sqrt{-1}F_j(x)}$ where necessary, make each F_j positive for large x.

1. *Compute a scale for the set of functions $F_1, \ldots, F_q$ as in Section 5.1. Use this to determine the order relations between $\gamma_0(F_1), \ldots, \gamma_0(F_q)$.*
2. *If the $\gamma_0(F_j)$ are all different, go directly to the next step. Otherwise choose a pair $j < k$ for which $\gamma_0(F_j) = \gamma_0(F_k)$ is maximal. Let $\lambda_{j,\alpha} = \lim\{F_k/F_j\}$, with α the next available index starting from 1. Then replace F_k by $\lambda_{j,\alpha}F_j + F_k^1$ where $\gamma_0(F_k^1) < \gamma_0(F_k)$, and make the consequent changes to the function Φ so as to keep Ψ unchanged.*
   We note that the replacement of F_k results in no increase in the number of F_i and a decrease in the number of them sharing the highest multiple γ_0 value. If F_k^1 tends to a finite limit, we do not need to add F_k^1 to the list under consideration and we can treat $e^{\sqrt{-1}F_k^1}$ as in Section 5.3. So it is clear that a finite number of repetitions of this process gives at most q F_is, all with different γ_0 values.
3. *Now rewrite the various $\lambda_{j,\alpha}$ to give linearly independent sets. More specifically, test each set $\{\lambda_{i,1}, \ldots, \lambda_{i,n_i}\}$ for linear dependence. If a linear relation is found it may be assumed to be of the form*

$$A_1\lambda_{i,1} = A_2\lambda_{i,2} + \cdots + A_{n_i}\lambda_{i,n_i}$$

   with $A_1, \ldots, A_{n_i} \in \mathbb{Z}$ and $A_1 > 0$. Then for each $j > 1$, replace $\lambda_{i,j}$ by $\lambda'_{i,j}/A_1$ and $e^{\sqrt{-1}\lambda_{i,j}F_i}$ by $(e^{\sqrt{-1}\lambda'_{i,j}F_i})^{A_1}$. Next replace $e^{\sqrt{-1}\lambda_{i,1}F_i}$ by

$$(e^{\sqrt{-1}\lambda'_{i,2}F_i})^{A_2}(e^{\sqrt{-1}\lambda'_{i,3}F_i})^{A_3}\cdots(e^{\sqrt{-1}\lambda'_{i,n_i}F_i})^{A_{n_i}}.$$

   Then search for a linear relation among $\lambda'_{i,2}, \ldots, \lambda'_{i,N_i}$, and so on.
4. *At the final stage the lim-sup of $\Psi(x)$ is computed. By Theorem 29 this is the sup of Φ on $[0,1]^n$ with $n = n_1 + \cdots + n_q$. To find it, compute the set of stationary points of the gradient of Φ on $[0,1]^n$. Φ will be constant on each connected component of this set and its sup is the largest of the finite number of values found. The zero set and the connected components can be computed using cylindrical decomposition, see [27, 36].*

As a corollary, one can also treat expressions of the form

$$\Phi(\sin F_1(x), \cos F_1(x), \ldots, \sin F_q(x), \cos F_q(x)).$$

10.3 Wandering Functions

The idea of a wandering function is that it oscillates, not just in the sense of not tending to a limit, but with no limit point being favoured. In particular the set where it is 'close to infinity' will be small and so the function will be bounded on most of its range. The fact that a function does not spend an abnormal amount of time close to any particular limit value can be interpreted as meaning that it does not represent any asymptotic trend. This is completely orthogonal to the behaviour of a Hardy–field element, and so wandering functions should be suitable candidates for coefficients in expansions with Hardy–field base elements.

It is not yet totally clear what is the optimal definition of a wandering function. The present treatment is based on [89]. Joint research in progress, by Bruno Salvy and the author, suggests that a slightly different approach may be more satisfactory. However the results obtained there are not in final form as this book goes to press.

The main motive in introducing wandering functions is to use them as coefficients in expansions, for example of elementary functions. So the definition needs to embrace the functions likely to be required there. In addition, one would like a reasonable level of generality and ideally some nice properties of closure. Obviously we cannot ask too much by way of closure properties. For example constants should not be expected to satisfy the definition. Then we cannot get closure under addition, since if f is a wandering function, $-f$ will surely also be one, but not $f + (-f)$. See also Lemma 47 below in this connection.

The following definition is from [89].

DEFINITION **26** *Let $\mathcal{S}$ be the set of C^∞ functions, f, defined on some interval $[\alpha, +\infty) \subset \mathbb{R}$ except perhaps at a countable number of points such that the following holds.*

> *For any $\varepsilon \in \mathbb{R}^+$ and any $l \in \mathbb{R}^+$ there exists $m = m(\varepsilon, l)$ and $M = M(\varepsilon, l) \in \mathbb{R}^+$ such that for any finite interval $I \subset [a, +\infty)$ with $|I| \geq l$, and any $v \in \mathbb{R}$ there are sub-intervals $S_1, \ldots, S_N$ with $\sum_{j=1}^N |S_j| < \varepsilon |I|$ and*
>
> $$m < |f(x) - v| < M \quad \forall x \in I \setminus \cup_{j=1}^N S_j.$$

There are a number of points to note.

1. The definition of $\mathcal{S}$ has some parallel with the notion of convergence in measure, see [38].
2. The elements of $\mathcal{S}$ are called *wandering functions.*
3. The dependence of m and M on l is needed in order to cater for small intervals containing a zero of $f - v$. However m and M must not depend on I itself or f could tend to zero and still satisfy the definition.
4. If we are thinking of using elements of $\mathcal{S}$ as coefficients in an expansion we shall need to add appropriate constants, e.g. use $\mathcal{S} \cup \mathbb{R}$.

5. One virtue of $\mathcal{S}$ is that its elements are well behaved except on relatively small sub-intervals, and the same is true of any finite number of elements. This means that when the elements of $\mathcal{S} \cup \mathbb{R}$ are used as coefficients, any finite part of the expansion will make sense on a large subset of $[a, \infty)$.
6. The definition of $\mathcal{S}$ is unfortunately too restrictive for $\mathcal{S}$ to contain all the functions we would wish.

The last point needs some elaboration. We would like $\sin h(x)$ to be a wandering function if h is a Hardy–field function which tends to infinity. This seems right intuitively and is needed for our most general result about expansions. But if h grows sufficiently slowly, $\sin h(x)$ can linger near zero for say half the time since the start at $x = a$, before moving away for a far longer time. An example is given by $h(x) = \log_2 x$, [89]. Suppose that $0 < \varepsilon < 1/2$. Then for any $N \in \mathbb{N}$, $|\sin(\log_2 x)| < \varepsilon$ if $\exp_2(\pi N - \varepsilon) < x < \exp_2(\pi N)$ and this interval has relative length

$$\frac{\exp_2(\pi N) - \exp_2(\pi N - \varepsilon)}{\exp_2(\pi N))}$$

which tends to 1 as $N \to \infty$.

We therefore make another definition as follows.

DEFINITION **27** *Let $\overline{\mathcal{S}}$ denote the set of C^∞ functions, f, defined on some interval $[\alpha, +\infty) \subset \mathbb{R}$ except perhaps at a countable number of points, with the following property.*

For any $\varepsilon \in \mathbb{R}^+$ there are $m = m(\varepsilon)$ and $M = M(\varepsilon) \in \mathbb{R}^+$ so that for any finite interval $I \subset [a, +\infty)$ and any $v \in \mathbb{R}$, there exists a finite interval J with $I \subseteq J \subset [a, \infty)$ and sub-intervals $S_1, \ldots, S_N$ of J such that $\sum_{j=1}^{N} |S_j| < \varepsilon |J|$ and

$$m < |f(x) - v| < M \quad \forall x \in J \setminus \cup_{j=1}^{N} S_j.$$

Here also m and M must not depend on I. The class $\overline{\mathcal{S}}$ contains functions like $\sin(\log_2 x)$, but also some functions which intuition suggests should not be there, like $g(x) = (-1)^{[\log_2 x]}$. Indeed this function only takes two values. However if we set $T_n = \exp(e^{2n+1} - 1)$ and $t_n = \exp(e^{2n})$ then $g = 2$ on $[t_n, T_n]$, and so obviously $1/3 < |g + 1| < 3$ there. Thus the subset of $[e, T_n]$ where this inequality fails is certainly contained in $[e, t_n]$, whose length relative to $[e, T_n]$ is less than $t_n/(T_n - e)$. This tends to zero as $n \to \infty$ because

$$\frac{T_n}{t_n} = \frac{\exp(e^{2n+1} - 1)}{\exp(e^{2n})} = \exp\{(e - 1)e^{2n} - 1\} \to \infty.$$

Similarly if we set $\tilde{T}_n = \exp(e^{2n} - 1)$ and $\tilde{t}_n = \exp(e^{2n-1})$, then $1/3 < |g - 1| < 3$ on $[\tilde{t}_n, \tilde{T}_n]$ and the relative length of this compared to $[e, \tilde{T}_n]$ is greater than $1 - \tilde{t}_n \tilde{T}_n^{-1}$, which tends to 1. Since any interval is contained in $[e, T_n]$ for sufficiently large n and similarly for $[e, \tilde{T}_n]$, it follows that $g \in \overline{\mathcal{S}}$.

Moreover this example shows that we cannot apply scaling in $\overline{\mathcal{S}}$ in the way we would like, any more than we can in $\mathcal{S}$. Specifically if $f \in \overline{\mathcal{S}}$ and h is a Hardy–field element which tends to infinity, we might like $f \circ h$ to belong to $\overline{\mathcal{S}}$. However with g as above, $g(\exp_2 x) = (-1)^{[x]}$, which certainly is not in $\overline{\mathcal{S}}$.

The example g gives us a reason for not referring to arbitrary elements of $\overline{\mathcal{S}}$ as wandering functions, and shows that it is worth trying to come up with a better definition of these if possible. However the example is not a disaster for us. The function g itself could reasonably serve as a coefficient in an expansion, and the bad scaling property does not arise with the trigonometric examples we shall be considering below.

Since it rapidly becomes tiresome to keep account of the relatively small intervals on which properties are allowed to fail, we introduce some special notation. We refer to the sets $S_1, \ldots, S_N$ above as the *exceptional sets.* If for each $\varepsilon > 0$ there is an $m = m(\varepsilon, l)$, as in Definition 26, such that $|f| > m$ except on exceptional sets of relative total length less than ε, we say that f is *mainly bounded away from zero* . Similarly if $M(\varepsilon, l)$ exists such that $|f| < M$ except on exceptional sets, we say that f is *mainly bounded.*

We now introduce our coefficient fields.

DEFINITION **28** *Let $b_1(x), \ldots, b_k(x)$ be elements of a Hardy field which tend to infinity such that $\gamma_0(b_1) < \gamma_0(b_2) < \cdots < \gamma_0(b_k)$. Let $J_i \in \mathbb{N}$ and let $\lambda_{i,j}, \in \mathbb{R}$ for $j = 1, \ldots, J_i$. Write*

$$\mathcal{R} = \mathbb{R}(\sin(\lambda_{1,1}b_1(x)), \cos(\lambda_{1,1}b_1(x)), \ldots, \sin(\lambda_{1,J_1}b_1(x)), \cos(\lambda_{1,J_1}b_1(x)), \ldots \\ \ldots, \sin(\lambda_{k,J_k}b_k(x)), \cos(\lambda_{k,J_k}b_k(x))).$$

To stress the dependence of $b_1, \ldots, b_k$, we sometimes write $\mathcal{R}(b_1, \ldots, b_k)$ instead of just $\mathcal{R}$.

Our first result was effectively given in [89], although not explicitly stated there. We give the main points of the proof and refer the reader to the preprint for further details.

THEOREM **31** *Suppose that there is a positive constant, $c \in \mathbb{R}^+$, such that $|b_1(x)| \geq c|x|$ for x sufficiently large. Then $\mathcal{R}(b_1, \ldots, b_k) \subset \mathcal{S} \cup \mathbb{R}$.*

We also have the following, again from [89].

THEOREM **32** *For any $b_1, \ldots, b_k$ as in Definition 28, $\mathcal{R}(b_1, \ldots, b_k) \subset \overline{\mathcal{S}}$.*

The nub of the proof of Theorem 31 is to show that a polynomial in

$$\mathbb{R}[\sin(\lambda_{1,1}b_1(x)), \cos(\lambda_{1,1}b_1(x)), \ldots, \sin(\lambda_{k,J_k}b_k(x)), \cos(\lambda_{k,J_k}b_k(x))]$$

is mainly bounded away from zero. We employ several lemmas. The first two are quite straightforward.

LEMMA **47** *Let $f, g \in \mathcal{C}^\infty(\mathbb{R}^+)$. If $f \in \mathcal{S}$ and $g \to 0$ as $x \to \infty$ then $f + g \in \mathcal{S}$.*

LEMMA **48** *A product of sines, $\prod_{j=1}^{J} \sin v_j$, may be written in the form*

$$\prod_{j=1}^{J} \sin v_j = \sum_{k=1}^{K} a_k \sin w_k,$$

where $a_1, \ldots, a_k \in \mathbb{R}$ and each w_k is of the form

$$w_k = \alpha_1 v_1 + \alpha_2 v_2 + \cdots + \alpha_j v_j + \beta,$$

with $\alpha_1, \ldots, \alpha_J, \beta \in \mathbb{R}$.

The next lemma is a key tool in the proofs of Theorem 31 and Theorem 32.

LEMMA **49** *Let $x_0 \in (a, \infty)$ and suppose that there exist M, $m_0 \in \mathbb{R}^+$ such that $|f''| \leq M$ on (a, ∞), $|f'(x_0)| > m_1$ and $x_0 - m_1/(2M) > a$. Then $|f| > m_1^3$ on $(x_0 - m_1/(2M), x_0 + m_1/(2M))$ except perhaps on a sub-interval of length at most $8m_1^2$.*

Intuitively this holds because if f is very small at x_0 and f' is not, then f will soon move away from zero as x moves away from x_0, and it cannot return quickly because of the upper bound on f''. The proof employs two applications of the First Mean Value Theorem.

As a corollary, we get the following.

LEMMA **50** *Let $\alpha \in \mathbb{R}$ and let $f \in C^2(\alpha, \infty)$. Suppose that f' is mainly bounded away from zero on (α, ∞) and f'' is bounded there. Then f is mainly bounded away from zero on (α, ∞).*

It follows that if such an f is bounded, it must belong to $\mathcal{S}$.

For the proof of Theorem 31, we assume inductively that the result holds for smaller values of k. The case $k = 0$ is a triviality. We suppose initially that $b_1(x) = x$.

We may assume the $\lambda_{1,i}$ to be pairwise distinct. Then the Wronskian $W(f, f', \ldots, f^{(2J_1)})$ is a polynomial in the quantities

$$\sin(\lambda_{2,1} b_2(x)), \cos(\lambda_{2,1} b_2(x)), \ldots, \sin(\lambda_{k,J_k} b_k(x)), \cos(\lambda_{k,J_k} b_k(x))),$$

and so by induction is bounded away from zero except on exceptional sets. Hence at any point outside these intervals, we obtain a lower bound on one of $|f|, |f'|, \ldots, |f^{(2J_1)}|$. Then Lemma 50 can be used repeatedly to get a lower bound on $|f|$ except on exceptional sets. So in this case f is mainly bounded away from zero.

Now it remains to apply scaling in order to complete the induction step. Transition to the slightly more general case when $\gamma_0(b_1(x)) = \gamma_0(x)$ is straightforward. The crux of the proof of Theorem 31 is now contained in the following lemma.

LEMMA **51** *Suppose that $f \in \mathcal{S}$ and that b is an element of a Hardy field such that $b(x)$ tends to infinity and $b'(x)$ is eventually increasing. Then $f \circ b \in \mathcal{S}$.*

Proof of Lemma 51

The behaviour of $f \circ b$ on an interval I is given by the behaviour of f on $b(I)$. We may take v to be zero. Then since $f \in \mathcal{S}$ there will be $m, M \in \mathbb{R}^+$ and exceptional sets $S_1, \ldots, S_N \subset b(I)$ such that $m < |f| < M$ on the set $b(I) \setminus \cup_{j=1}^N S_j$. On transferring back to the interval I we obtain exceptional sets $b^{-1}(S_1), \ldots, b^{-1}(S_N)$, and it is a matter of showing that these are suitably small.

We may assume without loss of generality that b' is everywhere positive. Then b has an inverse function h which tends to infinity and belongs to a Hardy field. Now the gist is that h' cannot tend to zero too rapidly (since h tends to infinity) and so provided that $b(I)$ is not too large, the shrinkage on applying h to $b(I)$ is not much more than the shrinkage on applying h to one of the S_j. Thus $\sum |h(S_j)|$ remains small compared with $|I| = |h(b(I))|$.

In terms of inequalities, the argument goes like this. Since $\gamma_0(h(x)) > \gamma_0(x^{-1})$, it follows that $\gamma_0(h'(x)) > \gamma_0(x^{-2})$. Hence for $c, d \in \mathbb{R}$ with $d > c$ and c sufficiently large,

$$\log\left(\frac{h'(d)}{h'(c)}\right) = \int_c^d (\log h'(t))' dt \geq -2 \int_c^d (\log t)' dt = -2 \log(d/c).$$

So

$$h'(d) > h'(c)(c/d)^2. \tag{10.7}$$

Now with $b(I) = [c, d]$, we may subdivide $b(I)$ if necessary to ensure that $c/d \geq 1/2$. Then (10.7) gives $h'(c) < 4h'(d)$. If $S_j = [c_j, d_j]$ for $1 \leq j \leq N$, we have

$$\frac{|h(S_j)|}{|I|} = \frac{\int_{c_j}^{d_j} h'}{\int_c^d h'} \leq \frac{h'(c_j)|S_j|}{h'(d)|b(I)|} \leq \frac{h'(c)|S_j|}{h'(d)|b(I)|} \leq 4 \frac{|S_j|}{|b(I)|}.$$

Therefore

$$\frac{\sum_{j=1}^N |h(S_j)|}{|I|} \leq 4 \frac{\sum_{j=1}^N |S_j|}{|b(I)|},$$

which is less than 4ε if $\sum_{j=1}^N |S_j| < \varepsilon |b(I)|$. Thus $h(S_1), \ldots, h(S_N)$ serve as exceptional sets for $f \circ b$ on I, and are of appropriate length. This proves Lemma 51.

Under the hypotheses of Theorem 31, b_1' increases. But $b_2 \circ b_1^{-1}$ will also have an increasing derivative and so by inductive assumption the result holds for elements of $\mathcal{R}(b_2 \circ b_1^{-1}, \ldots, b_k \circ b_1^{-1})$. The Wronskian argument above shows that it holds for $\mathcal{R}(x, b_2 \circ b_1^{-1}, \ldots, b_k \circ b_1^{-1})$ also. But then Lemma 51 allows us to scale by b_1, which proves Theorem 31.

Proof of Theorem 32

In the case when b_1' decreases, the above argument gets us as far as the conclusion for $\mathcal{R}(x, b_2 \circ b_1^{-1}, \ldots, b_k \circ b_1^{-1})$. We again need to scale by b_1 and now we need the following result.

LEMMA **52** *Suppose that $f \in \mathcal{S}$ and that b is an element of a Hardy field such that $b(x)$ tends to infinity and $b'(x)$ is eventually decreasing. Then $f \circ b \in \overline{\mathcal{S}} \cup \mathbb{R}$.*

Proof of Lemma 52
The problem now is that h, the inverse of b, might tend to infinity arbitrarily rapidly. So if there is an exceptional interval, S_N, at the extreme right–hand end of $b(I)$, the stretching effect could make $|h(S_N)|$ a substantial fraction of $h(b(I))$. In that case we should arrange for $b(I)$ to continue more to the right, but then we might hit a new S_j at the right–hand end. So we have to show that the S_js cannot keep winning this game. The details are rather delicate.

We may confine our attention to the case when f is not constant and $v = 0$. Let ε, $l \in \mathbb{R}^+$ as in the definition of $\mathcal{S}$ with $\varepsilon < 1/4$. Write $I = (a, a+l)$ and $I' = (a, a+2l)$. Let $S_1, \ldots, S_N$ be the exceptional sets of I'; so we have $\sum_{j=1}^N |S_j| < \varepsilon|I'|$ and $m < |f| < M$ on $I' \setminus \cup_{j=1}^N S_j$.

For any finite set of intervals $\mathcal{J}$ with $I \subset \cup\mathcal{J} \subset I'$, and any point $p \in \cup\mathcal{J}$, we say that property $P(\mathcal{J}, p)$ holds if

$$\sum_{j=1}^{K} |S_j \cap (\cup\mathcal{J}) \cap (p, a+2l)| < 2\varepsilon|(\cup\mathcal{J}) \cap (p, a+2l)|.$$

Often $\mathcal{J}$ will consist of a single interval J, and we will then refer to the property $P(J,p)$ rather than $P(\{J\},p)$.

Since $\sum |S_j| < \varepsilon|I'|$ and $a+l$ is the mid-point of I', it is clear that $P(I', p)$ holds for all $p \le a+l$. We seek a single interval J such that $P(J,p)$ holds for all $p \in J$. If this is not the case for $J = I'$, let q be the smallest value for which $P(I', q)$ fails, and take $J = (a, q)$. Then $q \ge a+l$, and we claim that $P(J,p)$ holds for all $p \in J$. For if there exists $p < q$ for which $P(J,p)$ fails then

$$\sum_{j=1}^{K} |S_j \cap (p,q)| \ge 2\varepsilon|J \cap (p,q)| = 2\varepsilon|I' \cap (p,q)|.$$

However since $P(I', q)$ fails, we already have

$$\sum_{j=1}^{K} |S_j \cap (q, a+2l)| \ge 2\varepsilon|I' \cap (q, a+2l)|,$$

which gives

$$\sum_{j=1}^{K} |S_j \cap (p, a+2l)| \ge 2\varepsilon|I' \cap (p, a+2l)|,$$

contrary to the definition of q.

Now with this J, replace each S_j by $S_j \cap J$. We are going to define, for $j = K, K-1, \ldots, 1$, a set of sub-intervals, $J_{1,j}, \ldots, J_{s(j),j}$, each to the right of S_j and contained in $J \setminus \cup_{j=1}^K S_j$ such that

(i) The $J_{i,j}$ are pairwise disjoint for $i = 1, \ldots, s(j), j = 1, \ldots, K$.
(ii)

$$|S_j| = \frac{2\varepsilon}{1 - 2\varepsilon} \sum_{i=1}^{s(j)} |J_{i,j}|.$$

We write $\eta = 2\varepsilon(1 - 2\varepsilon)^{-1}$. Since $P(J, p)$ holds for p equal to the left-hand end-point of S_K, this can be done for $j = K$ just by taking $s(K) = 1$ and $J_{1,K}$ to be an interval of the correct length to the right of S_K.

We call a sub-interval of J 'good' if it does not meet $\cup_{j=1}^K S_j$. So J is the union of the good intervals and the exceptional sets.

Now suppose that we have the required $J_{i,j}$ for $j = K, \ldots, r+1$. If we remove $S_K, , \ldots, S_{r+1}$ and $J_{1,r+1}, \ldots, J_{(s(K),K)}$ from J, then because (ii) holds, the property P remains true of the set of remaining intervals at the left-hand endpoint of S_r. That is to say $|S_r|$ is less than or equal to η times the sum of the lengths of the remaining good intervals to the right of S_r. We therefore have sufficient good intervals to the right of S_r to define $J_{1,r}, \ldots, J_{s(r),r}$ as required.

Now let j be between 1 and K, let S_j have end-points $c_j < d_j$ and let the end-points of $J_{i,j}$ be $\alpha_{i,j} < \beta_{i,j}$. Since h' is increasing,

$$|h(S_j)| = \int_{c_j}^{d_j} h'(t)dt \le (d_j - c_j)h'(d_j)$$

and for each $i = 1, \ldots, s(j)$

$$|h(J_{i,j})| = \int_{\alpha_{i,j}}^{\beta_{i,j}} h'(t)dt \ge (\beta_{i,j} - \alpha_{i,j})h'(\alpha_{i,j}) \ge (\beta_{i,j} - \alpha_{i,j})h'(d_j),$$

since each $J_{i,j}$ is to the right of S_j. Hence

$$\sum_{i=1}^{s(j)} |h(J_{i,j})| \ge h'(d_j) \sum_{i=1}^{s(j)} (\beta_{i,j} - \alpha_{i,j}) = h'(d_j)\frac{1}{\eta}(d_j - c_j) \ge \frac{1}{\eta}|h(S_j)|,$$

and thus

$$\sum_{j=1}^{K} |h(S_j)| \le \eta|h(J)| < 4\varepsilon|h(J)|.$$

Moreover $m < |f(b(x))| < M$ on $h(J) \setminus \cup_{j=1}^K h(S_j)$ because $m < |f(y)| < M$ for $y \in b(h(J) \setminus \cup_{j=1}^K h(S_j)) = J \setminus \cup_{j=1}^K S_j$.

Now take $\delta \in \mathbb{R}^+$ such that $\delta < m$ and $\delta^{-1} > M$, and let $h(J) = [\alpha, T]$. Then $\delta < |f \circ b| < \delta^{-1}$ on $[\alpha, T]$ except on the intervals $h(S_j)$, which are of length no more than $4\varepsilon(T - \alpha)$.

This completes the proof of Lemma 52, and the conclusion of Theorem 32 is now clear.

10.4 Wandering Expansions

These are expansions where the coefficients belong to $\overline{\mathcal{S}} \cup \mathbb{R}$.

DEFINITION **29** *let $\mathcal{H}$ be a Hardy field, let* $\mathbf{t}$ *be an element of $\mathcal{H}$ which tends to zero and let $\{r_n\}$ be a strictly increasing sequence of real numbers which tends to infinity. Suppose that $\{w_n\} \subset \overline{\mathcal{S}} \cup \mathbb{R}$ and that $F \in \mathcal{X}$. Then we say that $\sum_{n=1}^{\infty} w_n \mathbf{t}^{r_n}$ is a* wandering $\mathbf{t}$-expansion *of F if it is an asymptotic expansion of F in some interval (a, ∞), i.e. for each $N \in \mathbb{N}$*

$$\left| F(x) - \sum_{n=1}^{N} w_n \mathbf{t}^{r_n} \right| = O(\mathbf{t}^{r_{N+1}}),$$

as $x \to \infty$.

It should be noted that unlike the coefficients of normal asymptotic power series, the coefficients of a wandering expansion are not uniquely determined by the function. This is a consequence of the following very simple observation, c.f. Lemma 47.

LEMMA **53** *If $w \in \overline{\mathcal{S}}$ and $u(x) \to 0$ as $x \to \infty$ then $w + u \in \overline{\mathcal{S}}$.*

The Lemma is evident from the definition of $\overline{\mathcal{S}}$ since for b sufficiently large,

$$\delta < |F(x) - v| < \delta^{-1} \qquad \Rightarrow \qquad \frac{\delta}{2} < |F(x) + u(x) - v| < \frac{2}{\delta}$$

on $[b, \infty)$.

Thus $\sin x + (x-1)^{-1}$ has a wandering expansion $\sin x + \sum_1^{\infty} x^{-n}$, and another $\sum_0^{\infty} w_n x^{-n}$, where $w_0 = \sin x + 1/2x$, $w_1 = 1/2$ and $w_n = 1$ for $n \geq 2$. Of course there are infinitely many others, but the first one given seems the most natural. In this spirit, we insist that the coefficients in the wandering expansions we use shall belong to some $\mathcal{R} = \mathcal{R}(b_1, \ldots, b_k)$ where $b_1, \ldots, b_k$ have multiseries expansions with all terms tending to infinity. This is sufficient to ensure uniqueness of the coefficients for given $\mathbf{t}$. For suppose F has two such expansions. By subtracting terms as necessary, we may suppose that the two expansions differ in the first coefficient. Let the two expansions be $F \sim \sum_0^{\infty} a_n \mathbf{t}^{r_n}$ and $F \sim \sum_0^{\infty} b_n \mathbf{t}^{s_n}$. If $r_0 < s_0$ then

$$|a_0 \mathbf{t}^{r_0} - b_0 \mathbf{t}^{s_0}| \leq |F - a_0 \mathbf{t}^{r_0}| + |F - b_0 \mathbf{t}^{s_0}| = o(\mathbf{t}^{r_0}). \tag{10.8}$$

Hence $a_0 = o(1)$, which is impossible since $a_0 \in \mathcal{R} \setminus \{0\}$. A similar contradiction is obtained if $r_0 > s_0$, and so $r_0 = s_0$. But now (10.8) gives that $a_0 - b_0 = o(1)$, which implies that $a_0 = b_0$ since $a_0 - b_0 \in \mathcal{R} \setminus \{0\}$. Thus the coefficients are unique as asserted.

Now let $\{\mathbf{t}_1, \ldots, \mathbf{t}_n\}$ be an asymptotic scale with $\gamma_1(\mathbf{t}_1) < \cdots < \gamma_1(\mathbf{t}_n)$, and let $F \in \mathcal{X}$.

DEFINITION **30** *We say that* F *has a* wandering multiseries expansion *in the scale* $\{\mathbf{t}_1, \ldots, \mathbf{t}_n\}$ *if it has a* $\mathbf{t}_n$*-expansion*

$$F \sim \sum_{m=1}^{\infty} g_m \mathbf{t}_n^{r_m},$$

where $\{r_m\}$ *strictly increases to infinity and each* g_m *has a wandering multiseries expansion in the scale* $\{\mathbf{t}_1, \ldots, \mathbf{t}_{n-1}\}$*. In the case* $n = 1$ *a wandering multiseries expansion is just a wandering* $\mathbf{t}_1$*-expansion as in Definition 29.*

Let $\mathcal{F}$ be an asymptotic field. We write $\mathcal{O}(\mathcal{F})$ for the field of functions defined by rational expressions in elements of the form g, $\sin g$ or $\cos g$ with $g \in \mathcal{F}$. We give an algorithm which when supplied with an element $f \in \mathcal{O}(\mathcal{F})$ computes a wandering multiseries expansion for f with coefficients in a field $\mathcal{R}(b_1, \ldots, b_n)$ where $b_1, \ldots, b_n \in \mathcal{F}$.

The first stage is the determination of $\mathbf{t}_1, \ldots, \mathbf{t}_n$. Let

$$f = \frac{P(f_1, \ldots, f_N, \sin g_1, \cos g_1, \ldots, \sin g_M, \cos g_M)}{Q(f_1, \ldots, f_N, \sin g_1, \cos g_1, \ldots, \sin g_M, \cos g_M)}, \tag{10.9}$$

where P and Q are polynomials and the f_i and g_i all belong to $\mathcal{F}$. We construct a tower of function fields with the top field, $\mathcal{F}_n$ containing these quantities and then, as in Section 5.2, we obtain a scale, $\{\mathbf{t}_1, \ldots, \mathbf{t}_n\}$, in which they can be expanded. Then as in Section 4.3.3, we write each g_i in the form

$$g_i = h_i + c_i + \eta_i,$$

where c_i is a constant, $\eta_i \to 0$ and the multiseries of h_i contains only terms which tend to plus or minus infinity. Now we can add the sines and cosines of the c_i and η_i to the asymptotic field $\mathcal{F}$ as in Section 5.3 and apply the trigonometric expansion formulae to $\sin g_i$ and $\cos g_i$. Thus we reduce to the case when every term in the multiseries of each g_i tends to $\pm\infty$.

The next step is to obtain $b_1, \ldots, b_n$. Our method generates these in reverse order, and so we shall need to correct this at the end. We take b_1 equal to $|g_1|$ and examine the limit of g_2/b_1. If this is zero or infinite, we take $b_2 = |g_2|$ and interchange subscripts if $b_1/b_2 \to 0$. Otherwise, let the limit be k. Then $|g_2 - kb_1|$ is either identically zero or tends to infinity. For if the $\mathbf{t}_n$-expansions of g_2 and kb_2 first differ at the term in $\mathbf{t}_n^{r_i}$, then $r_i \leq 0$. If $r_i < 0$ then $|g_2 - kb_1| = \mathbf{t}_n^{r_i + o(1)}$, and if $r_i = 0$ we apply similar consideration to the $\mathbf{t}_{n-1}$-expansions of the two coefficients of $\mathbf{t}_n^{r_i}$. Eventually we must reach a case where the exponent corresponding to r_i above is negative because all terms in the two multiseries tend to $\pm\infty$.

We take $b_2 = |g_2 - kb_1|$ if this is not zero. On the other hand if $g_2 = kb_1$ we repeat the above with g_2 replaced by g_3, and if $g_3 = b_1 \lim(g_3/b_1)$ we continue similarly with g_4, and so on.

Suppose we have reached a stage when $b_1, \ldots, b_{j-1}$ have been defined and $g_1, \ldots, g_{m-1}$ are each linear combinations of these with constant coefficients.

We calculate the limit of g_m/b_i for successive values of $i = 1, \ldots, j-1$. If the answer is always zero or infinite, we take $b_j = |g_m|$. Otherwise suppose that α is the first index for which $g_m/b_\alpha \to k$ with $k \in \mathbb{R} \setminus \{0\}$. We write $g_{m,1} = g_m - kb_\alpha$ and if this is not zero, we compute the limit of $g_{m,1}/b_i$ for $i = \alpha+1, \ldots, j-1$ until we get a finite non–zero answer; if this never happens, we set $b_m = |g_{m,1}|$ and re–order $b_1, \ldots, b_m$ as necessary. Otherwise if $k_1 = \lim(g_{m.1}/b_\beta) \in \mathbb{R} \setminus \{0\}$, we put $g_{m,2} = g_{m,1} - k_1 b_\beta$ and, if this is non–zero, continue as above with $g_{m,2}$ replacing $g_{m,1}$ and β replacing α. It is clear that at some stage we must obtain g_m as a linear combination of b_is with constant coefficients. Continuing in this way we obtain $b_1, \ldots, b_n$ with $b_i/b_{i-1} \to 0$ for $i = 2, \ldots, n$ and each g_j a linear combination of the b_i.

Now we reverse the ordering of $\{b_1, \ldots, b_n\}$ to comply with Definition 28. We can then use the trigonometric expansion formulae to rewrite each $\sin g_j$ and $\cos g_j$ in terms of the sines and cosines of the b_i.

It is then just a matter of computing the expansion of our expression as in the asymptotic field $\mathcal{F}$ with elements of $\mathcal{R}(b_1, \ldots, b_n)$ being treated as constants. That is to say we perform the $\mathbf{t}_n$–expansion first keeping coefficients in closed form, then we calculate the $\mathbf{t}_{n-1}$–expansions of these, and so on, eventually arriving at the $\mathbf{t}_1$–expansions whose coefficients will be in $\mathcal{R}(b_1, \ldots, b_n)$.

Example
Let

$$f(x) = \frac{e^x \cos(x + e^{-x}) - \sin\left(\frac{e^x + x^2}{x-1} - 1\right)}{x \sin e^x + e^{-x} \cos x}. \tag{10.10}$$

Initially $g_1 = e^x$, $g_2 = x$, $g_3 = x + e^{-x}$ and

$$g_4 = \frac{e^x + x^2}{x-1} - 1 = \frac{e^x}{x-1} + x + \sum_{r=1}^{\infty} \frac{1}{x^n}.$$

We rewrite $\cos(x + e^{-x})$ as $\cos(x + e^{-x}) = \cos x \cos e^{-x} - \sin x \sin e^{-x}$ and similarly

$$\sin\left(\frac{e^x + x^2}{x-1} - 1\right) = \sin\left(\frac{e^x}{x-1} + x\right) \cos\left(\sum_{r=1}^{\infty} \frac{1}{x^r}\right) + \cos\left(\frac{e^x}{x-1} + x\right) \sin\left(\sum_{r=1}^{\infty} \frac{1}{x^r}\right).$$

On substituting into (10.10) we obtain the following expansion for the numerator of f

$$N = e^x \cos x + \left\{\sin x - \sin\left(\frac{e^x}{x-1} + x\right) - \frac{1}{x} \cos\left(\frac{e^x}{x-1} + x\right) + O\left(\frac{1}{x^2}\right)\right\} - \frac{\cos x}{2} e^{-x} - \frac{\sin x}{3!} e^{-2x} + \cdots.$$

We then write the denominator as

$$D = x \sin e^x \left(1 + \frac{\cos x}{\sin e^x} \cdot \frac{e^{-x}}{x}\right).$$

On expanding N/D we obtain

$$\begin{aligned}
f = \frac{\cos x}{\sin e^x}\frac{e^x}{x} &+ \left\{\left(\frac{\sin x}{\sin e^x} - \frac{\sin\left(\frac{e^x}{x-1} + x\right)}{\sin e^x}\right)\frac{1}{x}\right. \\
&\qquad \left. - \left(\frac{\cos\left(\frac{e^x}{x-1} + x\right)}{\sin e^x} - \frac{\cos^2 x}{\sin^2 e^x}\right)\frac{1}{x^2} + O\left(\frac{1}{x^3}\right)\right\} \\
&- \left\{\frac{\cos x}{2\sin e^x} + \left(\frac{\cos x \sin x}{\sin e^x} - \frac{\cos x \sin\left(\frac{e^x}{x-1} + x\right)}{\sin e^x}\right)\frac{1}{x} + O\left(\frac{1}{x^2}\right)\right\}\frac{e^{-x}}{x} \\
&- \left\{\frac{\sin x}{3!} - \frac{\cos^2 x}{2\sin e^x} + \frac{\cos x}{\sin e^x}\left(\sin x - \sin\left(\frac{e^x}{x-1} + x\right)\right) + O\left(\frac{1}{x}\right)\right\}e^{-2x} \\
&+ \cdots .
\end{aligned}$$

References

1. W.W. Adams and P. Loustaunau. *An Introduction to Gröbner bases*. American Mathematical Society, 1994.
2. N. Apostolatus. Allgemeine Intervallarithmetiken und Anwendungen. *Bull. Soc. Math. Grece (NS)*, 10:136–180, 1969.
3. J. Ax. On Schanuel's conjecture. *Ann. of Math.*, 93:252–268, 1971.
4. J. Baddoura. A conjecture on integration in finite terms with elementary functions and polylogarithms. In M. Bronstein, editor, *ISSAC '94 Proceedings*, pages 158–162, Kiev, Ukraine, 1994. A.C.M. Press.
5. R. Bellman. *Stability Theory of Differential Equations*. McGraw-Hill, 1953.
6. M. Berry and C.Howls. Hyperasymptotics. *Proc. Roy. Soc: Math. & Phys.*, A430:653–667, 1990.
7. L. Bertrand. Computing a hyperelliptic integral using arithmatic in the Jacobian of the curve. *Applicable Algebra in Engineering, Communication and Computing*, 6:275–298, 1995.
8. G. Birkhoff and G.C. Rota. *Ordinary Differential Equations*. Ginn & Co., 1962.
9. M. Boshernitzan. An extension of Hardy's class L of 'Orders of infinity'. *J. Analyse Math*, 39:235–255, 1981.
10. M. Boshernitzan. New 'Orders of Infinity'. *J. Analyse Math.*, 41:130–167, 1982.
11. M. Boshernitzan. Discrete orders of infinity. *Amer. J. Math.*, 106:1147–1198, 1984.
12. M. Boshernitzan. 'Orders of infinity' generated by difference equations. *Amer. J. Math.*, 106:1067–1089, 1984.
13. M. Boshernitzan. Hardy fields, existence of transexponential functions and the hypertranscendence of solution to $g(g(x)) = e^x$. *Aequationes Math.*, 30:258–280, 1986.
14. M. Boshernitzan. Second-order differential equations over Hardy fields. *J. London Math. Soc.*, 35:109–120, 1987.
15. M. Boshernitzan. Uniform distribution and Hardy fields. *J. D'Analyse Math.*, 62:225–240, 1994.
16. F. Boulier. *Étude et implantation de quelques algorithmes en algèbre différentielle*. PhD thesis, Université de Lille, 1994.
17. F. Boulier, D. Lazard, F. Ollivier, and M. Petitot. Representation for the radical of a finitely generated differential ideal. In *ISSAC95 Proceedings*. A.C.M. Press, 1995.

18. N. Bourbaki. *Éléments de Mathématiques. Ch. V: Fonctions d'une variable réelle. Appendice, pp. 36–55.* Hermann, Paris, Second edition, 1961.
19. T.J.I'A Bromwich. *An Introduction to the Theory of Infinite Series, 2nd ed.* Macmillan, 1926.
20. M. Bronstein. Simplification of real elementary functions. In G. Gonnet, editor, *ISSAC '89 Proceedings*, pages 207–211, Portland, Oregon, 1989. A.C.M. Press.
21. M. Bronstein. Symbolic integration: towards practical algorithms. In E. Tournier, editor, *Computer algebra and differential equations*, pages 59–85. Academic Press, 1989.
22. M. Bronstein. On the integration of elementary functions. *J. Symbolic Comp.*, 9:117–173, 1990.
23. B. Buchberger. Gröbner bases: An algorithmic method in polynomial ideal theory. In N.K. Bose, editor, *Multidimensional Systems Theory*, pages 184–232. Reidel, 1985.
24. B.F. Caviness and M.J. Prelle. A note on algebraic independence of logarithmic and exponential constants. *SIGSAM Bull.*, 12/2:18–20, 1978.
25. D. V. Chudnovsky and G. V. Chudnovsky. On expansion of algebraic functions in power and Puiseux series, I. *Journal of Complexity*, 2:271–294, 1986.
26. D. V. Chudnovsky and G. V. Chudnovsky. On expansion of algebraic functions in power and Puiseux series, II. *Journal of Complexity*, 3:1–25, 1987.
27. G.E. Collins. Quantifier elimination for real closed fields by cylindrical algebraic decomposition. In *Proc. 2nd Conf on Automata Theory and Formal Languages, Lecture Notes in Computer Science vol. 33*, pages 134–183. Springer Verlag, 1975.
28. B.I. Dahn and P. Göring. Notes on exponential-logarithmic terms. *Fundamenta Math.*, 127:45–50, 1986.
29. J.H. Davenport. *On the Integration of Algebraic Functions.* Springer Verlag, Berlin, Heidelberg and New York, 1981.
30. N.G. de Bruijn. *Asymptotic Methods in Analysis.* North Holland, 1958.
31. P. du Bois-Reymond. Ueber asymptotische Werthe, infinitäre Approximationen und infinitäre Auflösung von Gleichungen. *Math. Ann.*, 8:362–414, 1875.
32. J. Ecalle. *Introduction aux fontions analysables et preuve constructive de la conjecture de Dulac.* Hermann, Paris, 1992.
33. S. Eilenberg. *Automata, Languages and Machines.* Academic Press, 1974.
34. M. Fliess, J. Lévine, P. Martin, and P. Rouchon. Sur les systèmes non linéaires différentiellement plats. *C.R. Acad. Sci. Paris*, 315:619–624, 1992.
35. M. Fliess, J. Lévine, P. Martin, and P. Rouchon. Défaut d'un système non linéaire et commande haute fréquence. *C.R. Acad. Sci. Paris*, 315:513–518, 1993.
36. K.O. Geddes, S.R. Czapor, and G. Labahn. *Algorithms for Computer Algebra.* Kluwer, Boston, 1992.
37. D. Gruntz. *On Computing Limits in a Symbolic Manipulation System.* PhD thesis, E.T.H., Zurich., 1996.
38. P.R. Halmos. *Measure Theory.* Springer-Verlag, New York, Heidelberg, Berlin, 1974.
39. G.H. Hardy. *Orders of Infinity.* Cambridge Univ. Press, Cambridge, England, 1910.
40. G.H. Hardy. Properties of logarithmico-exponential functions. *Proc. London Math. Soc.*, 10:54–90, 1912.

41. G.H. Hardy. *Divergent Series.* Oxford Univ. Press, 1949.
42. W.K. Hayman. *Meromorphic Functions.* Oxford Univ. Press, Oxford, England, 1964.
43. O. Hölder. Über die Eigenschaft der Γ-Funktion, keiner algebraischen Differentialgleichung zu genügen. *Math. Ann.*, 28:1–13, 1887.
44. W. Kahan. A more complete interval arithmetic. Technical report, University of Toronto, 1968.
45. I. Kaplansky. *An Introduction to Differential Algebra.* Hermann, Paris, 1957.
46. A.G. Khovanskii. Fewnomials and Pfaff manifolds. In *Proc. Int. Congress Math., Warsaw*, pages 549–564, 1983.
47. D.E. Knuth. *Seminumerical Algorithms, 2nd ed.* Addison-Wesley Inc., Reading. Mass., 1981.
48. S. Lang. *Algebra.* Addison-Wesley Inc., Reading. Mass., 1965.
49. S. Lang. Transcendental numbers and diophantine approximation. *Bull. Amer. Math. Soc.*, 77/5:635–677, 1971.
50. A. Lightstone and A. Robinson. *Nonarchimedean Fields and Asymptotic Expansions.* Elsevier, New York, 1975.
51. F. Lindemann. Über die Zahl π. *Math. Ann.*, 20:213–225, 1882.
52. J. Liouville. Mémoire sur la classification des transcendents et sur l'impossibilité d'exprimer les racines de certaines équations en fonction finie explicite des coefficients. *J. Math. Pure et App.*, 2:56–104, 1837.
53. J. Liouville. Suite du mémoire sur la classification des transcendents et sur l'impossibilité d'exprimer les racines de certaines équations en fonction finie explicite des coefficients. *J. Math. Pure et App.*, 3:523–546, 1838.
54. A.J. Macintyre. The laws of exponentiation. In C. Berline, K. McAloon, and J.P. Ressayre, editors, *Model Theory and Arithmetic*, pages 185–197. Springer Verlag, 1981.
55. A.J. Macintyre and A.J. Wilkie. On the decidability of the real exponential field. In P. Odifreddi, editor, *Kreiseliana. About and around Georg Kreisel*, pages 441–468, Wellesley, Massachusetts, 1996. A. K. Peters.
56. K. Mahler. On the approximation of π. *Indagationes Math.*, 15:30–42, 1953.
57. A. Maignan. Solving one and two–dimensional exponential polynomial systems. In O. Gloor, editor, *ISSAC '98 Proceedings*, pages 215–221, Rostock, Germany, 1998. A.C.M. Press.
58. E.L. Mansfield. The differential algebra package diffgrob2. *MapleTech*, 3:33–37, 1996.
59. V. Marić. Asymptotic behaviour of solutions of a nonlinear differential equation of the first order. *J. Math. Anal. Appl.*, 38:187–192, 1972.
60. D. Marker, M. Messmer, and A. Pillay. *Model Theory of Fields.* Springer-Verlag, Berlin. Heidelberg., 1996.
61. R.E. Moore. *Methods and Application of Interval Analysis.* SIAM studies in Applied Mathematics, Philadelphia. USA, 1979.
62. F. Olver. *Asymptotics and Special Functions.* Academic Press, 1974.
63. A. Péladan-Germa. Testing identities of series defined by algebraic partial differential equations. In Gérard Cohen, Marc Giusti, and Teo Mora, editors, *Applied Algebra, Algebraic Algorithms and Error-Correcting Codes*, pages 393–407. Springer-Verlag, 1995. Proceedings of the 11th International Symposium, AAECC-11, Paris, France, July 1995.

64. J.P. Ramis and J. Martinet. Théorie de Galois Différentielle et resommation. In E. Tournier, editor, *Computer Algebra and Differential Equations*, pages 117–214. Academic Press, 1989.
65. D. Richardson. Some undecidable problems involving elementary functions of a real variable. *J. Symbolic Logic*, 33:514–520, 1968.
66. D. Richardson. Solution of identity problem for integral exponential functions. *Z. Math. Logik. Grundlag. Math.*, 15:333–340, 1969.
67. D. Richardson. Finding roots of equations involving functions defined by algebraic differential equations. Technical report, University of Bath, England, 1989.
68. D. Richardson. Towards computing non-algebraic cylindrical decompositions. In *Proceedings ISSAC91*, pages 247–255, Bonn, 1991. A.C.M. Press.
69. D. Richardson. Computing the topology of a bounded non algebraic curve in the plane. *J. Symbolic Comput.*, 14:619–643, 1992.
70. D. Richardson. A zero structure theorem for exponential polynomials. In *Proceedings ISSAC93*, pages 144–151, Kiev, Ukraine, 1993. A.C.M. Press.
71. D. Richardson. How to recognise zero. *J. Symbol. Comput.*, 24(6):627–646, 1994.
72. D. Richardson. The uniformity conjecture. pages 253–272. Springer Verlag, 2001. Lecture Notes in Computer Science.
73. D. Richardson, B. Salvy, J. Shackell, and J. Van der Hoeven. Asymptotic expansions of exp-log functions. In Y. N. Lakshman, editor, *ISSAC'96*, pages 309–313. ACM Press, 1996. Proceedings of the 1996 International Symposium on Symbolic and Algebraic Computation. July 24–26, 1996. Zurich, Switzerland.
74. R.H. Risch. The problem of integration in finite terms. *Trans. Amer. Math.Soc.*, 139:167–189, 1969.
75. R.H. Risch. The solution of the problem of integration in finite terms. *Bull. Amer. Math.Soc.*, 76:605–608, 1970.
76. J.F. Ritt. *Differential Algebra.* American Math. Soc., 1950.
77. A. Robinson. On the real closure of a Hardy field. In G. Asser et al., editor, *Theory of Sets and Topology*, Berlin, 1972. Deut. Verlag Wissenschaften.
78. M. Rosenlicht. Hardy fields. *J. Math. Anal. App.*, 93/2:297–311, 1983.
79. M. Rosenlicht. The rank of a Hardy field. *Trans. Amer. Math. Soc.*, 280/2:659–671, 1983.
80. M. Rosenlicht. Rank change on adjoining real powers to Hardy fields. *Trans. Amer. Math. Soc.*, 284/2:829–836, 1984.
81. M. Rosenlicht. Growth properties of functions in Hardy fields. *Trans. Amer. Math. Soc.*, 299/1:261–272, 1987.
82. M. Rosenlicht. Asymptotic solutions of $y'' = f(x)y$. *J. Math. Anal. App.*, 189:640–650, 1995.
83. M. Rothstein and B.F. Caviness. A structure theorem for exponential and primitive functions. *SIAM J. Comput.*, 8/3:357–367, 1979.
84. L. A. Rubel. Some research problems about algebraic differential equations. *Trans. Amer. Math. Soc.*, 280:43–52, 1983.
85. B. Salvy. *Asymptotique automatique et fonctions génératrices.* PhD thesis, École Polytechnique, INRIA Rocqencourt, 78153 Le Chesnay Cedex, France, 1991.
86. B. Salvy. Fast computation of some functional inverses. *J. Symbolic Comp.*, 17:227–236, 1994.

87. B. Salvy and J. Shackell. Symbolic asymptotics: Functions of two variables, implicit functions. *Journal of Symbolic Computation*, 25(3):329–349, March 1998.
88. B. Salvy and J. Shackell. Symbolic asymptotics: Multiseries of inverse functions. *Journal of Symbolic Computation*, 27(6):543–563, June 1999.
89. B. Salvy and J. Shackell. Asymptotic expansion with oscillating coefficients. Technical report, University of Kent at Canterbury, England, 2002.
90. B. Salvy and J.R. Shackell. Asymptotic expansions of functional inverses. In *ISSAC92 proceedings, Berkeley, California*, pages 130–137. ACM Press, 1992.
91. P. Seckington. *Zero Equivalence of Functional Expressions*. PhD thesis, University of Kent at Canterbury., 1996.
92. P. Seckington and J. Shackell. Computing zero equivalence and symmetry. Technical report, University of Kent at Canterbury, England, 1994.
93. R. Sedgewick and Ph. Flajolet. *An Introduction to the Analysis of Algorithms*. Addison–Wesley Publishing Company, 1996.
94. J. Shackell. The exponential identity problem and the Schanuel conjecture. Technical report, University of Kent at Canterbury, England, 1986. Unpublished.
95. J. Shackell. Asymptotic estimation of oscillating functions using an interval calculus. In P. Gianni, editor, *ISSAC '88 Proceedings*, pages 481–489, Rome, 1988. Springer Verlag.
96. J. Shackell. A differential-equations approach to functional equivalence. In G. Gonnet, editor, *ISSAC '89 Proceedings*, pages 7–10, Portland, Oregon, 1989. A.C.M. Press.
97. J. Shackell. Growth estimates for exp-log functions. *J. Symbolic Comp.*, 10/6:611–632, 1990.
98. J. Shackell. Inverses of Hardy L-functions. *Bull. London Math. Soc.*, 25:150–156, 1993.
99. J. Shackell. Rosenlicht fields. *Trans. Amer. Math. Soc.*, 335/2:579–595, 1993.
100. J. Shackell. Zero-equivalence in function fields defined by algebraic differential equations. *Trans. Amer. Math. Soc.*, 336/1:151–172, 1993.
101. J. Shackell. Extensions of asymptotic fields via meromorphic functions. *J. London Math. Soc.*, 52:356–374, 1995.
102. J. Shackell. Growth orders occurring in expansions of Hardy-field solutions of algebraic differential equations. *Annales de l'Institut Fourier*, 45:183–221, 1995.
103. J. Shackell. Limits of Liouvillian functions. *Proc. London Math. Soc.*, 72:124–156, 1996.
104. J. Shackell. Star products and the representation of asymptotic growth. In *ISSAC99 Proceedings*. ACM Press, 1999.
105. J. Shackell and B. Salvy. Asymptotic forms and algebraic differential equations. *J. Symbolic Computation*, 20:169–177, 1995.
106. M. Singer. Liouvillian first integrals of differential equations (extended abstract). In P. Gianni, editor, *ISSAC '88 Proceedings*, pages 57–63, Rome, 1988. Springer Verlag.
107. M. Singer. An outline of differential galois theory. In E. Tournier, editor, *Computer Algebra and Differential Equations*, pages 3–57. Academic Press, 1989.
108. M. Singer. Formal solutions of differential equations. *J. Symbolic Comp.*, 10:59–94, 1990.

109. D. Stoutemyer. Qualitative analysis of mathematical expressions using computer symbolic mathematics. In R.D. Jenks, editor, *Proceedings SYMSAC76*, pages 97–104, Yorktown Heights, New York, 1976. A.C.M.
110. W. Strodt and R.K. Wright. Asymptotic behaviour of solutions and adjunction fields for nonlinear first order differential equations. *Memoirs of the American Mathematical Society*, 109, 1971. 284 pages.
111. F. Ulmer. Representing functions by differential equations. Private Communication, 1989.
112. L. van den Dries. *Tame topology and o-minimal structure.* Cambridge University Press, The Edinburgh building, Cambridge, CB2 2RU, U.K., 1998.
113. L. van den Dries, A. Macintyre, and D. Marker. Logarithmic-exponential power series. *J. London Math. Soc. (2)*, 56:417–434, 1995.
114. J. van der Hoeven. *Asymptotique automatique.* PhD thesis, École Polytechnique, Laboratoire d'Informatique, École Polytechnique, Paris, France, 1997.
115. J. van der Hoeven. On the computation of limsups. *Journal of Pure and Applied Algebra*, 117 & 118:381–394, 1997.
116. J. van der Hoeven. Generic asymptotic expansions. *AAECC*, 9:25–44, 1998.
117. J. van der Hoeven. Zero–testing, witness conjectures and differential diophantine approximation. Technical Report 2001-62, Université de Paris–Sud, Orsay, 2001.
118. J. van der Hoeven. Relax, but don't be too lazy. *J. Symbolic Comp.*, 34:479–542, 2002.
119. M. van der Put. Symbolic analysis of differential equations. In H. Cuypers & H. Sterk A.M. Cohen, editor, *Some Tapas of Computer Algebra*, pages 208–236. Springer Verlag, 1999.
120. B.J. van der Waerden. *Algebra, vol.1.* Frederick Ungar Pub. Co., 1964.
121. W. Wasow. *Asymptotic Expansions for Ordinary Differential Equations.* Dover Publications Inc., 1965.
122. J.-A. Weil. *Constantes et polynômes de Darboux en algèbre différentielle: applications aux systèmes différentiels linéaires.* PhD thesis, École Polytechnique, Paris, France, 1995.
123. E. T. Whittaker and G. N. Watson. *A course of modern analysis, 4ed.* Cambridge University Press, 1962.
124. A.J. Wilkie. Some model completeness results for expansions of the ordered field of real numbers by Pfaffian functions and the exponential function. *J. Amer. Math. Soc.*, 9:1051–1094, 1996.
125. O. Zariski and P. Samuel. *Commutative Algebra vols I & II.* Van Nostrand, 1960.
126. D. Zwillinger. *Handbook of Differential Equations.* Academic Press, 1989.

Index

GPSR Compliance

The European Union's (EU) General Product Safety Regulation (GPSR) is a set of rules that requires consumer products to be safe and our obligations to ensure this.

If you have any concerns about our products, you can contact us on ProductSafety@springernature.com

In case Publisher is established outside the EU, the EU authorized representative is:

Springer Nature Customer Service Center GmbH
Europaplatz 3
69115 Heidelberg, Germany

Batch number: 10682213

Printed by Printforce, the Netherlands